DATE DUE

			PRINTED IN U.S.A.

THE HANDBOOK OF ENVIRONMENTALLY CONSCIOUS MANUFACTURING

THE HANDBOOK OF ENVIRONMENTALLY CONSCIOUS MANUFACTURING

From Design & Production to Labeling & Recycling

Robert E. Cattanach

Jake M. Holdreith

Daniel P. Reinke

Larry K. Sibik

IRWIN
Professional Publishing

Chicago • Bogotá • Boston • Buenos Aires • Caracas
London • Madrid • Mexico City • Sydney • Toronto

The Handbook of
environmentally conscious **nment**

...ompany is a large end-user of
fragile yet replenishable resources, we at IRWIN can assure you
that every effort is made to meet or exceed Environmental Protection Agency
(EPA) recommendations and requirements for a "greener" workplace.

To preserve these natural assets, a number of environmental policies, both
companywide and department-specific, have been implemented. From the
use of 50% recycled paper in our textbooks to the printing of promotional
materials with recycled stock and soy inks to our office paper recycling program, we are committed to reducing waste and replacing environmentally
unsafe products with safer alternatives.

© RICHARD D. IRWIN, INC., 1995

Editor in chief: Jeffrey A. Krames
Marketing manager: Angela Wells
Project editor: Rebecca Dodson
Production manager: Pat Frederickson
Designer: Larry J. Cope
Art manager: Kim Meriwether
Compositor: Precision Typographers
Typeface: 10/12 Times Roman
Printer: The Maple-Vail Book Manufacturing Group

Library of Congress Cataloging-in-Publication Data

The Handbook of environmentally conscious manufacturing: from design
& production to labeling & recycling / Robert E. Cattanach . . . [et
al.]
     p.   cm.
   Includes index.
   ISBN 0-7863-0147-3
     1. Environmental protection—Handbooks, manuals, etc.
  2. Technology—Environmental aspects—Handbooks, manuals, etc.
  I. Cattanach, Robert E.
  TD170.H39   1995
  658.4′08—dc20   94–33053

Printed in the United States of America
1 2 3 4 5 6 7 8 9 0 MV 1 0 9 8 7 6 5 4

Preface
Environmentally Conscious Manufacturing and Marketing

Manufacturers today have the good fortune to live in an age in which they need not harm the environment to succeed in business. Indeed, market incentives and legal controls increasingly offer rewards for those able to develop cleaner processes and products and penalties for those unable to eliminate waste. Moreover, this evolution is being helped along by a realization that industry is not the problem, but rather in many cases the solution.

Development of cleaner processes and elimination of waste make good business sense and good common sense. Environmental hazards and wastes represent a cost and a liability risk that are unnecessary in many cases. Eliminate the hazard or the waste, and you eliminate a cost and risk. Moreover, increasing consumer demand for green products offers a market advantage to cleaner products.

The costs and liabilities associated with hazards and wastes are not restricted to the threat of legal liability. To be sure, the risk of legal liability under environmental laws is substantial. Additional costs associated with hazards and waste include: (1) the increasing cost of purchasing hazardous materials as inputs, which reflect the internalization of environmental concerns; (2) the costs of managing and disposing of process waste, as waste disposal becomes more and more expensive; (3) the cost of stigmatization or market resistance to products perceived as harmful to the environment; and (4) the cost of public and regulatory hostility toward businesses that are perceived as unwilling to be part of the solution.

Not only are markets forcing corporate environmentalism, but also traditional "command and control" regulation is giving way to more flexible, market-based regulation that incorporates market rewards for environmental excellence. Emissions trading, for example, has allowed companies that can reduce their air emissions to make a profit on sales of emissions credits to companies with less foresight. On the other side of the equation, environmental fines, penalties, and even criminal sentences may be raised or lowered depending on the steps a company has taken—or failed to take—to prevent environmental problems.

Environmentally conscious manufacturing and marketing means capitalizing on the opportunity to lower costs, increase profit, and manage or eliminate environmental liability risks by incorporating environmental considerations into every aspect of business, from design of a product to production, distribution, marketing, use by consumers, and final disposition after use. Corporate environmentalism of this kind does not happen by itself. It requires deliberate integration of specific steps for environmental improvement into your company's management at every level.

This book provides a foundation for understanding and implementing principles of environmentally conscious manufacturing and marketing in your operation. It offers an explanation of environmentally conscious manufacturing and marketing techniques and provides a step-by-step plan for integrating these techniques into your business. Part I of the book focuses on design and manufacturing techniques, and Part II focuses on marketing and packaging techniques.

The environmental design and manufacturing techniques in this book start with accepted management principles, such as continuous improvement, and integrate environmental considerations, such as waste reduction and materials substitution. The techniques are explained with illustrative examples. The discussion includes both systems for the identification of opportunities to assess and improve your products environmentally as well as solutions to specific problems based on the examples.

The environmental marketing techniques in this book are introduced with an overview of challenges and opportunities that are unique to the environmental market. The complex arena of legal restrictions on specific environmental terms and techniques is discussed with respect to U.S. law and the law of other countries. Finally, detailed steps are suggested for the identification of proper environmental marketing claims, the evaluation of the effectiveness of the claims, and the analysis of legal restrictions that may apply to the claims, as well as strategies for dealing with logistical problems of making claims in markets with differing regulations.

As used in this book, environmentally conscious management does not mean simply meeting current legal standards. This book encourages you to adopt environmental excellence as a goal for your processes and products. Setting a goal of minimum compliance simply puts you in a trap of reacting to continually changing legal requirements and disregards the opportunities and demands of a market that can recognize and reward environmental excellence.

The market has already demonstrated that it will reward corporate environmentalism. Consumers report that they are willing to pay more for greener products, and actual consumer behavior supports this claim. Increasing internationalism has also increased the incentives for environmentally conscious manufacturing. Emerging standards such as ISO 14000, the environmental counterpart to ISO 9000, and BSI 7750 will require incorporation of environmental considerations into design and manufacturing processes as a condition of certification. As purchasers in developed countries increasingly require compliance with these or similar standards, those with a foundation of environmentally conscious manufacturing will have better access to the most attractive markets.

Although environmentalism is a relatively young aspect of the market, it has a firm beachhead. As efficiency, product safety, and advertising have done, environmental responsibility inevitably is becoming a fundamental requirement for effective competition in the market.

Acknowledgments

Many talented people assisted in the creation of this book, which would not have been possible without them. Jennifer Muenchrath provided invaluable assistance, especially with respect to Chapter Six. Kevin Saville's help in updating the constantly changing international requirements was particularly important. Tim Kenny assisted in the overview of U.S. regulations on marketing and advertising. Karen Timmerman provided superior editing and organizational support. Oppenheimer's library staff, especially Trudi Busch and Norma Knudson, provided excellent support. Jeanie Thielen deserves special credit for long hours of information-processing support. Carol DeWolf provided helpful editing support on early drafts of the manuscript. The assistance of Joanne Sims and Linda Voss-berry of the Dorsey firm was essential in the final stages of the editing.

R.E.C.

J.M.H.

D.P.R.

L.K.S.

Table of Contents

Chapter One

An Overview of Environmentally Conscious Manufacturing and Marketing

E veryone accepts the concept of increasing efficiencies by decreasing waste. The reason it does not always happen lies in the details. This book explains how to make it happen and gain the marketing advantages that go with it.

1.1   ESTABLISH A BASELINE

The first step in the process requires that you assess where you are and then where you want to be. Manufacturers must be aware of the environmental regulations that apply to their operations. Chapter Two contains an outline of the key regulations likely to apply to most manufacturing processes, such as the federal "Clean" acts (Air, Water), waste-oriented legislation (Resource Conservation and Recovery Act), as well as Superfund (or CERCLA). It goes on to outline a simple process for establishing compliance with these regulations.

Compliance alone, however, is not enough. Fundamental to the concept of environmentally conscious manufacturing and marketing (which we will refer to as ECM for short) is the notion of a proactive approach to the environmental consequences of production activities. The trend in regulations toward voluntary initiatives, also addressed in Chapter Two, mirrors this approach. Government has begun to recognize that significant advances are not likely through more command and control types of regulation, and market incentives are expected to play a major role in the evolving regulatory scheme.

Those companies positioned to take advantage of market opportunities by anticipating regulatory developments through proactive environmental programs will reap rewards. The new emissions trading provisions of the 1990 Clean Air Act are a classic example. Those utilities that adopted more stringent control mechanisms to minimize emissions not only provided cleaner power to their customers but also gained a windfall when they were able to sell excess emission credits to other utilities that had not had the same foresight. Similar trading in volatile organic credits in California's South Coast Air Quality Management District demonstrates a more localized approach to the same concept, with much broader applicability.

1.2   ENVIRONMENTAL MANAGEMENT

Once a company has mastered the basics of Chapter Two's step-by-step approach to environmental compliance, the real gains of a total quality environmental management program can be realized. Chapter Three provides helpful examples of alternative ways that companies can go "beyond compliance" to integrate an ongoing environmental improvement process into all aspects of their operations. Continuous environmental improvement becomes part of the company's culture. The management standards being discussed today, exemplified by the British Standards Institute (BSI) 7750, provide process-oriented approaches that can be adapted to various management styles and cultures in virtually any company.

The important issue in environmental management is not the precise way in which environmental excellence is achieved, but rather that a process exists that makes it possible. Without an ongoing improvement program in the manufacturing process itself, the most exhaustive of efforts to identify and comply with all environmental regulations will inevitably become inefficient and outdated. Whether the international system ultimately adopted is BSI 7750 or some variation on that system, there is no question that significant standardization of environmental management is upon us. Growing pressures for a level playing field among trading partners are causing environmentally advanced nations to look for ways to require importers from less developed countries to internalize all environmental costs of production. GATT purists may disagree, but significant disparities in internal environmental programs cannot exist for long.

Meeting minimum standards may suffice for today, but this book is for companies that want to prepare for tomorrow. Simply articulating an environmental management system, or even adopting it at the production level, will not provide the full advantages that are available under ECM. Environmental objectives must be integrated into a company's business plan and all levels and divisions made accountable for achieving them. Chapter Three provides a process for accomplishing that objective, including insightful case studies to illustrate the specifics of how environmental management systems can be made to work in real-life situations.

1.3   MAKING IT HAPPEN

A sound environmental theory is great, but results are even better. Chapter Four provides the practical steps necessary to establish an environmental baseline for any particular product, division, or function and then measure the production process in a way that can provide immediate efficiency gains as well as improved environmental performance. The process is disarmingly simple: measure the current status, set new goals, establish a plan to meet them, implement the plan, measure its effectiveness—and then start all over again with new goals, plans, and implementations.

While some of the documentation required of ISO 9000 and similar programs can be daunting, a well-devised plan, particularly aided by the revolution in information-processing technology now under way, can make that documentation become second nature. And, if the documentation gets in the way, it can always be modified in a way that still makes significant improvements possible, albeit perhaps not with the added benefit of formal ISO certification.

It sounds easy, but once again, companies have been discouraged by the lack of a detailed road map on how to get it done. Chapter Five contains specific examples

of how to identify ECM process options for improvement, how to evaluate those options, and then how to document and implement the plan to make them a reality. Obviously space does not permit discussions of every conceivable situation, but the discipline and methodology provided in this chapter can be easily adapted to apply to almost any manufacturing situation.

1.4   PRODUCT LIFE CYCLE: THE NEXT FRONTIER FOR ECM

Talking about evaluating and changing manufacturing operations, however, means we are still playing catch-up. The real gains to be made from ECM can be found in the design stage. Product life cycle assessments (LCA) are the first step toward a truly integrated environmental awareness. From raw materials to production to packaging, transportation, use, and disposal, product life cycle assessments permit us to look at the full extent of environmental consequences. Several countries in Europe have already adopted aggressive product take-back legislation for automobiles and electronics, and other products are sure to follow. Even ignoring the marketing and free trade consequences, these initiatives herald a new era of environmental awareness.

Product take-back and its first cousin, packaging take-back, will create challenges for all manufacturers, distributors, providers of raw materials, and of course marketers. Those companies prepared to meet the aggressive demands of this new body of legislation will have market advantages over competitors that have been slow to anticipate the trend, and those that have failed to prepare for these requirements will suffer the consequences of having to react to the advances of their competitors.

Chapter Six provides a concise explanation of the concepts of product life cycle and sets out a business guide to implementing LCA. Like the earlier chapters, this guide provides a methodology for evaluating your current situation and plotting the strategy necessary for incorporating LCA into your design, manufacturing, and ultimately product return or disposal process.

1.5   MARKETING THE ENVIRONMENTALLY CONSCIOUS PRODUCT

Beyond improving product quality and eliminating waste, ECM provides potentially significant advantages in marketing the product itself. Over half of all consumers report that they make purchasing decisions based on environmental factors. The key to environmentally conscious marketing is product differentiation. So much has been said and written about environmental performance of consumer products that the currency is in danger of losing its value. Rigorous new regulations prevent empty or misleading claims, however, and products with genuine environmental advantages should be able to gain an edge through carefully crafted (and documented) marketing.

Exploiting that market edge is the focus of Chapters Seven through Ten. Chapter Seven starts by explaining the framework of Environmentally Conscious Marketing, including media campaigns, traditional advertising claims, private and government certifications, and packaging and take-back requirements. Critical to navigating this maze of alternatives is the process of developing market data and other facts to substantiate ECM claims as well as ways to deal with inconclusive science and overcoming consumer skepticism through education.

Once the context for making the claims is established, the environmentally conscious marketer must fit the claims into the applicable regulations and guidelines. Chapter Eight sets forth the American model (adopted by the United States, Canada, and Australia), which approaches ECM from a consumer protection prospective, primarily using truth-in-advertising laws. Those engaging in ECM will have to be aware of issues concerning outright deception, advertising substantiation, and general fairness in advertising. The Federal Trade Commission's guidelines are set forth in some detail as the most likely criteria that regulate such claims.

The book goes on to discuss specific ECM terms and their appropriate usage, such as recycled, recyclable, reusable, compostable, degradable, as well as how key states are also regulating such matters (including California's Proposition 65), so that manufacturers can evaluate how their products can be marketed using common ECM parlance. Specific discussions on commonly encountered products, such as plastics, ozone depleting substances, and typically recycled products, as well as the complex area of packaging, should provide helpful insight into exactly how to maximize the use of the product's environmental advantages. In addition, the book explains how manufacturers can participate in more broadly based emblems and certifications such as Green Seal, Scientific Certification Systems, and the Canadian National Emblem.

1.6   INTERNATIONAL STANDARDS

ECM regulation in other parts of the world tends to focus more on protection of the environment than the consumer. Specific environmental performance standards are likely to be encountered, particularly in Europe. Standards unique to each nation, such as Germany's Blue Angel, abound. Chapter Nine sets forth in summary fashion the various competing international schemes and provides enough detail for the reader to determine how to apply and meet various international ECM standards.

The European Commission has initiated the Eco-Label program in an effort to minimize the potential for confusion in ECM among its member states. Based on a life cycle assessment of the product, the Eco-Label has the potential to standardize ECM throughout Europe and perhaps worldwide. Limited to specific product groups that have undergone rigorous LCA scrutiny, the Eco-Label will not likely have an immediate impact on a broad range of products, but the process used to award the Eco-Label must be kept in mind as a likely future standard.

1.7   IMPLEMENTING AN ENVIRONMENTALLY CONSCIOUS MARKETING PROGRAM

Faced with a dizzying array of ECM standards, especially for products marketed worldwide, manufacturers need to be able to sift through the standards and develop a marketing process that not only complies with the law but also does so in a cost-effective manner. Chapter Ten provides the strategy for meeting the challenges of marketing ECM products. A step-by-step process is established to help identify opportunities and challenges, both in the context of marketing specific products as well as communicating the company's environmental commitment generally. The chapter will help you identify areas of distribution and determine the need for collection and perhaps generation of market research.

Once the data have been gathered, the manufacturer needs to identify applicable rules and regulations, both as to products and various jurisdictions in which they will be marketed. The chapter on implementation then walks the manufacturer through the process of evaluating options, risks, and finally developing a specific marketing plan. The same continuous improvement process provided in the manufacturing section of the book applies equally to the marketing process.

1.8   SUMMARY AND OVERVIEW

Manufacturers with virtually no prior experience in ECM can use the processes summarized in this chapter, and contained in more detail in the chapters to follow, to evaluate the environmental performance of their products and create a system for improving that performance under a total quality approach. Using the documentation generated by that very same process, manufacturers can then achieve not only direct cost savings and product improvement but also marketing advantages.

I

DESIGN AND MANUFACTURING

Regulatory Framework

C ompliance with environmental regulations is assumed as the baseline goal for all companies. However, a company's management should not rely on compliance as a given fact. The complexity of today's environmental regulations virtually assures that a company will experience periods of unintended noncompliance.

The risks and liabilities associated with manufacturing and service industries are significant enough to require that industry take steps to limit the potential effects of noncompliance. The specter of plant closings due to environmental noncompliance has been held over the head of industry, but relatively few plants are actually closed by regulatory agencies. The more likely effects of regulatory noncompliance are fines, stipulation agreements, and in rare cases prison sentences.

The effects of noncompliance listed above are tangible and as a result relatively easy to understand. Other less tangible effects also have to be considered in assessing the risk associated with a company's noncompliance with environmental regulations. Adverse public relations can quickly snowball with media sensationalism. Often the resources and time spent on reacting to an accident, spill, or notice of noncompliance far outweigh the cost of proper management.

Companies that are committed to developing environmentally conscious products and services need to start with the basics and assure themselves that their environmental program is being handled properly. The long-term payback expected from investing the resources of time and capital needed to develop environmentally conscious products and processes can quickly dissipate with one unfortunate incident. This section covers the regulations that need to be addressed by a facility, the development of a written regulatory compliance program, compliance assessments, and training.

The subject of environmental regulations becomes ex-

ponentially more complex as the regulations of more than one country are considered. For purposes of illustration, we will limit the discussion to U.S. regulations.

2.1   ENVIRONMENTAL REGULATIONS

Not intended to be a comprehensive discussion of the environmental regulations, this section highlights the major acts, points out potential areas of concern, and discusses the common areas where noncompliance occurs.

Clean Air Act.   The Clean Air Act (CAA) controls air emissions from manufacturing facilities. Originally enacted in 1955, the CAA has been amended in 1967, 1970, 1977, and 1990. The 1977 and 1990 laws are the main portions of the CAA affecting manufacturing today. The 1977 amendments included provisions for national ambient air quality standards, new source performance standards, hazardous air pollution standards, motor vehicle standards, fuel and fuel additive provisions, and aircraft emission standards.

The 1990 amendments, which will be implemented over the next several decades, contain provisions to regulate emissions of 189 hazardous air pollutants (HAPs). Major sources, those with the "potential to emit" 10 tons per year of a single HAP or 25 tons per year of a combination of HAPs, will be required to meet minimum control criteria and apply for an air emissions permit. The definition of "potential to emit" can be interpreted to be very broad. In some cases, such as with spray painting operations, the definition of potential to emit assumes that a spray gun is triggered continuously for a year. Using this definition, a conventional spray gun applying a paint containing just 7 percent xylene has the potential to emit over 10 tons of xylene in a year.

Other provisions of the 1990 amendments that will

affect manufacturing include the stratospheric ozone protection provisions: Under these provisions and a subsequent Executive Order, the production of common cleaning chemicals such as CFCs and 1,1,1-trichloroethane will be phased out by the end of 1995. While the use of recycled solvents will still be allowed, their availability and cost will likely be prohibitive for most manufacturers. Finally, the administrative burdens of applying for and complying with a federal permit will come as a shock to those companies previously able to avoid the need for a federal permit.

Clean Water Act.   The Clean Water Act (CWA) controls wastewater discharges from manufacturing facilities. Originally enacted in 1948, the act initially focused on providing federal aid to municipalities for the construction of sewage treatment systems. The 1972 revision set up effluent discharge limitations for specific pollutants for both municipalities and industry. Industry is either regulated directly through the National Pollutant Discharge Elimination System (NPDES) permits (for discharges directly to the environment) or through pretreatment standards imposed by the municipal authority that receives the plant's wastewater and in turn discharges to the environment under that municipality's NPDES permit.

The Water Quality Act of 1987 addressed the major weakness of the previous portions of the CWA—nonpoint discharges. Nonpoint discharges such as agricultural drainage or stormwater runoff must be identified, permitted, and tested under these new provisions. Industries must develop and implement plans to reduce or eliminate pollutant discharge via stormwater runoff.

The Resource Conservation and Recovery Act.   The Resource Conservation and Recovery Act (RCRA) regulates the generation, storage, transportation, and treatment or disposal of hazardous waste. Enacted in 1976 to replace the Resource Recovery Act of 1970, RCRA provided the first cradle-to-grave approach to hazardous wastes management. Amendments to RCRA in 1984 called for a ban on landfilling untreated hazardous wastes. The new amendment also closed loopholes that allowed hazardous wastes to be burned for energy recovery or disposal in industrial furnaces.

RCRA represents the first regulatory attempt at the life cycle process. For example, RCRA discouraged disposal through landfilling even though it was initially less expensive, because of the perceived greater long term costs to the environment. Other disposal options considered less of a long-term risk, such as incineration, aboveground vaults, and chemical destruction, however, proved significantly more expensive and difficult to construct because of neighborhood opposition. The net result was a drastic increase in the cost of producing hazardous waste by-

products, which in turn created the first real need for the ultimate long-term option, pollution prevention.

Comprehensive   Environmental   Response, Compensation, and Liability Act.   The Comprehensive Environmental Response, Compensation, and Liability Act (CERCLA), also known as Superfund, was a logical extension of RCRA. RCRA governed the current disposal of hazardous wastes, and CERCLA provided the funding and the guidelines to clean up those sites contaminated from prior unregulated disposal.

Generally viewed as horrendously inefficient, Superfund drove home the need for considering and internalizing all costs of production.

Toxic Substances Control Act.   With other legislation protecting the quality of air, water, and land resources, the Toxic Substances Control Act (TSCA) was intended to determine the risks associated with the use of raw materials. Under TSCA, the potential toxic effect of new substances must be fully evaluated before use in production and production of chemicals that pose a significant risk can be banned. Chemicals regulated under TSCA include polychlorinated biphenyls and asbestos.

Emergency Planning and Community Right-to-Know Act.   The Emergency Planning and Right-to-Know Act (EPCRA) was enacted in 1986 largely a result of the accidental release of methyl isocyanate in Bhopal, India, in 1984. Concern over the potential for a similar occurrence here in the U.S. led to legislation requiring companies to report the release and storage of specific chemicals and chemical compounds above threshold limits. Release reporting, commonly know as Form R reports, has provided the public with information on the types of chemicals used and the amount and nature of releases to the environment. Citizen groups often are able to use the Form R data as a cross-reference to determine environmental compliance in other areas.

2.2   REGULATORY EVOLUTION

The compliance programs developed by the federal government began with a command and control approach, with the Environmental Protection Agency developing regulations that mainly focused on pollution control techniques. As the EPA has evolved, so have the programs it has developed. Pollution prevention language now is included in virtually all environmental regulations developed since the enactment of RCRA in 1984. Specifically, RCRA requires a company to certify that it has a program in place to reduce the volume and/or toxicity of hazardous waste generated. The Community Right to Know (EPCRA) Form R emission report includes a section for reporting the planned emission reductions. Chemical Pro-

Case Study

Government Computer Purchases

An executive order by the Clinton Administration requires all federal government procurement agencies to purchase computer equipment that includes energy-saving features. Computer usage accounts for approximately 5 percent of the energy used in the United States. Computers are idle a large portion of the time they are turned on during a business day. The ''Energy Star'' computer equipment specified by the Clinton administration goes into a sleep mode, or low power mode, if inactive for a preset time period. A touch of the keyboard quickly powers up the entire computer, bringing the user back to the previous activity. According to EPA estimates, the by-products of this wasted energy consumption include 20 million tons of carbon dioxide, 140,000 tons of sulfur dioxide, and 75,000 tons of nitrogen oxides per year.

cess Safety Analysis, Spill Prevention Control and Countermeasure planning and the stormwater provisions of the Water Quality Act all seek to use pollution prevention planning to prevent chemical releases to the environment.

Programs such as the EPA's Industrial Toxics Project (ITP), also known as the 33/50 program, seek to offer industry incentives to reduce pollution. In the case of the ITP, the EPA's carrot was positive publicity for those companies that were able to reduce their Form R reported emissions of 17 targeted chemicals by 33 percent by 1992 and 50 percent by 1995. The Clean Air Act Amendments of 1990 also provided industry with an incentive to reduce air emissions—a six-year reprieve from meeting applicable standards for those companies willing to commit to reduce toxic emissions by 90 percent.

In addition to the federal government, states have enacted their own pollution prevention measures. At least 27 states have some form of pollution, waste, or emission reduction legislation. While state requirements vary in the specifics, generally they require industry to develop pollution prevention plans that identify technically and economically feasible options to reduce the generation of hazardous wastes or chemical emissions.

The EPA and the Clinton Administration indicate the next evolution in environmental regulation will come in the form of life cycle analysis. Life cycle analysis involves determining the complete environmental impacts of all activities involved in the life of a product. The analysis goes far beyond what occurs in the factory, to include the extraction of raw materials, transportation and distribution of raw materials, the product, and waste byproducts, product use, packaging waste, and the ultimate reuse, recycling, or disposal of the product.

Life cycle analyses weigh the total effects a manufacturing activity will have, over the product's life, on all receptors and media. For example, a portion of a life cycle analysis would consider energy consumption, including the energy used during manufacturing as well as the energy consumed or saved by the product during

its expected life. This energy consumption can then be translated into the amount of carbon dioxide, sulfur oxides, and other pollutants released into the atmosphere.

The EPA released a guidance document on life cycle analysis in 1992 titled, ''Life Cycle Assessment: Inventory Guidelines and Principles.'' This guide describes the first segment of the life cycle analysis, the assessment. Two other segments, impact analysis and improvement analysis, will be dealt with in subsequent guidance documents.

2.3   MANAGEMENT PLAN FOR HAZARDOUS MATERIAL AND WASTE

The baseline for environmentally conscious manufacturing is compliance with applicable regulations. Similarly, the basis of compliance is developing and implementing a management plan for hazardous material and waste. There are two reasons for developing such plans: They are required under various regulations, and they are an important part of a company's strategic business plan.

The requirements of RCRA, CERCLA, EPCRA, and CWA all include some form of planning, with considerable overlap of the planning components among the various regulations. To eliminate some of the overlap, a composite of the various plans can be developed into a ''super plan.'' For the purposes of this discussion, this composite plan will be referred to as the management plan.

Unfortunately, management plans are often viewed as little more than regulatory busywork. As a result, these plans frequently suffer from three main problems: they are incomplete; they are not kept up to date or reviewed periodically; or, they are prepared but never implemented. During a regulatory inspection, agency personnel typically will not understand all the intricacies of a manufacturing operation, but they do understand what paperwork and plans are required. Not having an up-to-date management plan is easy to spot and easy to cite in a report or notice of violation. Citing a facility for an outdated plan also allows the agency to show it is enforcing the regulations.

Just meeting regulatory requirements, however, should not be the only reason for preparing a management plan. Having a plan helps limit the potential for a disaster and reduce the effects of a disaster if one does occur. Conversely, the effects of not having an adequate management plan are well understood by companies that have had to deal with the aftermath of a disaster. What would Union Carbide have done differently if it could have turned back the clock on Bhopal? Would a comprehensive management plan have made a difference? Could the leak of methyl isocyanate have been prevented? If not, could the surrounding residents have been notified and evacuated quicker? Hindsight notwithstanding, many of the consequences of this disaster could have been avoided with a comprehensive management plan.

A management plan also reflects a company's corporate philosophy. If the philosophy is to meet minimum requirements, just to get by, then that is what a management plan will reflect. A company must always balance the current cost of "overcompliance" with the risk of catastrophic losses later. Even a potentially well-reasoned balance can prove wrong. A major oil company chose to provide the minimum spill containment imposed by regulation. However, the release of 3.8 million gallons of oil was more than the containment could hold. The 700,000 gallons that escaped triggered a $2.25 million federal fine and a $4.6 million settlement with the state. In addition to the fine and settlement, the company incurred over $11 million in cleanup costs, $5.25 million in legal fees, and civil damages and penalties of nearly $10 million.

If a company's ideology is to go beyond just meeting the requirements and reduce a company's overall effects on the environment, then that ideology should be embodied in the management plan. In the case of such companies, the management plan should be considered as an integral part of the company's business plan. As in any business plan, the environmental management plan investigates the requirements of doing business; determines the cost, risk, and effects of various options; and then makes decisions that help the company realize its goals. These goals can include maximizing market share, improving customer satisfaction, achieving higher profits, as well as the more traditional objective of reducing environmental impact.

The components of the management plan, discussed below, include the following:

Contingency plan.

Preparedness and prevention.

Facility inspections.

Operating log.

Waste management.

Security.

Personnel training.

Spill Prevention and countermeasures plan.

The regulatory element of each component, which provides the minimum requirements, is discussed first followed by a strategy for each component that can be adopted by a company seeking to develop an environmentally conscious posture.

Contingency plan.   Those facilities that manage hazardous materials or wastes are required to prepare and implement emergency preparedness and response procedures. The management plan is intended to provide information relating to the immediate measures and procedures that will be implemented during an emergency to minimize the potential hazards to human health and the environment. Such emergencies could include: accidental chemical releases to the land, air, or water; floods; fires; tornadoes; or explosions.

A management plan includes the following information about a facility: the facility description, emergency notification procedures, telephone numbers of agencies and response personnel, and lists and locations of equipment that would be useful during emergency response activities. The facility map should show exits, gates, processing and storage areas, and emergency equipment (such as fire extinguishers, eye washes, and showers).

Facility description.   The facility description provides general information on the facility layout and operations for use by the facility's emergency response team, regulatory personnel, and outside emergency response groups including police, fire, and ambulance personnel. The brief description should include the facility location, types of manufacturing operations in use, general information on the hazardous materials and wastes typically stored on site, and utility providers. Because the user of the management plan may not be familiar with the facility, the description should be written with this frame of reference in mind; acronyms, slang, and generally unfamiliar terms should be avoided.

The facility layout and site maps should show major points of interest that would be important to locate during an emergency. The facility layout should include the locations of the following items:

- Exits, gates, and emergency evacuation routes.
- Emergency communication equipment.
- Major production areas.
- Raw material and waste storage areas.
- Fire extinguishers, hydrants, and hoses.
- Eye washes, showers, and first aid stations.
- Spill containment, cleanup materials, and salvage drums.

- Electrical transformers and major switch boxes.
- Natural gas, propane, gasoline, and fuel oil lines and shutoff valves.
- Stormwater and sanitary sewer lines and drains.
- Water lines and shutoff valves.

The site map should include the locations of the following items:

- All buildings and floors on the site.
- Roads and railroad tracks.
- Property and fence lines.
- Stormwater sewers and drainage ditches.
- Nearby bodies of water.
- Fire hydrants.

Immediate response. Immediate response information should include immediate response procedures, emergency phone numbers, and an emergency response notification form. In the event of an emergency involving hazardous materials, the immediate response procedure provides the individuals involved with a simple, step-by-step plan for the initial response. Ideally the emergency coordinator should be responsible for implementing the immediate response procedure, but other individuals such as security, facility management, and/or supervisors should be familiar with the details of the procedure. A sample immediate response procedure is included in Appendix C.

During an emergency, time may be precious. The management plan should include a list of phone numbers of both internal and external resources that may be required in an emergency. Because release reporting is required under CERCLA, the Oil Pollution Prevention Act, SARA, and the CAA, a list of the appropriate agencies and guidelines as to when to call should be included in the management plan. Blank forms showing the different resources that should be included in a phone list are shown in Appendix C at the back of the book.

The emergency response notification form provides a framework for recording the details of an emergency. The information recorded can be useful after the fact in reconstructing the events leading up to the emergency and the response taken by the facility.

Emergency response procedures. Unlike the immediate response procedure, emergency response procedures are developed for specific events. Procedures need to be developed in the event of fire or explosions, floods, severe weather (tornadoes, hurricanes, high winds, etc.), spills or material releases to the environment, and spills contained within the facility. The emergency response procedures are intended to provide specific information to be used to quickly and safely affect the

control of an emergency situation. The items contained in the emergency response procedures include:

- Evacuation routes.
- Procedures for evacuation of and accounting for employees.
- Listings of safety equipment and communication devices.
- Emergency shutdown procedures.
- Specific contingency plans.

Personnel responsibilities. RCRA requires that one person, the emergency coordinator, assume charge during an emergency involving hazardous waste. The emergency coordinator must be familiar with both the facility and the management plan and must be given the authority to make decisions and carry them out under emergency conditions. The emergency coordinator can delegate responsibilities but still retains responsibility for the execution of any delegated tasks. Other facility personnel that are identified as part of an emergency response team are also required to be familiar with emergency response plans and have the necessary authority and resources. The responsibilities of the emergency coordinator are outlined in Appendix C.

Preparedness and prevention. The preparedness and prevention section of a management plan can also be considered as a risk reduction plan. The goals of this section are to identify potential risks to human health and the environment, minimize the potential occurrence of accidents or emergencies, and have plans and procedures available to minimize the impact should an event or emergency occur. The implementation of a preparedness and prevention plan will also assist the facility in maintaining compliance with regulations relating to protection of the groundwater. The topics covered under this section include:

- Role of the environmental coordinator and others.
- Setting up agreements with emergency response contractors.
- Verification of local agreements.
- Conducting and documenting periodic emergency drills.
- Listing of incompatible materials.
- Listing of hazardous materials.
- Security plan.
- General housekeeping.

The environmental coordinator is responsible for developing and implementing the facility preparedness and prevention plan, which encompasses a number of areas and may require the involvement of several facility staff and employees. For this reason, it is recommended that the environmental coordinator assemble a preparedness, prevention and emergency (PPE) committee whose primary responsibility would be to develop, implement, and maintain the preparedness and prevention plans.

The PPE committee is an integral part of successful hazardous materials management. This committee, under the guidance of the emergency coordinator, develops and ensures the management plan is correct and up to date. First and foremost, the PPE committee, in conjunction with the emergency coordinator, is responsible for raising the awareness of facility employees about the dangers of hazardous materials and wastes, the importance of environmentally safe work habits, and the company's commitment to protection of the environment.

The PPE committee will develop procedures to avoid potential releases of hazardous materials and also ensure that the facility is prepared to respond to such incidents when they do occur. The committee also will develop contingency plans for those areas where the potential for fire, explosion, or spills of hazardous materials exists and where releases could pose a threat to human health or the environment. Also the committee is responsible for ensuring that the materials management plan is accurate and complete at all times. Topics the committee might address include, but are not limited to:

1. Production and raw material changes where hazardous material handling procedures will require modifications or additions.

2. Training for new employees whose job descriptions require that they be exposed to or handle hazardous materials.

3. Upgrading of training programs and annual refresher training for existing employees.

4. Inspection and security measures for sensitive plant areas.

5. Adequacy of communication, safety, and emergency response equipment, effectiveness of practice drills, and current personnel responsibilities.

6. New federal, state, or local regulations, revised facility policies and procedures, and any other issues that may have a bearing on the effectiveness of the plan.

Members of the PPE committee should meet after any spill, fire, explosion, or other accidental discharge to evaluate the incident and countermeasures used and to make appropriate changes to the plan, if necessary or if required by regulation.

Other information addressed in the preparedness and prevention plan includes a description of prior arrangements with outside contractors and responding agencies, as well as the results of emergency response drills and equipment testing and maintenance, which should be reviewed periodically by the PPE committee. The documentation of prior arrangements with outside contractors should include the services contracted for, the contract period, response time, and any contract limitations. Prior arrangements with local community services such as fire, police, and hospitals, should also be verified and documented. Documentation and verification can be accomplished by sending a copy of the management plan to the outside contractors and agencies either as a registered letter or by including a letter of verification. The letter of verification requests that the specified sections of the management plan involving the contractor or agency be reviewed and that a copy of the letter be signed and returned acknowledging receipt of the management plan. The verification letter should also offer to meet to further discuss the management plan and to conduct a facility tour. A sample letter of verification is included in Appendix C.

In preparing a preparedness and prevention plan, one area that is often overlooked is housekeeping. Sloppy working conditions lead to small spills, which eventually add up and become indistinguishable from a large spill. In some instances, regulators have questioned the nature of an area based on visual staining or stressed vegetation. Subsequent sampling and analysis can indicate high contamination levels, triggering costly remediation. Attention to housekeeping should extend throughout the property. Large quantities of empty barrels, scrap materials, and general trash on facility grounds can indicate to regulators or neighbors the type of housekeeping that goes on within the plant.

Facility inspections. Routine inspection procedures are recommended as an aid to monitor for equipment deterioration or malfunction, errors, leaks, accidental spills, or other situations that could result in the release of hazardous material into the workplace or environment. Scheduled and documented inspections should be conducted for areas where hazardous wastes are generated, accumulated, stored, treated, or disposed of, and also for hazardous raw material storage and usage areas. Some examples of areas requiring inspection include:

- Container storage areas.
- Operating and structural equipment.
- Security devices.
- Safety and emergency response equipment.

The frequency of inspection of each area depends upon several factors including the type, age, history, and use of the item being inspected, as well as applicable federal, state, and local regulations. Areas that have a high potential for the release of a hazardous material may require weekly or even daily inspection. Other areas, such as a loading dock, may require inspection only after usage. Personnel who are assigned inspection tasks must be properly trained in inspection procedures and applicable regulations. They must also know who to contact if any emergency response, corrective actions, or repairs are necessary.

Typical problems encountered with each item of inspection are included on the inspection form to aid the inspector and to ensure a complete and thorough inspection. The inspector is required to check the status of each item and indicate the condition of the equipment. When problems such as the deterioration of equipment or spills are noted, the inspector must take appropriate actions to promptly correct the situation. If this is not possible, the inspector must immediately contact the individual who has the related responsibility. All corrective actions will be recorded on the appropriate inspection record and in the operating log. Following completion of an inspection, the inspector must submit the form to the appropriate area as designated on the form. The completed form is then entered into the inspection file. Written records of all inspections must be retained a minimum of three years; however, it is recommended that such documentation be retained at the facility indefinitely so that it is available for review by authorized regulatory agencies as required.

Operating log.   The operating log is a written record of a facility's hazardous waste and other environmental activities to assist the environmental coordinator in managing and monitoring these activities. These logs should be maintained on site until facility closure. The specific type of information to be included in these logs will depend upon the facility's operations but may include the following:

- Quantities of hazardous wastes sent off site for recycling, reclamation, storage, treatment, and/or disposal.
- Results of waste analyses.
- Descriptions of incidents that require implementation of the emergency plan, including any summary reports.
- A description of other environmental incidents.
- Generator reports.
- Inspection records.
- Permit summaries.
- Any citations, notices of violation (NOVs), and resulting corrective actions taken.
- Monitoring requirements established to prevent future violations.
- Any other activities related to environmental regulations and compliance.

The information summarized should be reviewed periodically to ensure compliance with regulatory requirements applicable to facility operations. Information provided in an operating log also can be useful in completing biennial or other reporting requirements to EPA, state, and local authorities. For records not physically located in this plan, notations may be made on the subsequent operating logs regarding their locations.

Waste management.   The waste management section of the management plan is designed to provide procedures for correctly categorizing, handling, and dis-posing of waste materials, including solid or containerized wastes, in a manner that is consistent with corporate environmental guidelines and in compliance with all applicable federal, state, and local regulations. Procedures should additionally take into account methods that will aid in reducing potential short-term and long-term liabilities that can occur from improper or careless management of both hazardous and nonhazardous waste materials.

Because the law imposes ultimate responsibility on the generator of wastes, facilities should strive to employ the following waste management practices and to include these practices as part of the Waste Management section of the management plan:

- Compliance at all times with applicable federal, state, and local regulations pertaining to the handling and disposal of industrial wastes.
- Proper analyses of waste materials before on-site or off-site treatment, storage, or disposal for the protection of human health and the environment and the reduction of potential liabilities associated with their final disposition.
- Implementation of improved materials and manufacturing processes wherever possible to reduce and/or eliminate the use of toxic materials and generation of wastes.
- The use of only permitted, reliable transporters and treatment and disposal facilities to transport, handle, process, and dispose of all waste materials.

These guidelines should be revised as necessary to reflect changes in government regulations and facility operations, procedures, and conditions.

Security.   Security measures should be outlined in the management plan that will prevent unauthorized persons and livestock from entering areas where hazardous materials and wastes are located. Examples of such security measures include:

- Physical barriers such as fences and gates and buildings that are locked or otherwise secured.
- Communication systems.
- Guards and/or electronic security systems

Personnel training.   A complete and comprehensive training program for all employees is a critical step in both preventing and effectively responding to an emergency situation at a facility. It is the responsibility of each employee to successfully complete each training program and understand the importance of performing each job function in a safe and environmentally responsible manner. The type and intensity of worker training will vary depending on the position and work responsibilities of the employee. However, at a minimum, each employee *must* understand safe handling techniques for hazardous materials and proper response procedures to emergency situations.

A number of training requirements are mandated under several federal regulations; however, it is also important to remember that, in many cases, individual state programs can require training under additional or more stringent regulations. Therefore, it is recommended that employee training be provided, at a minimum, at the time of initial employment and on an annual basis thereafter. Documentation of each training session, including a log of the training session conducted, the date and time of training, the information discussed, and the personnel in attendance, should be retained in facility files permanently. A copy of any certificates of completion issued to employees should also be retained on file at the facility. Further, it is recommended that the facility prepare a list of job descriptions and associated required training programs to assist in monitoring employee training activities. Also, personnel training records should accompany employees transferred within the company.

All training programs should be conducted by individuals knowledgeable of the training program requirements and current regulations. Each program must be designed and taught to ensure that employees understand their responsibilities in each area. The types of training programs that should be provided include, but are not limited to:

- Hazardous materials and wastes that employees may be exposed to, the associated physical and health hazards, and proper handling and disposal procedures.
- Procedures for using, inspecting, repairing, and replacing emergency equipment.
- Preparedness and prevention plans and procedures.
- Critical operation and facility shutdown procedures.
- Proper response to spills, fires, explosions, and other emergencies.
- Proper use of safety and emergency equipment.
- General information to increase individual awareness of health and safety and environmental regulations pertaining to work environments.

Note that the above requirements should not be considered complete; depending on specific facility operations and individual state regulations, additional training requirements may be applicable. Also, pending and future federal and state regulations may require additional training for proper employee awareness and performance of duties.

Spill prevention and countermeasures plan. In accordance with Title 40 CFR, Part 112 of the Clean Water Act, owners or operators of facilities that have discharged or could reasonably be expected to discharge oil in harmful quantities into navigable waters of the United States or adjoining shoreline are required to prepare a spill prevention, control and countermeasures (SPCC) plan. These regulations were revised with the passage of the 1990 Oil Spill Prevention Act. This act was enacted in response to the Exxon Valdez oil spill in Alaska. Facilities are subject to these regulations if:

- The facility's underground storage capacity of oil is greater than 42,000 gallons, OR
- The cumulative aboveground storage of oil is greater than 1,320 gallons, OR
- A single aboveground container of oil has a capacity greater than 660 gallons.

An SPCC plan must be certified by a professional engineer and must include, at a minimum, a discussion of the following:

- Facility drainage.
- Bulk storage tanks.
- Facility transfer operations.
- Tank can and tank truck loading/unloading.
- Inspections and inspection records.
- Security.
- Personnel training and spill prevention procedures.

Some states have additional SPCC requirements that are more stringent than federal requirements. Refer to your local environmental control authority if you have questions.

2.4   COMPLIANCE ASSESSMENTS

So far the discussion of compliance with environmental regulations has centered on actions at the facility level. However, some compliance programs are best focused at the corporate level. Compliance assessments provide a function similar to financial audits. Financial audits determine if the company is being put at risk through mismanagement or fraud. Before embarking on a compliance assessment program, it is important that upper management has committed the resources necessary to resolving those problems found. Once areas of noncompliance are identified, the company is obligated to resolve them. While ignorance has never been a defense, indifference and cover-up of environmental noncompliance easily could lead to more vigorous enforcement, including criminal prosecution.

2.4.1   *Staffing*

A compliance assessment program can be staffed in three ways, each method with its own strengths and weaknesses. Regardless of the makeup of the assessment staff, the depth of the review and the size of the facility will determine the size of the team and the amount of time spent conducting the assessment. As a rule of thumb, a 250,000-square-foot manufacturing facility should take three people between two and four days to conduct a thorough assessment.

The first assessment staffing method is to use the environmental coordinators and staff of each facility as an assessment pool, with a team assembled from the pool for each assessment. The main advantage with the assessor pool is that it provides a cross-training avenue for each environmental coordinator by conducting assessments of outside facilities. This arrangement also exposes inexperienced or smaller facility personnel to a wider background, allowing them to move into positions with additional responsibility at a faster rate.

The second type of staffing is to use outside resources such as consultants or nonregulatory state programs. Using outside staff is helpful when the environmental coordinator's schedule does not allow for his or her full participation. Outside resources can also provide an unbiased and sometimes brutally honest view of a facility's compliance program. In cases where it is suspected that noncompliance is being hidden, an outside consultant may provide the most accurate assessment of the situation. In addition, some states offer regulatory compliance assistance through technical assistance programs (TAPs). These programs are available either free or at a nominal charge, but they usually focus more on providing waste reduction assistance rather than information on environmental compliance.

The third staffing option is a hybrid of the previous two options. With this method, a consultant or other outside staff person is used to perform the bulk of the time-consuming assessment activities such as regulatory research, confirming any calculations or assumptions, and report writing. Internal personnel conduct the less time-consuming tasks such as facility walk-throughs and inspections and follow-up on the concerns identified. This option provides many of the benefits of the previous options at a reasonable cost. Depending on the arrangements made with a consultant, training for the environmental coordinators can also be included in the assessment.

Finally, careful thought should be given to conducting the assessment under the protection of the attorney privilege, especially if any significant areas of noncompliance are suspected. Despite its best efforts, a company may simply lack the time or the resources to correct all deficiencies immediately. Using outside counsel to provide legal advice on the status of compliance, under whose supervision the assessment would be conducted, may provide critical protection to the results of the corrections being implemented. Absent that protection, there is always the risk that the regulators may learn of the noncompliance before the correction, and use the assessment as "smoking gun" proof of the company's knowing violation of the law in subsequent criminal enforcement proceedings.

2.4.2 Assessment Process

Preparation for an assessment starts several weeks before the assessment visit and includes scheduling, background research, and preparation of the assessment survey form.

Scheduling includes working out the actual logistics of the assessment and notifying the facility's environmental coordinator about what the assessment will entail, which records should be reviewed, and a tentative schedule of events. The assessment team reviews state and local regulatory agency records on the facility before the assessment, including Form R filings, notices of violation, past inspection reports, and operating permits. Preparation of the assessment survey form involves updates to reflect regulatory changes and tailoring questions to reflect the results of the background search or known conditions at the facility.

The on-site portion of the assessment consists of a preassessment meeting, facility tour, records review, and a closing meeting. The preassessment meeting includes assessment staff, facility personnel, and upper management, if possible, and covers the assignments of each of the participants, the schedule, and the goals of the assessment. The facility tour should follow a logical progression through all areas of the facility, including receiving, storage, manufacturing, waste handling and disposal, and shipping. Where possible, the supervisors or engineering of a department or production area should be included in the tour to answer specific questions. The records review is the most time-consuming portion of the visit as assessment personnel examine the facility's management plan, waste shipping manifests, operating permits, analytical records, and emission reporting forms and backup data.

The closing meeting provides facility personnel with a summary of the findings and the next steps to be completed in the assessment process. This meeting also allows the facility to respond to any problems discovered during the assessment. The goal of the closing meeting is to ensure that there are no surprises in the final report. If agreement is reached on areas of noncompliance, a schedule should be developed at this time for bringing the facility into compliance. This schedule should be included in the final report.

Since many areas of environmental compliance are not black and white, some further regulatory research may have to be conducted to confirm the findings of the assessment and to clarify any questionable findings. Such interpretations underscore the desirability of protecting the entire process through attorney-client privilege. Further, interpretations used by the state and local regulators may need to be clarified or even challenged by legal counsel in order to make sure that the company fully understands all requirements.

Based on the findings of the on-site assessment and the regulatory research, a report is prepared and issued. A section of the report should be reserved for the facility's response to the concerns raised in the assessment and a timetable for bringing the facility into compliance with the items noted. Based on the compliance timetable, periodic follow-ups will confirm that the facility is meeting the agreed upon remedial actions.

Chapter Three

Integrating Environmentally Conscious Manufacturing within the Company

Historically, environmental management at a manufacturing facility was considered as an afterthought. Waste and emission treatment technology was incorporated as an end-of-pipe design, rarely as an integral part of the manufacturing operation. The end-of-pipe reaction to environmental control only encouraged more environmental control as each new treatment method failed to provide a permanent solution to harmful wastes and emissions. As an example, industry installed scrubbers to meet Clean Air Act requirements that specified emissions be controlled. However, the scrubber created a water discharge that was regulated under the Clean Water Act. Again, industry reacted and installed wastewater treatment, the sludge from which was regulated under RCRA. Industry complied by disposing of the sludge in a landfill. However, when the landfill eventually leaked, CERCLA was enacted to regulate the cleanup. And the cleanup method selected, incineration, started the whole process over again.

At the plant level, the environmental manager of the 80s could also be found to wear the hat of safety coordinator, personnel director, facilities maintenance, or production line supervisor, with additional environmental responsibilities tacked on to other duties. On-the-job training was the normal course for developing an environmental manager. Overworked and undertrained managers led to interdepartment feuds and missed communication, which contributed to the problem of inadequate environmental management.

Despite these barriers in the past, today the costs and liabilities associated with environmental compliance necessitate a proactive approach to environmental management. This approach requires that the concept of environmentally conscious manufacturing be integrated throughout a company's management structure. Integrating ECM into a company's operations involves a two-step process. The first step is to create a supportive management structure within the company for proactive environmental management. The second step is to use that supportive environment to establish and communicate an environmentally conscious business plan.

3.1 ESTABLISHING A PROACTIVE CORPORATE MANAGEMENT STRUCTURE

Much of the discussions involving how to implement pollution prevention, waste reduction, or design for the environment include the use of "cross-functional resources." What are cross-functional resources? What can someone not trained in environmental management or engineering do to improve the environmental performance of the company?

The use of cross-functional resources assumes that no individual has the knowledge or resources available to implement changes that cut across all levels of a company. Therefore, the company uses people with talents in specific areas to most efficiently achieve the desired results. A cross-functional environmental management team is a dynamic group. Not all individuals' talents are needed throughout a project. Therefore, people come into the group and leave as necessary. This keeps the number of people involved to a minimum, making the group more efficient and freeing up group members for other tasks.

3.1.1 Management Resources

With the past emphasis on pollution control, the responsibility of environment management and problem solving was placed at the facility level. The facility's environmental performance was judged on compliance, with qualitative gauges being the ability of the facility to stay out of the regulators' sights. Quantitative compliance goals were limited to the cost of compliance, treatment costs, disposal costs, and fines and penalties. The resources needed to maintain compliance were adequate funding for wastewater treatment systems, air scrubbers, sludge driers, and other treatment technologies.

As pollution prevention became more important, measures of a facility's environmental performance were expanded to include the amount of waste generated, pollutants emitted, and risk reduction activities. Much of the earliest gains in pollution prevention involved relatively easy changes in procedures, housekeeping, and waste segregation. More aggressive pollution prevention efforts involved process updates with existing technology. The resources required were available again at the facility level.

New concepts in pollution prevention, design for the environment (DFE), total quality environmental management (TQEM), and life cycle analysis (LCA) invoke a much larger scope, a scope that cannot be addressed at the facility level. These new concepts require that companies evaluate the environmental friendliness of the product. Product "enviro-soul searching" can consider the raw materials used, the manufacturing process, energy requirements, packaging issues, recyclability, and the environmental value of the product itself. These issues cannot be addressed at the facility level; they require input from marketing, research and development, and process engineering, as well as manufacturing. The resources and talent necessary to address all these issues are typically found not at the facility level but at the divisional or corporate levels of a company. To truly adopt an environmentally friendly approach to manufacturing requires not only developing the goals to reach that level of environmental performance, but also providing the tools and resources necessary to achieve these goals, which is the responsibility of the senior management.

3.1.2 Corporate Environmental Staff

Everyone agrees that an environmental coordinator is needed at the facility level. However, facility coordinators are often too busy with their day-to-day functions to keep abreast of all areas within the "environmental universe." Corporate environmental staff can weed through the regulations and guidance documents and direct a "predigested" package to the affected facilities. In addition to information dissemination, corporate environmental staff members also serve a planning function within an environmental program. The planning function of the corporate staff allows facilities to proactively and efficiently plan for regulatory and other changes rather than constantly reacting. The corporate staff also provides consistency within a company. Because they are not under the thumb of manufacturing, short-term crises do not lead to long-term mistakes.

3.1.3 Purchasing

The purchasing department is often used as a checkpoint to ensure that new chemicals are not allowed into a facility without proper documentation. This role can be expanded to include a review of the chemical constituents in new products purchased by a company. The purchasing staff does not necessarily need to become experts on chemistry or toxicology, but the staff should serve as the gatekeeper, passing a new chemical on to a review group or checking the constituents against a company's toxic hit list. Purchasing can also develop guidelines to encourage the purchase of recycled materials, to control inventories, and minimize non-returnable packaging including drums.

3.1.4 Accounting

Pollution prevention and cost savings are often interrelated. However, because environmental management costs are buried in a lump sum budget, the potential cost-

Case Study

3M

In addition to the goals of production, sales, and product development, 3M's divisional managements also have environmental goals. These goals are reviewed along with the more typical goals and are used to determine the division's performance. Many of the early pollution prevention gains achieved through relatively easy changes such as improved operations, housekeeping, and modest process modifications have already been implemented. 3M now realizes that the next improvements in environmental performance will come through process and product redesign. For management's performance to be fairly judged, resources must be made available to make the necessary changes. Developing environmental goals at the divisional level places the necessary resources from marketing, research and development, and manufacturing at a manager's disposal.

---

savings are not easily identified. Breaking out environmental management costs, including those associated with permit modifications, hazardous material storage, or other environmental services, and charging these costs back to the departments responsible for generating the waste, educates managers on the financial impacts of their operations.

3.1.5 *Finance*

As a company begins to investigate more far-reaching pollution prevention projects that require intensive capital investment, the financial arm of the company will play a larger role. All companies have some method of determining the financial attractiveness of a proposed project. Larger corporations typically use an involved process of capital requests that includes the preparation of worksheets containing the information required for upper management to make a decision. This capital request process needs to include an environmental worksheet to demonstrate to upper management that the environmental ramifications have been considered and minimized. This environmental justification worksheet should include:

- The raw materials (highlighting recycled materials) to be used.
- Energy usage and energy saving gains made.
- Waste and emissions generated.
- Anticipated environmental management costs including disposal, permitting, training, and storage costs.
- Future regulatory changes that may affect the project such as threshold limit changes or restrictions on disposal methods.
- Environmental risk and steps taken to reduce these risks.
- Justification as to how this project meets the company's environmental goals.
- Alternatives and their environmental impact and costs.

A sustainable pollution prevention program has to be cost-effective. The environmental justification worksheet tries to quantify the environmental management costs associated with a project; however, environmental risk is difficult to quantify. A less rigorous method of accounting for risk is to assign a lower standard for capital approval for projects that meet environmental management goals.

As an example, a company requires a two-year payback on capital expenditures over $50,000. For projects that meet specified goals such as the elimination of a hazardous waste stream, an additional six months of payback is allowed. The final project justification would have to meet a 2.5-year payback. The less rigorous payback is a method of quantitatively accounting for the risk associated with off-site disposal of a hazardous waste.

The goals used to offer a lower capital justification should require important milestones in environmental management. The example above requires that a waste stream be eliminated, not just reduced. Any hazardous waste stream, even a relatively small waste stream, carries a potentially high environmental risk.

3.1.6 *Research and Development*

As stated at the beginning of Part II of this book, the fourth and final tier in an environmental management program requires the development and use of innovative technology to achieve a company's long-term goals of environmentally friendly products and processes. (This subject will be dealt with more thoroughly in Chapter Six.) Successful implementation of new technology requires access to new developments and changes in technology. Companies such as Dow and IBM provide access to technology in different ways. IBM maintains a company "expert list." The names of the top individuals within each field are distributed throughout the company. If a facility needs help or advice outside of its

Case Study

Frost Paint

Many experts disagree whether monetary awards provide long-term incentives for various types of company programs. Frost Paint has used two types of structured monetary awards programs with vastly different results. The first program was designed to reduce manufacturing costs. The company offered to share a portion of the savings from each idea that was implemented with the employee that submitted the idea. While the program had several good success stories, it also created several problems. Some ideas were rejected from the incentive program because they already were in the process of planning and implementation. Communication between employees was reduced over the fear that a fellow employee would "steal" an idea.

Despite these problems, the company again decided to add a monetary incentive to its waste reduction program, only this time with a twist. The mone-

tary incentive for the waste reduction program did not provide a direct payback. Any savings was added to a waste reduction pool, and two-thirds of the pool was shared among all employees. The change was dramatic. At the end of the program's first year, the company had reduced hazardous waste generation by 55 percent. Savings associated with the project amounted to over $25,000. But the real benefits came in the form of employee involvement. Instead of jealously hoarding ideas, recommendations were freely shared among employees. Often one idea would spark an improvement or a whole new idea. In addition to the monetary rewards, employee recognition was also given for individual contributions. The biggest recognition however came from the State of Minnesota, which presented Frost Paint with the state's annual Governor's Award for Pollution Prevention.

---

own pool of talent, a company expert can be contacted. Dow takes a more formal approach through its technology centers. These technology centers focus on the specific manufacturing process technologies used within their divisions. All process changes and improvements are directed through these technology centers for testing and approval.

3.1.7 *Incentives*

Changing a company's focus toward environmentally friendly products and processes requires the commitment and participation of the entire company. To sustain this activity, employee contributions need to be rewarded. While the improvement in a company's environmental management program provides satisfaction for everyone involved, outstanding individual efforts should be recognized. Such recognition can be given through awards, praise at special functions, or a simple thanks from an employee's manager. Monetary awards tend to provide short-term gains but can be effective if structured properly, as the accompanying case study illustrates.

3.1.8 *The CERES Principles*

In partial response to the Valdez oil spill in Alaska, the Coalition for Environmentally Responsible Economics (CERES) established what is now known as the CERES

Principles. The principles were designed to be used as the foundation upon which a company's environmental program could be built. Signatories to the principles are required to complete and submit an annual audit to CERES. This audit is subject to review by outside authorities.

Because of the constraints placed upon signatories to the principles, industry has not warmly embraced the movement. However, the individual points provided by the principles can be used as a starting place for the development of a company's own set of principles.

Introduction

By adopting these principles, we publicly affirm our belief that corporations have a responsibility for the environment and must conduct all aspects of their business as responsible stewards of the environment by operating in a manner that protects the ability of future generations to sustain themselves.

We will update our practices constantly in light of advances in technology and new understandings in health and environmental science. In collaboration with CERES, we will promote a dynamic process to ensure that the principles are interpreted in a way that accommodates changing technologies and environmental realities. We intend to make consistent, measurable progress in implementing these principles and to apply them to all aspects of our operations throughout the world.

Protection of the biosphere.   We will reduce and make continual progress toward eliminating the release of any substance that may cause environmental damage to the air, water, or the Earth or its inhabitants. We will safeguard all habitats affected by our operations and will protect open spaces and wilderness, while preserving biodiversity.

Sustainable use of natural resources.   We will make sustainable use of renewable natural resources, such as water, soils, and forests. We will conserve nonrenewable natural resources through efficient use and careful planning.

Reduction and disposal of waste.   We will reduce and where possible eliminate the waste through source reduction and recycling. All waste will be handled and disposed of through safe and responsible methods.

Energy conservation.   We will conserve energy and improve the energy efficiency of our internal operations and the goods and services we sell. We will make every effort to use environmentally safe and sustainable energy sources.

Risk reduction.   We will strive to minimize the environmental, health, and safety risks to our employees and the communities in which we operate through safe technologies, facilities, and operating procedures, and by being prepared for emergencies.

Safe products and services.   We will reduce and where possible eliminate the use, manufacture, or sale of products and services that cause environmental damage or health or safety hazards. We will inform our customers of the environmental impacts of our products or services and try to correct unsafe use.

Environmental restoration.   We will promptly and responsibly correct conditions we have caused that endanger health, safety, or the environment. To the extent feasible, we will redress injuries we have caused to persons or damage we have caused to the environment and will restore the environment.

Informing the public.   We will inform in a timely manner everyone who may be affected by conditions caused by our company that might endanger health, safety, or the environment. We will regularly seek advice and counsel through dialogue with persons in communities near our facilities. We will not take any action against employees for reporting dangerous accidents or conditions to management or to appropriate authorities.

Management commitment.   We will implement these principles and sustain a process that ensures the board of directors and chief executive officer are fully responsible for environmental policy. In selecting our board of directors, we will consider demonstrated environmental commitment as a factor.

Audits and reports.   We will conduct an annual self-evaluation of our progress in implementing these principles. We will support the timely creation of generally accepted environmental audit procedures. We will annually complete the CERES Report, which will be made available to the public.

Disclaimer.   These principles establish an environmental ethic with criteria by which investors and others can assess the environmental performance of companies. Companies that sign these principles pledge to go voluntarily beyond the requirements of the law. These principles are not intended to create new legal liabilities, expand existing rights or obligations, waive legal defenses, or otherwise affect the legal position of any signatory company, and are not intended to be used against a signatory in any legal proceeding, for any purpose.

3.2   DEVELOPING AN ENVIRONMENTAL BUSINESS PLAN

Once a proactive corporate structure is in place, business planning based on environmentally conscious manufacturing can begin. Most successful companies rely on a long-term business plan to ensure their future growth and success. Developing a business plan requires that a company identify potential markets, the competition's vulnerability, and the company's strengths. Specific performance goals are set based on the potential market identified. This information is then used to devise marketing, R&D, and manufacturing strategies that will achieve the company's goals.

The same approach can be used to develop an environmental management program using an environmental business plan. Such a plan focuses on developing and capitalizing on a company's good environmental performance. Good environmental performance can be defined as complying with environmental regulations, practicing pollution prevention, or using low-impact product and/or packaging design.

Of these, pollution prevention activity is coming into its own as an indicator of a company's environmental business sense. The federal government has become a strong promoter of pollution prevention, as seen in the latest amendments and new legislative packages coming out of Washington. At the state level, pollution prevention has taken on a more stand-alone guise. Twenty-seven states have enacted some form of pollution prevention legislation, with the most effective legislative packages requiring manufacturing facilities to prepare pollution prevention plans. In preparing a plan, a company develops and investigates options for preventing the generation of wastes or the release of pollutants. The plan must address a facility's operations during a specified time period, and it must be updated periodically. Common elements of a pollution prevention plan include the following:

- Policy statement.
- Description of current waste- or emission-generating processes.
- Description of current and past practices used to eliminate or reduce the generation of toxic pollutants.
- Identification of pollution prevention options that are both economically and technically practical.
- Pollution prevention objectives with numeric goals.
- Rationale used to develop each goal.
- Options that are not economically or technically feasible.

The pollution prevention plans required by the states suffer from one major flaw—they address only a specific or limited number of wastes or emissions. State planning requirements may focus only on hazardous wastes or Form R emissions, but rarely both. Nonhazardous or industrial waste streams are also conspicuously absent. The pollution prevention plan of an environmentally conscious company addresses all facets of the company's operations and wastes. To avoid confusion with the pollution prevention plans, the phrase "environmental business plan" was coined to differentiate the scope of the two plans. While the scope is different, the components are essentially the same.

3.2.1 *Policy Statement*

The policy statement required by most state pollution prevention plans can be satisfied by a weak, "feel good" statement like the following:

> It is the policy of the Minnesota Beargrease Company to eliminate or reduce at the source of generation, recycle or reuse on site all materials that result in the generation or release of hazardous materials, toxic pollutants, and industrial wastes to the environment, where such activities are judged to be both technically and economically feasible.

Great, but how is this going to get done? What are the responsibilities of each department in implementing this policy statement? How are the day-to-day actions of individual employees impacted by this policy statement? How is an environmentally conscious consumer of beargrease going to interpret this policy statement?

The above statement is merely a good start. In addition to the vague goals stated, the policy should encourage individual divisions, departments, and groups within the company to develop their own versions of this statement as it relates to their own areas. For example, the package engineering department might develop its own policy statement:

> It is the policy of the packaging engineering group of the Minnesota Beargrease Company to reduce the environmental impact of the packaging through the use of recycled/recyclable raw materials and to ensure that any packaging does not pose an unnecessary threat to the environment when managed or disposed. Specifications developed to date to meet this policy include:
>
> - Use nontoxic inks and pigments within its packaging.
>   —Toxic inks are defined as those that include the use of heavy metal (cadmium, mercury, chromium, or lead) or have a volatile content in excess of 2.8 pounds per gallon.
> - Use packaging material with the following specifications:
>   —Paper products: minimum of 50 percent, post-consumer, recycled content.
>   —Shipping pallets: use only collapsible, returnable plastic pallets.
>   —Product containers: recyclable aluminum.

Upper management, in developing a policy statement, sets the guidelines but cannot develop the specifics. Individual divisions, departments, and groups then have to take the corporate, fledgling policy and flesh it out. The specifics help employees in conducting their day-to-day tasks. The packaging engineer in charge of launching a new improved product, beargrease with aloe, is not going to ask, "Does the packaging meet the corporate environmental policy?" But the department's policy requiring nontoxic inks and recycled paper products will be incorporated in the final packaging design. These individual, specific policies and guidelines should be collected and consolidated into the environmental business plan.

3.2.2 *Generating Processes*

The description of the generating processes is a quick discussion of all the facility operations. This description includes the raw materials used, the end product, any by-products, and the waste and emissions generated. The description of the manufacturing operations helps to establish a baseline. Much of the waste and emissions data gathered for regulatory reporting is done on a facilitywide basis. While facilitywide data are useful in setting priorities or when looking at the overall performance, the details of a manufacturing operation are required in making operational, engineering, and process changes when dealing with the reduction of environmental impacts.

In many cases, baseline data have already been established under the state pollution prevention guidelines, such as reported volumes of RCRA wastes generated, SARA 313 emissions, or state-listed wastes generated or emitted. Since this information was compiled from specific operations, it is just a matter of breaking it back down into the individual components and including it in the description of the generating process. Where no baseline data exist, the facility must develop the data through an intensive process audit.

A process audit clearly shows how an operation really works. There is a surprising lack of understanding between the engineer that designs an operation and the operators that actually run it. Operators may have made minor changes to an operation that fixes one problem but may affect operations downstream. Conversely, the changes on one line may directly transfer to another line to solve the same problem. The process audit is simply a tool that makes engineers, supervisors, management, and operators sit down and discuss their manufacturing process so everyone is working with the same complete set of information.

3.2.3   *Current and Past Practices*

A description of current and past pollution prevention efforts is important for two reasons. First, such a description documents past activities in the event that conditions change or an idea resurfaces in the future. This helps prevent work from being repeated unnecessarily and provides the groundwork for feasibility studies of similar work. Too many times an idea is brought up by a new player only to be rebuffed with, "We tried that before and it didn't work." More often than not, what was tried and why it failed was not documented. With changing technology, an idea that did not work in the past might now work. The increasing cost of environmental management makes the economically infeasible options of the past more attractive today.

Second, this documentation is useful to facilities that have adopted a proactive approach to pollution prevention. One of industry's greatest fears of mandated pollution prevention is that if industry reduces pollution today, tomorrow the regulatory agencies will mandate an across-the-board cut regardless of past efforts. This has not been the case. When pollution prevention legislation at the state level has mandated a cut, a historical baseline is selected. The company's current waste and emission generation rates are then compared to this baseline. If a company instituted pollution prevention efforts before this baseline year, credit is given if the company has the documentation to prove its claims.

In documenting a company's pollution prevention efforts, make sure all efforts are recorded, not just efforts that deal with hazardous or toxic materials. Regulations change, and it is difficult to accurately predict where future regulation will go. Documenting all a facility's pollution prevention activities provides some protection for the future.

3.2.4   *Investigation of Options*

Investigating the available options is very similar to a brainstorming session. Ideas should be solicited from as many sources as is practical for the size of the project. Judgment on the feasibility of each idea should be made only after considering all aspects. In developing options, several areas should be considered: raw material substitutions, product design, waste segregation, employee training, housekeeping, process modifications, on-site recycling and reuse. This topic is covered in greater detail in Chapter Five.

3.2.5   *Feasibility*

Both the technical and economic feasibility of the options presented in the previous section are addressed as part of a pollution prevention plan. State pollution prevention planning requirements allow an option to be rejected on either technical or economic grounds. In considering the technical feasibility of an option, a number of details need to be addressed, including pollution prevention, product quality, production rate, risk, and worker health and safety.

In considering the economic feasibility, most companies have some method of evaluating a project's economics. Typical methods are return on investment (ROI), payback years, net savings, or internal rate of return (IRR). This issue is covered in Chapter Five.

3.2.6   *Development of Goals*

Several state pollution prevention laws require that the facility set reduction or elimination goals. Where a facility is not in a position to set firm, numeric goals, it must state how it will proceed in developing these goals. Such goals should be related to the current wastes and emissions generated or released as either a percentage reduction or as a reduction in the total volume or weight.

In considering these goals, it is important to remember that the amount of waste or emissions generated is often closely related to production. Where possible, the plan's objectives should be tied to production to prevent under- or overreporting of the progress made. It is also important to be realistic in developing these objectives and not set goals based solely on manufacturer or vendor claims, or best-case conditions. Start-up problems and "the real world" have a habit of reducing the actual progress made; these should be considered in setting objectives. Any calculations, assumptions, or rationale used in setting these objectives should be documented in the plan.

3.2.7   *Measurement of Progress*

As difficult as it is to measure the volume of waste and emissions generated, measuring the amount of progress made is even more difficult. Variations in production can

Case Study

Custom Painting Job Shop

The manager of a painting shop had just finished a lengthy air permit for a new facility. The permitting process was complicated because the current facility had been out of compliance by operating for a number of years without an air permit. In addition to the time spent preparing the permit application, the facility also had incurred legal, stack testing, and consulting costs as well as a small fine. The new air permit restricted the solvent content of the paints applied, capped the total volume of solvent emissions, and required the facility to keep accurate and detailed records of paint and solvent use.

Interested in finding out how the competition had fared in the permitting process, the manager had requested a list of all facilities in the state operating with an air permit. None of the competition had an air permit. Thinking that this was now an unfair position, the manager asked how could he now compete with operations that were in noncompliance with environmental regulations.

The plan presented to the plant manager was to use the fact that the company was operating in an environmentally sound manner as a positive selling point. A number of the company's clients were larger corporations that had adopted just-in-time (JIT) inventory and manufacturing methods. Since this placed considerable pressure on the painting shop and other suppliers to react to the customer's needs, a letter was drafted outlining the changes that had occurred recently. The letter stated the company had adopted a policy of complying with all environmental regulations. The company also had added additional emphasis to its research and development efforts to stay ahead of any future regulations. The stated purpose of the letter was to inform the company's customers that the changes were intended to make the company a more reliable supplier.

The response from the company's largest customer was:

> [We are] currently formulating a corporate pollution prevention plan and it will have an effect on our suppliers. It is nice to know that some suppliers are proactively anticipating these changes.

---

mask or swell any actual gains made. While a number of measures of production exist, they were generally developed to gauge production for different reasons. Several generally available measures include: sales, units produced, square footage produced, raw materials used, number of employees, and direct labor hours.

Selecting a method of measuring production to include in an environment business plan requires that the individual circumstances of the facility be considered. Each of the measures of production listed above contain inherent strengths and weaknesses. Sales figures are readily available; however, they vary with inflation, and changes in inventory also will affect the results. Units produced, square footage produced, and raw materials used are probably the best methods of relating production to wastes and emissions. However, for plants that produce a large variety of products, it can be difficult to produce accurate comparisons if the product mix changes. The number of employees or direct labor hours can help smooth out any variation in product mix but are affected by changes in productivity or automation.

3.2.8 *Progress Reports*

Most states that require pollution prevention planning also require progress reports or annual summaries. Of the various state pollution prevention planning requirements,

this is the area with the greatest amount of variability among the states. Components of several of the most comprehensive progress reports are assembled below:

- Summary of each objective.
- Summary of the progress made during the year, if any.
- Methods used to achieve this progress.
- An explanation of why the objectives were not achieved, if necessary.

Any explanation of why the objectives contained in the plan were not achieved should include a discussion of the barriers encountered. Such barriers could include technical, economic, and regulatory hurdles.

RESOURCES

Baker, Rachel D., and John Warren. "Management of the Product Life Cycle to Promote Pollution Prevention," *Pollution Prevention Review*, New York, Autumn 1991.

Callenbach, Ernst, et al. *EcoManagement, The Elmwood Guide to Ecological Auditing and Sustainable Business* (San Francisco: Berret-Koehler Publishers, 1993).

Huttner, Claudia. *Management Plan for Hazardous Material and Waste* (St. Paul, MN: Capsule Environmental Engineering, 1991).

"Life-Cycle Assessment: Tracking Impacts from Cradle-To-Grave," *Pollution Prevention News* (Washington, D.C.: EPA, July–August 1993), p. 9.

McClain, Wallis E., Jr., ed.   *U.S. Environmental Laws* (Washington, D.C.: The Bureau of National Affairs, Inc., 1991).

McIntyre, Melissa. *Comprehensive Hazardous Materials Management Plan—U.S. Edition* (St. Paul, MN: Capsule Environmental Engineering, 1993).

Minnesota Guide to Pollution Prevention, Minnesota Office of Waste Management (St. Paul, MN: March 1, 1991).

Sibik, Larry K. *The Minnesota Pollution Prevention Act. Requirements for Electroplaters and Surface Finishers,* November 5, 1990.

Sibik, Larry K. *Preparing a Pollution Prevention Plan* (Minneapolis: American Institute of Plant Engineers Conference, September 30, 1991).

State Legislation Relating to Pollution Prevention, compiled by Waste Reduction Institute for Training and Applications Research, Inc. (Minneapolis, MN: April 1991).

Chapter Four

Evaluating Existing Products and Processes

In Chapter Three we learned how to refocus a management structure to increase the level of environmental consciousness. Following this plan installs a team of environmentally aware personnel representing all major areas of company operations. The next step is for this team to identify more specifically where the company is, where it needs/wants to go, and how to get there. This may be done on a division-by-division basis, product-by-product basis, process-by-process basis, or facility-by-facility basis—whatever works best within the organization.

This chapter presents methods to determine an environmental baseline for a facility. Developing this baseline will lay the groundwork for all future environmental activities, and the process must be well thought out to avoid having to repeat this step. Bypassing this assessment step of the ECM process most likely will result in serious repercussions for a company's environmental program. Without an accurate, properly measured snapshot in time, it is impossible to demonstrate progress. As discussed elsewhere in this book, making environmental claims will be impossible without the ability to defend them with accurate data.

Several examples in this chapter will show different techniques used by firms to establish their environmental baseline. While there will be no perfect solution for any company, the discussion should help identify a workable measurement system for any type of operation or product line.

4.1   ESTABLISHING A BASELINE

Developing an accurate baseline helps you identify opportunities for greatest improvement. By knowing the 20 percent of products, plants, or processes that are causing 80 percent of your problems, you can establish a system of triage that is critical to accomplishing your goals. As items are crossed off the list, additional opportunities can be addressed.

Another reason to develop a baseline is to track progress. In 1988, the EPA began requiring reporting of chemical emissions by manufacturers. The initial years' reports have been used by the EPA and a large number of companies in documenting the progress they have made. This national reporting of emissions perhaps accomplished more than any command and control regulations because it got image-conscious top managers to insist on action in reducing discharge levels. Likewise, a corporate annual environmental report can highlight the most pressing issues for decision makers.

Developing a baseline takes time and money, and you may be tempted to skip this step and get on with identifying and implementing your ECM opportunities. In essence, this would be a continuation of the status quo. Every organization that has competition naturally seeks ways to be better by improving efficiencies, expanding capacity, reducing waste, and improving quality. Without a baseline and the resulting plan, however, such efforts are most likely unfocused and counterproductive.

Many companies attempt to respond to regulatory violations without first establishing a baseline and developing a plan. For example, a metal finisher was hit with a notice of violation for exceeding its wastewater discharge permit limits. While the plant had a number of different processes that contributed contaminants to the wastewater, the facility team made significant (and expensive) changes in one area where it knew contaminant reductions were possible. After changing the process, the violations still occurred. Consultants were brought in, and they measured all the contaminant sources in the plant and developed a mass balance for the contaminants of concern. This mass balance showed that most of the contaminants came from a process that plant personnel had not considered a significant source.

Facilities have difficulty stepping back and taking a fresh look at their operations. Spending all of your time in one facility can make you blind to new ideas. The baseline assessment, when done properly, requires that you gather the facts on the status of your operations. This information may be revealing if processes, plants, and departments that were not viewed as having any problems emerge with key concerns.

Anyone who has tried to measure a corporation's status on sensitive issues understands some of the inherent difficulties in identifying environmental problem areas within an operation. Any corporation or group is made up of individuals, all with egos and objectives that often make it very hard, if not impossible, to conduct an unbiased environmental assessment without alienating them. One way to soften the blow is to use an outside third party to assist in developing the corporate baseline. A consultant with an environmental background and a product and process background can be used to rate all of a corporation's operations based on priorities and rating methods established before the assessment activity.

For example, one corporation hired a consultant to conduct environmental compliance and simple pollution prevention assessments at all its 70 U.S. manufacturing operations. The work identified some divisions, facilities, and processes of special concern for the corporation. Following completion of the work, corporate management used the information collected to strongly encourage certain groups to "clean up their act." The consultant conducted follow-up assessments to determine if the requested improvements were made. Additional pressure was applied to anyone who was responsible for facilities or processes that remained a high concern. The program was highly successful for the corporation because it greatly reduced the environmental impact of all operations. In return, however, representatives of the consulting group were viewed by many with an ire previously reserved for EPA and OSHA inspectors.

While the assessment process identifies the environmental problem areas of a facility or corporation, it will also find that many of the operations are staffed with individuals that have been working for years to reduce the environmental impact of their processes. These operations will have fewer environmental compliance problems, be implementing continual product and process improvements, have good housekeeping, and have a satisfied work force. These operations will invariably be headed by an individual committed to harnessing the creativity and power of the employees. The goal of your efforts should be to develop this attitude throughout all

your facilities, product lines, or processes by taking the models that work and encouraging others to apply them to their operations. An effective plan should utilize a carrot and stick approach to achieve continuous improvement.

4.1.1 *Determining Evaluation Methods*

A wide variety of evaluation techniques can be used to establish a baseline. Developing the best evaluation method for a particular operation requires an understanding of the purpose of the evaluation. Overall corporate environmental objectives might include those such as the following:

- Reducing packaging materials by 50 percent since 1991.
- Reducing toxic emissions by 90 percent since 1988.
- Reducing hazardous process waste generation by 50 percent since 1987.
- Reducing wastes and emissions by 8 percent annually.
- Producing a totally recyclable product.

A model self-assessment program was developed by the environmentally conscious manufacturing group at the National Center for Manufacturing Sciences (NCMS) in Ann Arbor, Michigan. This program, titled "Achieving Environmental Excellence," is included in Appendix N and contains over 150 yes-no questions addressing 12 areas of excellence:

1. Policy.
2. Management.
3. Planning.
4. Cost.
5. Stakeholders.
6. Human resources.
7. Operations and facilities.
8. Suppliers.
9. Compliance.
10. Technology and research.
11. Auditing.
12. Measurement and continuous improvement.

Environmental programs can be patterned after existing quality programs, allowing them to be more easily incorporated into manufacturing operations. In 1990, the President's Council on Environmental Quality was formed to develop and test ideas for merging the goals of environmental quality and economic growth. This council published a framework for pollution prevention that included eight steps. The progress along these steps can also be a measurement tool for the success of an environmental program. These steps are:

1. Management commitment.
2. Quality action team.
3. Training.
4. Determine environmental impact.
5. Select improvement projects.
6. Implement improvement projects.
7. Measure the results.
8. Standardize the improvements.

The Global Environmental Management Initiative has published an environmental self-assessment program that allows corporations to evaluate the status of their programs and procedures. This program was developed based on the 16 principles of the Business Charter for Sustainable Development. These topics include:

1. Recognize environmental management as a top corporate priority.
2. Integrate environmental programs into each business.
3. Continually improve environmental programs.
4. Educate employees.
5. Assess environmental impacts before starting projects.
6. Minimize the impact of products and services.
7. Advise customers in the safe handling of products.
8. Operate facilities with minimal impact.
9. Research the environmental impacts of operations and ways to reduce these impacts.
10. Change processes to prevent serious environmental harm.
11. Promote improved environmental activities of contractors.
12. Prepare for emergencies.
13. Transfer environmentally sound technologies.
14. Contribute to public education and policy development.
15. Foster openness with employees and the public.
16. Measure and report environmental performance.

This assessment program can help to identify the weakest areas in an organization's environmental program and to focus future efforts on those areas. Conducting annual evaluations against these standards can help you determine the impact any changes are having to the overall corporate program.

Members of the Chemical Manufacturers Association (CMA) are evaluated by the standards of the Responsible Care® program. As of 1992, the 180 members of the program which accounted for more than 90 percent of U.S. productive capacity for basic industrial chemicals. Part of this program is a code of management practices, which includes six codes that address:

Pollution prevention.

Community awareness and emergency response.

Distribution.

Process safety.

Employee health and safety.

Product stewardship.

Within the pollution prevention code, CMA has outlined 14 management practices that provide a framework for reducing waste and releases to the environment and for managing the remaining materials. These practices are comparable to the methods outlined in this book.

1. Commit the organization.
2. Inventory wastes and releases.
3. Evaluate potential impacts.
4. Educate and listen to employees and the public.
5. Establish a reduction plan, goal, and priorities.
6. Implement the reduction plan.
7. Measure progress.
8. Communicate progress.
9. Integrate reduction concepts in planning.
10. Outreach.
11. Facility evaluation.
12. Review, select, and retain contractors and toll manufacturers.
13. Protect the groundwater.
14. Address past sites.

The International Organisation for Standards (ISO) is developing standards for environmental performance that may become universally accepted. This standard for environmental programs will likely be accepted worldwide. The ISO standard for quality (ISO 9000) has become of great importance to any company seeking to do business within the European Economic Community. While not yet finalized, the ISO 14000 environmental standard will likely incorporate items from existing standards such as the British BS 7750 standard and the Canadian Standards Association environmental management system standards. Companies should be aware of these standards and incorporate relevant items into their surveys.

The proposed British BS 7750 environmental standards address the following general issues:

Preparatory environmental review.

Environmental policies.

Organization and personnel.

Environmental effect evaluations.

Environmental objectives.

Environmental management program.

Policy documentation.

Operational control.

Compliance recordkeeping.

Environmental management audits.

Periodic management reviews.

The general methods presented in this book, based on the authors' experience, provide a model that should allow companies to readily incorporate the documentation and recordkeeping components of the ISO environmental management system standard. Companies that have gone through the ISO 9000 quality process have learned to focus on good communication, written procedures, and measurement and review systems. These fundamentals are also key to a sound environmental program.

Also, structured life cycle assessments will likely emerge as a measurement tool to define the environmental impact of a product or process. The life cycle assessment process is described in Chapter Six. A highly publicized life cycle assessment compares the environmental impacts of disposable diapers versus cloth diapers.

In summary, many existing evaluation methods are available for use within an organization. These methods all generally support the same fundamental principles that are key to having a proactive environmental management system. Each corporation must decide whether to adopt one of these programs outright or to modify them to meet specific operations, goals, and objectives.

4.1.2 *Defining Evaluation Goals*

As with successful quality programs, the best environmental programs require an attitude of "zero environmental defects." If a company accepts wastes, emissions, or permit violations as inevitable by-products of its operations, those problems will occur. This attitude of zero environmental defects must be understood and held as a tenet for the environmental assessment team. Every waste, every emission, every violation must be seen as a problem that should be an opportunity to reduce cost and risk.

The tenet of zero environmental defects requires that priorities be set when dealing with environmental problems. In the evaluation and implementation stages, decisions will be made to prioritize issues and direct energies. Those decisions should not be made until after the assessment activities are completed, when the scope and magnitude of all problems are understood. For example, Ingersoll-Rand realized in the early 1980s that the corporation was spending an increasing amount on hazardous waste disposal. To determine the magnitude of the problem, the company surveyed each facility to determine the amount and type of wastes generated, the processes generating these wastes, and the costs associated with waste management. The information gathered was combined with other available information on risks and liabilities and used to prioritize corporate environmental concerns.

These results ultimately led to directives for eliminating the use of underground storage tanks, PCBs, chlorinated solvents, and cyanides.

Consumer products manufacturers will want to take a baseline evaluation approach that can be seen and understood by their customers. Procter & Gamble has taken the approach to minimize the amount of packaging materials in response to national concerns on the filling of landfills. Kimberly-Clark has addressed the same concerns with its Huggies diapers by developing lower volume absorbent materials.

Some manufacturers are aggressively working toward totally recyclable products. If this is the desired goal, the assessment should determine what product components are not recyclable and then work to modify these items. Xerox used a similar approach and came up with a toner cartridge for laser printers and copiers that can be easily separated and recycled. These advancements contributed to Xerox being awarded the World Environment Center's 1993 Gold Medal for International Corporate Environmental Achievement. According to Paul A. Allaire, Xerox chairman and CEO, "Proactive environmental practices are good for the Earth and are good for business. It's part of being a total quality company."

The automobile industry is also using recyclability as a potential measurement area, though efforts have been hampered by the use of various plastics and composites in cars. Presently each car that is recycled generates approximately 500 pounds of non-recycled fluff. To address this, Chrysler, Ford, and General Motors have formed the Vehicle Recycling Partnership to develop guidelines to help production engineers design for recycling. Mercedes and Volvo are also working to develop a totally recyclable car. These advancements are in part due to laws in Germany requiring total vehicle recycling.

4.1.3 *Considering Future Environmental Issues*

In looking to the future, other environmental issues may increase in importance, which will undoubtedly affect evaluation goals. Several issues are on the horizon, such as those listed below, that could emerge to affect manufacturing and product designs and ultimately how a company's environmental baseline should be established. Knowing your company's present status on these issues will make it easier to address them in the future.

- Timber harvesting in old growth forests and national parks and the rapid deforestation of tropical rain forests could affect lumber prices and availability, as well as image problems toward the use of certain wood products. In this light, it may be useful for companies such as furniture manufacturers to measure the amount of timber (possibly broken down by wood types) consumed and replaced during the baseline assessment process.

- The amount of carbon dioxide and other greenhouse gasses emitted could emerge as an important environmental scorecard item and should be measured as a part of your current baseline. Companies could be rated and compared by their emissions per unit of production.

- The advent of increased energy costs or a federal tax on energy consumption may further encourage energy conservation efforts among consumers and industry.

- Acid rain is a major issue throughout the world. If current regulatory controls are relaxed or if additional destruction from acid rain is noted, the largest sources of acid rain pollutants may be faced with consumer boycotts or political pressure.

- U.S. companies with Mexican operations should be prepared to be questioned on the use of hazardous chemicals in the *maquiladora* plants and the potential mismanagement of resulting waste products. Now may be the time to develop plans to eliminate the use of certain problem materials rather than to have to react to public outcry in the future.

- Concerns could arise over the use and ultimate disposal of products made in the United States and sold in countries with less developed environmental infrastructures. The manufacture of pesticides that are banned in the United States for use in other countries is a recent example of this growing problem.

In summary, serious thought must go into developing review criteria before initiating your environmental baseline work. For a broad-based manufacturing corporation, the following suggested criteria would present a thorough picture of a company's environmental status. While it takes a little more energy to develop this information now, it is impossible to go back and try to assemble it with any level of accuracy if it is needed in the years to come.

Raw material usage by type.

Energy consumption.

Toxics use amounts.

Hazardous waste generation.

Nonhazardous waste generation.

Annual emission levels (toxics, greenhouse gasses, ozone depleters, acid rain precursors).

Worker safety rates.

Annual fines.

Annual citations for noncompliance with environmental laws.

Number and size of spills.

4.2 MEASURING THE BASELINE

Once you have developed the parameters of interest, the assessment phase can begin. This section will present two general methods of establishing the baseline—the survey and the assessment.

4.2.1 Surveys

Environmental surveys are forms or reports that are typically developed in corporate headquarters and completed by the facility or process personnel. The information is submitted to corporate staff or a consultant for compilation and analysis. These surveys can be completed annually or quarterly. Depending on the data being collected, it may take a month or two before all the information for a reporting period is finalized. Results of sampling can take a month to obtain. This delay prevents a facility from having final compliance data immediately following the survey period. For this reason, it may be best to require that the results of complicated annual calendar year surveys be submitted in March.

A sample survey is included in Appendix B. This survey focuses on waste and emissions information, although it also includes some questions on environmental compliance status. This type of survey would be applicable for an organization that had a good understanding of the plant operations. For a corporation that did not have a thorough understanding of each facility's products and processes, it would be advisable to include additional questions on process flow diagrams, product specifications, and raw material and energy consumption. This information is also beneficial for international surveys where it may be impractical to obtain this process information through facility assessments.

Collecting process information using surveys can be difficult but can lead to interesting findings. At a company with several plants with flexographic printing operations, it was found that while the processes, products, and equipment were the same at each facility, waste generation levels were drastically different. The corporate environmental manager found that some facilities had found ways to reduce waste generation and recycle spent inks. Applying these methods to all the plants yielded annual savings of over $100,000 and increased product quality.

Environmental management information can also be collected using surveys. The Global Environmental Management Initiative self-assessment process described earlier is one tool for reviewing environmental policies and procedures.

4.2.2 Assessments

In an assessment, an individual or team is sent to each facility or process area to gather the necessary information. In a pollution prevention assessment, information is gathered on the environmental impacts of the manufacturing processes, either from existing data or from samples taken and flows measured to provide a snapshot of the present operations. Often a combination of approaches is required, with existing data used wherever possible and sampling conducted to fill in data gaps.

Assessments can be done by a single individual, a team

of people, or a series of specialized teams. They can take as little as a single day or as long as months, depending on the complexity of the facility and the amount and quality of information required. The assessment team should fill out a questionnaire in advance of the assessment to allow the facility team to gather all available information.

The objective of the assessment is to determine what hazardous materials are used in the plant, where they are used, and where they end up after the product is made. It is important to emphasize that the assessment operation is generally not an evaluation procedure—it is a data-gathering phase. Only after the process information is gathered can the identification and selection of pollution prevention options begin.

4.2.2.1 Assessment personnel.

Whether a single individual or a team, those conducting the facility assessment should possess or have access to knowledge of plant processes, product specifications, environmental considerations, maintenance issues, and manufacturing requirements. Typical assessor personnel include an environmental and health and safety representative, process engineers, facility or plant engineers, and a maintenance supervisor. The key personnel also should be strongly committed to the project, have an open mind, and most importantly be able to solicit and listen to the ideas of machine operators, product designers, and maintenance personnel—those who are most familiar with the particular operations. In addition, the one chosen as team leader must be given adequate time to devote to the project to ensure its success. This cannot be a fill-in task for the leader to work on only as time exists; it must be a high-priority, major responsibility.

Team members from outside the facility operations are often beneficial in providing new viewpoints. In a multi-facility company, sister plant employees or employees from the corporate groups work well to provide a fresh look at the operations.

4.2.2.2 Assessment process.

The following list outlines the steps to be taken before conducting the facility assessment.

Preparing for the assessment

1. Identify all potentially significant processes and key personnel for each. Identify contacts that will be available during the assessment to address questions regarding product and process specifications, waste generation amounts, and process costs including raw material usages and costs.

2. Collect copies of material safety data sheets (MSDSs) for all major materials used at the site. MSDSs should be recent and include a notation as to where the chemicals are stored and used.

3. List annual purchase rates for all major raw materials.

4. Collect copies of Form Rs, energy consumption reports, waste generation reports, wastewater reports, and air testing results. Also include information on applicable permit requirements. If no summary reports are available, have actual manifests and testing results ready for review.

5. Inform key personnel, including operators, of the assessment dates and objectives.

6. Make several copies of facility maps for use during the assessment.

7. Gather all available information on product and process specifications.

8. Assemble the production information necessary for allowing comparisons of environmental impacts on a constant production basis.

The project team begins the assessment by collecting a complete list of the types and quantities of manufacturing materials used in facility processes. This list must include not only purchased chemicals but also metal stocks and other raw materials that potentially may discharge to air, water, or ground. This inventory helps determine which processes use hazardous materials and have the potential to generate hazardous wastes and emissions.

Once the team has determined the quantities of materials used, it can develop the total amount of hazardous constituents using information contained on the MSDSs. Both the chemical inventory and a file of current MSDSs must be maintained on site under OSHA Worker Right-To-Know and EPA SARA emergency planning and emission reporting regulations. For MSDSs that appear incomplete or suspicious, it is worthwhile to contact the material suppliers to obtain additional information on any hazardous constituents in the goods.

To learn which of the facility's processes and wastes would benefit most from pollution prevention activities, the project team must determine the types and quantities of hazardous wastes resulting from each process. Information on hazardous waste inputs and outputs is most easily presented in a series of calculations called *mass balances*. These mass balances work under the premise that all hazardous materials entering the facility must exit the facility somewhere, whether in the finished product, the solid waste removed from the plant, the wastewater, or the air emissions. To determine pollution prevention possibilities for individual processes, team members must complete these mass balances for each hazardous constituent of each process. If the facility manufactures several very different products in one process, the project team should develop separate mass balances for each product.

Facilities have already accumulated much of the information needed to determine mass balances if they are required to submit Tier I or Tier II forms or Form R reports under the SARA regulations. Chemical processing facilities, refineries, paper mills, and plants producing high-value products also generally have mass balances available for every major process. When mass balances are available, they should be reviewed for completeness and accuracy. In addition to the mass balance information, the project team should also determine the unit cost of raw materials, products, and wastes for future economic evaluation of pollution prevention options.

To show how mass balances are calculated, consider as an example a solvent-based painting process in a facility that produces sheet metal cabinets. The inputs to the painting process would be the following:

- Sheet metal.
- Paint containing xylene as solvent.
- Xylene thinner and solvent.
- Clean booth filters.

Outputs from the process would include:

- Painted parts.
- Waste paint.
- Waste solvent/paint mix.
- Dirty booth filters.
- Air emissions.

For this operation, the hazardous material of concern is the xylene solvent. Purchase records, waste records, MSDSs, and waste analyses will provide the necessary xylene usage and waste information. Under the mass balances concept, it is assumed that the amount of xylene entering the facility equals the total amount of xylene exiting the facility in the finished product, the solid waste, the wastewater, and the air emissions. Air emissions information typically is not available, but team members can calculate such information using the following equations if they know all other inputs and outputs:

$$\text{xylene in} = \text{xylene out}$$

where

$$\begin{aligned}\text{xylene in} = &(\text{lb. paint used}) \times (\text{percent xylene in}\\ &\text{paint}) + (\text{lb. thinner used}) \times (\text{percent}\\ &\text{xylene in thinner})\end{aligned}$$

and

$$\begin{aligned}\text{xylene out} = &(\text{lb. paint waste}) \times (\text{percent xylene in}\\ &\text{paint waste}) + (\text{lb. solvent/paint mix}\\ &\text{waste}) \times (\text{percent xylene in solvent/}\\ &\text{paint mix waste}) + (\text{lb. booth filters})\\ &\times (\text{percent xylene in booth filters}) +\\ &(\text{lb. xylene in air emissions})\end{aligned}$$

Facility personnel know the pounds of paint purchased from purchasing records and the pounds of waste generated from waste manifests or billing information. These figures need to be adjusted to an annual basis. The amount of xylene in the raw material can be found on the MSDS. Personnel also should know the composition of the waste generated. If this information is not in facility records, the waste contractor can often supply the data for a small fee. Personnel can calculate air emissions if they know all other figures. Typically, the solvent concentration in the booth filters is negligible, leaving only the paint and waste information to be collected.

4.2.3 *Production Measure Parameters*

With either a survey or an assessment, a parameter should be determined for assessing the relative production levels at the facilities so future results can be compared on a constant production basis. Some production measurement parameters that reportedly have been used include dollar volume of product, number of items produced, and square feet of product coated.

For example, 3M uses a baseline that is easy to obtain and can be used for its wide variety of manufacturing operations. According to Tom Zosel, 3M's manager of pollution prevention, each manufacturing facility must report the pounds of waste, meaning all materials not converted into primary or secondary products, divided by the total pounds of output from the facility. Each plant is then evaluated on its ability to reduce this figure, taking into account that these figures are not comparable from facility to facility due to the different products made at each location. The simple formula is presented below.

$$\frac{\text{lb. waste}}{\text{lb. primary products} + \text{lb. secondary products} + \text{lb. waste}}$$

Ingersoll-Rand uses the number of direct labor worker-hours to compare production levels from one year to the next. This gives the corporation the pounds of waste generated per worker-hour for each facility. Again, this measure is not applicable for comparisons among different plants.

This assessment process leads the way to identifying opportunities within an organization that will reduce the environmental impact of the operations. Now that the information has been collected, it should be compiled to identify the divisions, product lines, facilities, or processes that pose the greatest environmental challenges and the greatest opportunities for making a positive impact. The next step is to identify product and process modifications to exploit these opportunities and obtain the environmental benefits.

RESOURCES

Global Environmental Management Initiative, *Environmental Self Assessment Program*, (Arlington, VA:1992).

Personal conversation with Thomas Zosel, pollution prevention manager, and Robert Bringer, Ph.D., vice president environmental engineering and pollution control, 3M Company, 1993.

Personal conversation with James O'Dell, senior vice president of technology, Ingersoll-Rand, 1993.

President's Commission on Environmental Quality, *Total Quality Management: A Framework for Pollution Prevention* (Washington, D.C.: 1992).

Preventing Pollution in the Chemical Industry (Washington, D.C.: Chemical Manufacturers Association, 1993).

Specification for Environmental Management Systems, British Standard 7750, 1992.

Chapter Five

Developing ECM Process Options

Once a company has determined the environmental status of its operations, management can take two approaches. The first is the passive approach—to sit back and wait for the employees to initiate the changes toward environmentally conscious manufacturing. The second approach is to take an active role in targeting projects to eliminate wastes and emissions. This chapter will concentrate on accelerated, active ECM programs that can show effective results for a corporation in one to four years.

Passive programs may be best for some operations. Passive programs conducted by some large companies have been very effective because of the supportive role of the corporate groups. In these programs, corporate staff set goals and monitor progress, while leaving the development and implementation of plans to meet these goals to the research and production staff who are most familiar with the processes. The success of passive environmental programs depends on plant staff having the technical expertise and available time to carry out the necessary responsibilities. Corporations that have the technical talent within their manufacturing environments can best function by setting general goals and letting the plants meet them.

There are several potential problems with passive programs, however. The importance of the facility "champion" cannot be understated. This initiative may not be taken by facility personnel for many reasons. If liability concerns for environmental issues are addressed at the corporate level, as is common in medium-sized organizations, then the facility staff may not be as motivated to reduce wastes or emissions. Also, if facility accounting procedures do not relate the comprehensive costs of waste generation to a production department, there will not be an economic motivator to reduce these wastes. Another factor is that often facility personnel are kept fully burdened fighting fires and attempting to keep operations in compliance with regulations and do not have the time, energy, and/or expertise to allocate to process modifications. In these instances where additional technical support is advantageous, the methods described in this chapter are effective to develop, prioritize, and implement ECM opportunities.

5.1   IDENTIFYING OPTIONS

The objective in identifying options is to develop as many realistic pollution prevention alternatives as possible for later evaluation and selection. Environmentally conscious manufacturing opportunities may be found through modifying basic customer specifications, product features, manufacturing or maintenance techniques, or packaging methods. Small group brainstorming to identify these opportunities can be highly effective, if expertise in the various areas is represented.

It is appropriate at this point to discuss the role of outside consultants in identifying pollution prevention opportunities. The use of technical experts to provide outside business assistance is a familiar practice for many companies and is promoted through state and federal technical assistance programs. President Clinton has expanded the funding of advanced manufacturing centers to provide technical assistance to businesses. In the field of pollution prevention, engineering students, experienced professionals, and retired engineers have all been used in various state programs to provide on-site and over-the-phone technical assistance to identify opportunities.

The use of experienced pollution prevention consultants can lead to the development of a comprehensive list of options since an outsider often is better able to question practices and techniques that are taken for granted within the manufacturing organization. A consultant with a sound engineering background can help evaluate these options and present them to a nontechnical audience. An outside viewpoint also can simplify the opportunity assessment process. Typically, a consultant will spend from one to five full days in a plant reviewing written procedures, observing operations, and questioning as many people as possible, especially operators and engineers. An outside consultant is more likely to be able to conduct such a thorough research effort and to obtain the unbiased answers from staff that are necessary to ensure progress toward ECM goals.

The brainstorming process for developing environmentally conscious manufacturing options is demonstrated below using several hypothetical examples. The key to the process is to always ask why—much like an inquisitive three-year-old—until you run out of ideas. Using brainstorming tools, such as the fishbone diagrams taught in total quality management programs, can be helpful to overcome any mental roadblocks.

5.1.1   *Example 1—A Machining Operation*

For this example, our facility is a machine shop making a variety of small parts that are then shipped to another plant for cleaning, plating or painting, and final assembly.

In one area, screw machines are used to make small components from bar stock steel. A sulfurized straight oil is used as a machining fluid. In a second section, five multispindle CNC machines are used to cut, mill, drill, and tap steel. Each machine uses a different machining fluid in its sump as specified by the machine operators. Individual 30- to 100-gallon sumps are located under each machine and are used to collect and recirculate the fluids that are sprayed on the tools to provide lubrication and dissipate heat. These sumps are emptied and cleaned by each operator when they go rancid, approximately every two weeks. In a third area, rough-and-finish grinding is done on select parts. This area has a 750-gallon central grinding fluid collection and recirculation system that is filled with a water soluble oil. This central system is dumped to the sewer every two weeks to prevent the buildup of grinding fines.

Finished parts are placed in lined cardboard boxes and sent to the assembly plant. The facility uses a just-in-time pull system to only make parts as they are ordered to reduce in-process inventory. The facility had used 1,1,1-trichloroethane in a vapor degreaser to clean parts. They tried to convert to an aqueous cleaner but found that some parts came out dirty and others were clean but soon rusted. In response to customer concerns about the dirty parts, the facility agreed to reduce costs and eliminate cleaning operations since the customers would clean the parts anyway.

While the facility had no outstanding environmental violations, there were several regulatory concerns. Review of the wastewater records indicated regular exceeding of city ordinances for oxygen demand (BOD and COD) and oil and grease loadings. The facility was paying $75,000 annually in surcharges for these loadings. In addition, the discharge from the facility was above proposed new city limitations for lead. Stormwater discharges were also a concern. Runoff from the chip storage area carries coolants and oils into a ditch that leads off of the property.

5.1.1.1   Identifying wastes.   From the assessment of facility operations, we tentatively identified the following wastes:

Screw machine area

Screw machine straight oil—not typically dumped.

Oil-soaked absorbent—25 gallons per day.

Oily chips—3 cubic yards per day.

Waste solvent used at the machines—10 gallons per week.

Oily rags—50 per day.

CNC area

Water-based machining fluids, mostly soluble oils—125 gallons per week.

Oil-soaked absorbent—10 gallons per day.

Oily chips—2 cubic yards per day.

Waste solvent used at the machines—10 gallons per week.

Oily rags—30 per day.

Grinding area

Grinding fluid—750 gallons every 2 weeks.

Oil-soaked absorbent—25 gallons per day.

Waste solvent used at the machines—10 gallons per week.

Oily rags—30 per day.

Spent grinding wheels—5 per day.

In addition, floor cleaning operations generate approximately 50 gallons of mop water per day that is dumped into the sewer. Stormwater flows to the ditch. Scrap metal is collected by a recycler. Packaging materials and other nonhazardous plant wastes are thrown in a dumpster for disposal at the local landfill.

The plant layout is shown in Exhibit 5–1.

5.1.1.2 Identifying options.
Screw machine area. The first process is the screw machine area. Questions to ask include:

Why are we machining the parts? Are other options available?

Why is so much absorbent needed? Can we prevent machines from leaking?

Why are we using a sulfurized straight oil? Are other fluids applicable that may be more environmentally friendly? Can the machining be done without the fluids?

Why are we using leaded steels? Are there options that would reduce the introduction of lead into the wastewater?

Why do we mop the floors, and why do we have to dump the mop water? Are there other methods to keep the shop clean and prevent slips and spills?

Why is the solvent used at the machines? Is cleaning necessary? Can another material be used to clean the parts and machines?

From this questioning we may find that:

Having the parts machined by a third party would reduce revenues and potentially impact quality with no obvious gain, environmental or otherwise. Using injection-molded parts, composites, or powder metallurgy products would not be cost-effective but could be used to effectively eliminate the need for the screw machines.

The machines are old and have leaky hydraulic systems that contribute to the mess in the department. The facility uses loose absorbent to keep the areas clean because of its low cost. They have never considered absorbent socks as an alternate method of controlling the oil.

The inherent leaky nature of the older screw machines dictates the use of straight oils, since other fluids would

quickly become contaminated with the hydraulic oils. Also, water-based fluids could damage the machines. Sulfurized oils have been used for as long as anyone could remember and have worked well.

Leaded steels are used due to their machinability. The facility was not aware of alternatives to leaded steels, such as bismuth-based steels.

Floor mopping is needed in this area because of the oils from the machines and from the chips that are stored in small bins at each machine and rolled to the chip hopper. Janitorial personnel are charged with keeping the manufacturing areas clean. Reducing the frequency of floor cleaning would not be acceptable to the safety coordinator or the plant manager. One of the coolant vendors recommended the floor cleaner and it has worked well to remove the oils. (Review of the MSDS for the cleaner found it to consist of triethanolamine and chelating agents.) The mop water is generally so dirty that facility personnel consider reuse or filtration to be impractical.

Based on these responses, the following options could have potential merit and will be listed for further review:

- Reducing machine leaks by rebuilding seals.
- Using absorbent socks, wringing out and collecting the oils and reusing the socks.
- Using nonsulfurized straight oils.
- Using nonleaded steels.
- Using a floor cleaner that may allow removal of oils through a simple treatment system.
- Eliminating cleaning at the individual machines and installing a small central cleaning system.

CNC area. The second area is the CNC department. Questions to ask include:

Why are parts machined? Are there design options to reduce the amount of machining?

Why is the coolant waste generated? Are there options to prevent this material from going bad?

Why are the different machining fluids used? Can these be consolidated to allow easier recycling?

Why are leaded steels used? Are nonleaded options available?

Why is the absorbent needed? Can machine leaks and drippings from chips be avoided? Are there other methods to contain the oils around the machines?

Why is the solvent used at the machines? Is cleaning necessary? Can another material be used to clean the parts and machines?

From this questioning we may find that:

The facility has not considered alternate machining methods because of the capital investment in the present procedures.

The machines are fully enclosed, but sometimes doors are left open, which allows coolants to mist onto the floor and contributes to the mess in the department. Cool-

EXHIBIT 5–1
Machine Shop

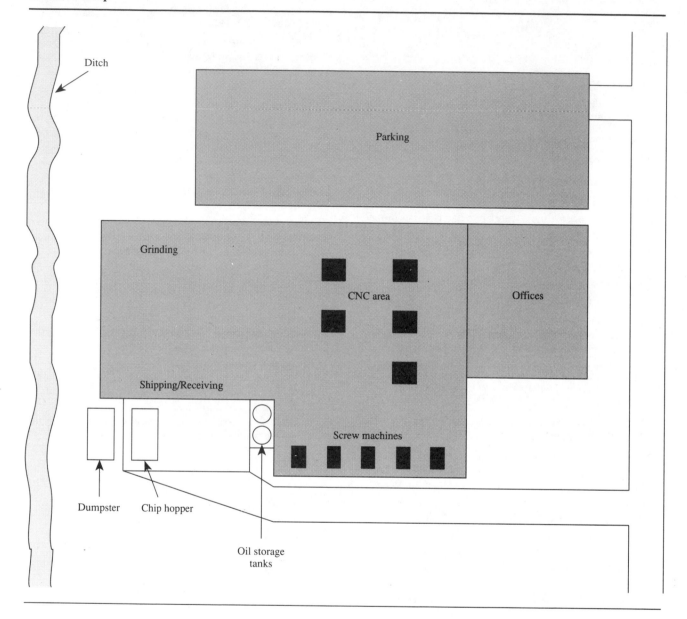

ant also drips off parts and chips that are stored next to the machines. Loose absorbent is used to keep the areas clean because of its low cost. Personnel have never looked at absorbent socks as an alternate method of controlling the oil.

Operators have always been in charge of "their" machines and they each have the authority to order whatever machining fluid they feel is appropriate to give them the highest production rates. The operators are also charged with maintaining the fluid concentrations, although lower wage, third shift janitorial personnel are used to empty and clean the sumps when requested by the operators.

Leaded steels are used due to their machinability. Personnel were not aware of alternatives to leaded steels, such as bismuth-based steels.

As in the screw machine area, floor mopping is needed in this area because of the oils from the machines and from the chips that are stored in small bins at each machine and rolled to the chip hopper.

From these responses, we can identify the following options worth further investigation:

- Consolidating the machining fluids to one solution.

- Identifying machining fluid cleanliness requirements and establishing a continuous filtration/contaminant removal system.

- Using absorbent socks, wringing out and collecting the oils, and reusing the socks.

- Investigating the use of nonleaded steels.

- Implementing procedural changes and conducting operator awareness training to keep machine doors closed to minimize splashing.

- Modifying responsibilities to increase the accountability of production personnel for the cleanliness of the area and the amount of waste generated.

- Reducing the amount of cleaning required and installing a small central cleaning system with less hazardous chemicals.

- Modifying floor washing procedures and methods, as in the screw machine area.

Grinding area. In the grinding department the following questions are asked:

Why is this operation performed? What are the customer requirements for surface finish? Can precision machining operations be performed to reduce the dependence on rough-and-finish grinding?

Why are solvent cleaners used at each grinder? Can cleaning be centralized and less hazardous materials be used?

Why is floor mopping needed? Can process modifications, such as improved equipment ventilation, be made to reduce the level of oil mists in the air (which end up on the floor)? Can floor cleaning be done less frequently with a less hazardous material?

Why is the grinding coolant used? Why is it disposed? Can the grinding fines and tramp oil be continuously removed to keep it from going bad?

Why is leaded steel used? Can an alternative steel be used that would not cause the waste grinding fluid to be classified as a hazardous waste?

Responses to these questions conclude the following:

Grinding is used as a finishing operation and cannot be reduced through the use of higher precision machining practices. It is not practical to have these activities performed by a third party.

Local ventilation on the grinders can be improved to reduce the oil mists that eventually condense on the shop floor and necessitate cleaning.

As in the other areas, reducing the frequency of floor cleaning is not an option unless the oil buildup is reduced.

Using a nonleaded steel is seen as a way to prevent the generation of a significant amount of hazardous waste.

The use of a grinding coolant purification system was investigated previously, but the department supervisor objected to it based on the additional work that would be required to maintain the equipment. As described previously, janitorial staff were presently used to clean out the sumps. Grinding department personnel were reviewed based on the number of parts produced per shift and had a pay incentive program to promote high productivity. This management structure did not encourage the use of a coolant recycling system, even though the system could help the department and the company by im-

proving product quality while reducing the waste generation and costs.

Based on this review, the following items were listed for further investigation:

- Improvement of local ventilation on grinders.
- Investigation of nonleaded steels.
- Investigation of coolant recycling systems.
- Identification of a less hazardous floor cleaner.

5.1.1.3 Other areas. There are a number of other opportunities to review at a manufacturing operation such as the one in this example. Through the assessment process, things like packaging wastes and shipping containers may also be identified as concerns. For these items, procedures such as the following can be effective:

- Controlling purchasing and inventory to eliminate ordering of unnecessary materials.
- Purchasing in appropriate size containers for the expected use requirements.
- Ordering materials in reusable or recyclable bulk containers.
- Controlling raw material use on a first-in, first-out basis to minimize shelf aging.

In addition, it is important to identify other processes or procedures that may be affected by potential changes. How are the present processes affecting product quality and cost, and are there certain operations that cause excessive quality or cost concerns? Typically a material does not instantly become "bad" once it is a waste. If a grinding fluid sump is emptied every two weeks, it probably isn't much better on the 13th day than the 14th day. Fines in the fluid could be reducing product quality by providing a rougher finish. Likewise, dirty coolants in CNC machines can reduce the life of the tools and the part production rates. Obviously, systems that can continually filter and remove contaminants will have paybacks in other areas besides waste reduction.

Process control can also be improved through some pollution prevention opportunities. This can be an important issue for companies seeking certification of their process quality controls methods, such as through the ISO 9000 certification process. As stated above, filtering a grinding fluid can reduce some of the variability in grinding operations, which could otherwise cause variations in the surface finish of the parts.

As discussed in other sections of this book, life cycle analyses can also be used to identify pollution prevention opportunities, especially as they relate to product design.

5.1.2 *Example 2—Cleaning, Assembly, and Painting Operations*

In our second facility, machined parts are received, cleaned, assembled into a final product and painted before being packaged and shipped to a customer. Parts are received with an oil coating to prevent rusting. They are stored until needed for assembly. Before assembly, the oil is removed through the use of a conveyorized vapor degreaser. Parts are hung on hooks or placed in baskets and hung on a conveyor. The conveyor carries the parts through an enclosed machine that contains solvent vapor zones created by boiling a nonflammable solvent, 1,1,1-trichloroethane (TCA).

While used extensively for its relative low toxicity and safety, TCA has been noted to contribute to the depletion of the stratospheric ozone layer. In the United States, the Clean Air Act Amendments of 1990 required products cleaned with TCA to be clearly labeled as being manufactured with an ozone-depleting compound. This has led to a marked increase in interest in other cleaning chemicals and methods by manufacturers.

Following cleaning, parts are assembled into the final product. In our sample facility, there is little waste in assembly operations. The assembled products are passed to a final cleaning and painting operation. A water-based, five-stage washer is used to provide a clean surface for good paint adhesion. An iron phosphate conversion coating with a hexavalent chrome sealer is used in the washer to increase the corrosion resistance of the final product.

The parts are then conveyed to the painting area where automatic spray guns are used to coat the parts with a variety of colors of solvent-based paints. Four different paint colors are kept on line in separate painting areas, with a final manual painting area for touch-up painting and the use of other custom colors. The four colors represent over 95 percent of the product. When custom colors are requested, they are mixed in 3-gallon paint pots and manually sprayed onto the parts. Spray guns, pots, and paint lines are cleaned with solvent once per day and between all color changes. Paint booth air filters are replaced when they become dirty.

Coated parts are dried in a low-temperature oven then removed from the conveyors, inspected, packaged, and shipped to customers.

5.1.2.1 Identifying wastes.
From our assessment we identified the following wastes and emissions:

Cleaning operations

TCA/oil waste—70 gallons per month.
TCA emissions—60,000 pounds per year.
Cooling water—3 gallons per minute.

Pretreatment operations

Wastewater—5 gallons per minute.
Spent cleaning baths—900 gallons every two weeks.

Painting operations

Spent paint and cleaning wastes—55 gallons per day.

EXHIBIT 5–2
Assembly and Painting Facility Block Process Flow Diagram

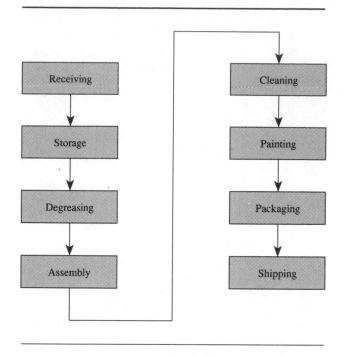

Solvent emissions—80,000 pounds per year.
Dirty paint booth filters—10 cubic yards per week.

The facility did not have any immediate regulatory concerns. Plant issues included the desire to eliminate the use of TCA and concern over future air pollution control costs and increasing wastewater treatment and waste disposal costs. The use of phosphate and chrome coatings placed the facility under stringent wastewater regulations and led to the generation of a wastewater treatment sludge that was regulated as a hazardous waste because of its chrome content.

5.1.2.2 Identifying options.

Cleaning operations. Questions asked to help identify opportunities in the cleaning operations include the following:

Why are the parts cleaned? What are the product cleanliness requirements? Can this stage of cleaning be eliminated? Can oils be removed by the suppliers? Can alternative rust inhibitors be used?

Why is vapor degreasing with TCA used? Can alternative methods or chemicals be used to meet the cleanliness requirements? What other methods have been investigated?

Can the cleaning system be enclosed or otherwise modified to reduce emissions?

Can the oil and other contaminants be removed from the cleaning agent to recover and reuse the material?

From these questions we learn the following:

Parts are cleaned to allow assembly. Oils would interfere with the assembly process. Alternative corrosion protectors may be viable to provide interim protection. Internal parts are actually electroplated with a corrosion-resistant coating (zinc) and are not subject to rusting. The cleaning requirement is that parts must be able to be easily handled in the assembly operations.

The vapor degreaser has been used for years as a preferred method for removing oils from parts. Facility personnel have estimated that a conveyorized aqueous cleaning system comparable to the existing system would cost $300,000, have higher energy requirements, and would exceed the capacity of the existing wastewater treatment system. Facility personnel have also estimated that it would take approximately $150,000 to modify the present degreaser to reduce emissions. These modifications include the installation of additional refrigeration capacity and restricting the openings to the machine.

Based on this review, the following options are selected for further evaluation:

- Investigating the elimination of the use of oils on plated parts.
- Investigating the effectiveness of oil-free corrosion inhibitors.
- Investigating alternative cleaning methods for any parts that still need cleaning.

Pretreatment operations.   Questions asked to help identify pretreatment pollution prevention opportunities include:

Why is pretreatment necessary? Are there other options to prepare the parts for painting?

Can the wastewater generation be reduced? Can the water be recycled?

Can alternatives to the phosphate coating and chrome sealant be used?

From these questions we learn the following:

The pretreatment process was developed by a consultant retained to ensure the highest practical corrosion resistance for the finished product. Environmental costs were addressed as a minor factor during process selection.

Based on this review, the following options are selected for further evaluation:

- Implementing rinsewater conservation techniques.
- Investigating the impact of using nonphosphate cleaners and nonchrome sealants.

Painting operations.   Questions for the painting operations include:

Why are the parts painted? Could prepainted components be used?

Why are so many colors necessary? Can custom colors be eliminated since they represent only 5 percent of the business?

Why are high-solvent paints used? What are the painting specifications? What are the customer requirements? Can high solids, water-based, or powder coatings be used? If not, why not? Can they be used with other process modifications? If the use of solvent paints was to be banned, effective in six months, what would you do?

From these questions we learn the following:

Painting is necessary for product aesthetics and corrosion protection. All that really needs to be painted is the outside case of the product, however, and prepainting the sheet metal cabinet parts is a potentially viable option.

Presently over 100 colors are used. Marketing staff state that reducing this availability will greatly affect product sales and portray an image of not being able to respond to customer needs.

No real efforts have been placed on investigating alternative coatings. The existing coatings have been used because of their fast drying characteristics. Major oven modifications would be necessary to allow higher temperature curing.

Based on this review, the following options are selected for further evaluation:

- Prepainting of sheet metal cabinet components.
- Eliminating all but the four primary product colors.
- Investigating alternative coatings (in conjunction with the investigation of the pretreatment alternatives discussed earlier).

The second option could meet severe resistance from an ''unenlightened'' marketing staff. In a marketing-based organization, this resistance can be hard to overcome, eliminating some very viable pollution prevention opportunities. Changing this mind-set requires a senior management commitment to becoming an environmentally conscious manufacturer.

5.1.3 *Identifying Research Needs*

In some processes, a need for pollution prevention can be defined but no technically feasible options exist. At this point, brainstorming efforts may become bogged down. When this occurs, several methods can be used to help think ''outside of the box.'' These methods can be used to identify options that then may require research and development.

One method is to restate the ultimate purpose of the product being made, and then discuss other ways in which this need would be fulfilled. For example, if a part is coated or electroplated for corrosion resistance, is it feasible to make the product using plastics, composites, or corrosion-resistant metals?

Another method is derived from an experience in assisting in a voluntary program to site and build a hazardous waste treatment and disposal facility. At a number of public meetings, citizens living near proposed disposal sites offered the suggestion that the hazardous wastes that were being generated "by industry" be disposed in the backyards of the CEOs that were profiting from these industries. Ignoring the philosophical issues of the responsibility of living in a cooperative society, this mindset can be a valuable tool to help spur the creativity of the brainstorming team.

The typical engineer is not trained to address these bigger picture issues, but instead is taught to develop processes to improve product quality and reduce manufactured costs. Developing this broader mind-set can be done only through extensive training and awareness building, such as has been used in U.S. companies in the "quality" programs of the 80s and the energy conservation programs of the 70s.

A third option is useful to help identify waste reduction and recycling options. This method entails getting facility personnel to imagine what they could do to reduce or recycle wastes and emissions if their product was made out of solid gold. Anyone who has ever worked in a facility using precious metals can understand the potential value of waste material.

5.2   EVALUATING OPTIONS

Once the potentially viable options are identified, the next step is to evaluate them, select those that will be implemented, and develop a plan and schedule for implementation. The evaluation of process modifications is a continual process in any successful manufacturing environment. However, evaluating modifications to reduce the environmental impact of manufacturing operations can be more involved than a typical cost-benefit review. The problem lies in attempting to quantify the environmental impacts and financial benefits of various options.

Both technical and economic reviews should be carried out to determine the impact of these options. Technical evaluations are conducted before financial evaluations for several reasons. First, if the process modification will not work, the economics does not matter. Second, through the technical evaluation you will identify enhancements that will affect the capital or operating costs of the process.

5.2.1   Technical Evaluation

The goal of the technical evaluation should be that all key individuals are comfortable with the probable success of the process change. This comfort level will vary for each project depending on the potential ramifications of failure.

Supporters of a process change can be blind to the potential avenues for failure. It seems, however, almost every manufacturing operation has at least one individual who points out all the reasons something should not work. "We tried something like that once and it didn't work" must be recited thousands of times a day around the world. Unfortunately, the people making these statements are often in a position to make their statements come true by withholding their support in addressing inevitable problems.

It therefore is important to obtain key employee support for a project. The technical evaluation process provides an excellent opportunity to allow these people to provide their input in a structured environment. Other methods of obtaining employee support are addressed later in the book.

The technical evaluation needs to be structured for the specific process or product changes being proposed. The questions listed below provide a good starting point for many changes. Answering these questions requires talking to as many knowledgeable sources as possible, preferably others who have implemented or tried to implement the proposed options. This requires some investigative skills. Good starting points include talking to equipment or chemical vendors, trade associations, competitors, authors of books or articles on the topic of interest, and others who may be able to provide you with valuable leads.

5.2.1.1   Evaluation questions.   Questions to ask in a technical review include the following:

Will this change work in my application?

Has it worked elsewhere? What is the development status? How much research and testing is required before it can be implemented in my facilities?

What impacts will this have on my products and my customers? Will customer approvals be necessary? What is required to obtain these approvals? What impact will the change have on product sales?

What impact will the change have on productivity and quality?

What are the environmental impacts of the proposed change? Will the change simply shift contaminants from one medium to another? What are the regulatory ramifications of the change? Will a permit or a permit modification be necessary? How will these items affect the facility?

What impacts will the change have on maintenance operations? Will complicated and frequent maintenance procedures actually lead to increased wastes and emissions?

How will the change impact worker safety? Will chemical exposures increase as a result of the change? Are there odor problems that need to be addressed?

Are there any material or product handling issues that need to be addressed?

Are the staff and resources available to implement the change? How will reallocating resources to the process change affect other areas of the operations?

Are there legal or contract issues impacted by the change? Does a license for use of the technology need to be obtained?

How will laborers be affected by the change? Are operator responsibilities increased? Will contracts and pay scales need to be adjusted to address changing responsibilities? Are these changes feasible?

What are the ramifications of a failure of the change? What could cause a system failure? Can parallel systems be run during a trial period before making major changes? What happens if the whole thing just doesn't work?

The answers to these questions will give a good indication of the likelihood of success for a proposed change. Obtaining the answers obviously requires a great deal of work, but can be simplified for most operations. Typically, you will be able to find other facilities willing to share their experiences with a product or process and give you enough information to determine the potential for success.

5.2.1.2 Evaluation example.

The following example illustrates how this thought process can be used. In this example, we will investigate the technical feasibility of converting a hexavalent chrome plating system to a less toxic trivalent chrome plating line. In our facility, steel moldings that are used in high quality office furniture are chrome plated for aesthetics.

The plating line is a series of tanks with an automatic hoist that carries the parts through the chemical solutions. The sequence of the tanks is shown in Exhibit 5–3.

The chrome plating is applied by flowing electrical current to the part while it is immersed in a solution containing hexavalent chromium. Hexavalent chromium is electrically converted to chromium metal on the surface of the part. The part is then removed from the tank and excess hexavalent chrome solution is rinsed from the part in the flowing water rinse tanks.

In the plating process, hexavalent chrome mists are generated. Hexavalent chrome is a carcinogen and uncontrolled mists can pose a threat to both employees and persons living near the facility. Typical plating lines have extensive fume capture hood systems, which then vent to an air pollution control system to remove the chrome mists before discharging the air.

One of the options that was identified was using a less toxic trivalent chrome system to apply the chrome plating to the parts. For this option we will step through the questions listed above.

Will this change work in my application? Has it worked elsewhere? What is the development status? How much research and testing is required before it can be implemented in my facilities?

The vendor of the trivalent chrome chemicals was able to identify six companies in the region that used the process on similar parts. Our facility talked to all six and found that while they had occasional problems, they all could make the system work reliably.

EXHIBIT 5–3

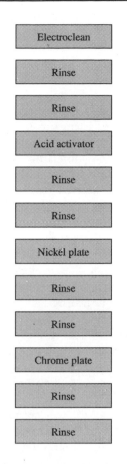

What impacts will this have on my products and my customers? Will customer approvals be necessary? What is required to obtain these approvals? What impact will the change have on product sales?

Parts from our shop were plated by the vendor in pilot tests. The trivalent chrome color was slightly different from the existing hexavalent chrome-plated parts. While there is no formal customer approval process for the company, the marketing manager was extremely concerned that the new office furniture may not exactly match other office components that are hexavalent chrome plated. The marketing manager noted that since the product was sold based on quality and appearance, changing the finish could have a disastrous impact on sales.

A second opinion was offered by the environmental manager. The change could actually be incorporated into the marketing efforts to promote the environmental stewardship of the company. This could match with the existing emphasis on quality in the marketing approach.

What impact will the change have on productivity and quality?

To answer this question we went back to the existing users of the process. All of them freely discussed their periodic problems in maintaining the plating bath compositions, which would lead to unacceptable product quality. Most firms had such severe quality problems that they had to dump and recharge the chrome bath to allow production to resume. Several of the firms had hired the chemical vendor to come into their shop weekly to check the bath composition and make sure they were maintaining it properly. One firm, which operated a just-in-time operation, had to temporarily lay off employees while the chrome problem was being resolved. While tempted to drop it, they stuck with the trivalent process.

What are the environmental impacts of the proposed change? Will the change simply shift contaminants from one medium to another? What are the regulatory ramifications of the change? Will a permit or a permit modification be necessary? How will these items affect the facility?

The impact of the change is to eliminate the use and release of a carcinogen. The trivalent chrome process did not appear to pose any of the airborne health risks that the hexavalent chrome process had. In addition, a portion of the wastewater treatment system dedicated to chemically converting hexavalent chrome in the wastewater to trivalent chrome could be eliminated, saving operation costs. The generation of hazardous waste from the wastewater treatment process was also expected to decrease by the conversion.

Based on a review, the only permit modification that would be necessary was the wastewater treatment permit, which had to be adjusted before elimination of the hexavalent chrome reduction system. A number of response and contingency plans would need to be modified, and operator training would have to address the potential hazards of working with the new material. It was expected that operators would be pleased to see the elimination of the use of hexavalent chrome.

What impacts will the change have on maintenance operations? Will complicated and frequent maintenance procedures actually lead to increased wastes and emissions?

Based on the feedback from the existing users, a dedicated chemical technician position was recommended to maintain the baths in addition to working in other areas of the plant. An alternate to adding this position was to hire the chemical vendor to come into the plant weekly to check the system operation. No significant change in maintenance wastes was noted by the other plants.

How will the change impact worker safety? Will chemical exposures increase as a result of the change? Are there odor problems that need to be addressed?

As noted above, the worker response was expected to be very favorable.

Are there any material or product handling issues that need to be addressed?

No special product handling concerns were noted. Spills and splashes of hexavalent chrome from the ex-

isting system had severely discolored and dissolved portions of the concrete floor in the plating area.

Are the staff and resources available to implement the change? How will reallocating resources to the process change affect other areas of the operations?

While the plating supervisor was busy, she offered to dedicate herself to seeing to the success of the conversion, provided she had an additional staff person made available as a chemical technician. It was estimated from the discussions with the other companies that it could take up to one month of dedicated efforts before the system would be up and running and producing continuous high-quality results.

Are there legal or contract issues impacted by the change? Does a license for use of the technology need to be obtained?

No licensing issues were affected by the proposed change.

How will laborers be affected by the change? Are operator responsibilities increased? Will contracts and pay scales need to be adjusted to address changing responsibilities? Are these changes feasible?

Provided the additional technician was supplied, the change would have no serious impact on operator responsibilities.

What are the ramifications of a failure of the change? What could cause a system failure? Can parallel systems be run during a trial period before making major changes? What happens if the whole thing just doesn't work?

The new chemicals could not be placed in the existing hexavalent chrome tank because of the potential for contamination of the new bath. Conversion would require tearing out the old tank, installing a new tank, and charging it with the new chemistry. If the system could not be made to work, the chemicals could be drained from the new tank and the old hexavalent chemicals be added back into the tank. Running parallel systems was not an option, but it was decided that in case of total failure, the old system could be easily brought back.

Based on this review there is a known potential for failure, but a commitment to make the system work by the plating supervisor. The one area that needs to be addressed through further review and an executive decision is the impact on any color differences on product sales.

5.2.2 Economic Evaluation

The pollution prevention opportunities that make it through the technical evaluation are then evaluated for their economic impact. The typical pollution prevention success stories touted by the big companies point out how simple changes reduced emissions by hundreds of thousands of pounds and paid for themselves in two days.

This may have caused more problems for industry as now regulators and the public point to these stories and tell companies, "Why can't you do that?"

Unfortunately, as those who are actively working in industry know, the payoff is not always that outstanding. The typical benefits are difficult to quantify. These can include incorporating process modifications that allow a facility to remain in compliance with its discharge permit. Without the modifications, the facility could have been forced to spend hundreds of thousands of dollars and waste valuable technical resources in upgrading an end-of-pipe treatment system. Worse yet, the facility could have been cited and fined for violation of its permits, requiring significant management and legal resources to resolve. Unfortunately, generally accepted accounting practices have difficulty recognizing these avoided costs.

One issue that is often overlooked is the opportunity cost of the resources spent on complying with regulations through end-of-pipe treatment systems. The federal Environmental Protection Agency has estimated that implementing the Clean Air Act Amendments of 1990 will cost industry $25 billion by the year 2005. What is missing, however, is that much of these costs will be spent working toward sampling, reporting, and installing air pollution control equipment. What would be the impact, however, if industry would spend this $25 billion on improving their process efficiencies and product quality? For one, a lot of consultants and pollution control equipment manufacturers would go out of business, but an equivalent amount of small manufacturers might be able to stay in business. In times of global competition, these issues cannot be overlooked.

In an ideal business, all investment options can be quantified in terms of costs, savings, and potential risks. Unfortunately, this is never the case, especially with projects that affect environmental issues. The liabilities associated with the disposal of hazardous waste are a good example. One company has used the figure that every dollar spent on waste disposal represents $10 of costs of a future cleanup. Using this figure in an economic evaluation helps justify projects that reduce waste generation.

The EPA has developed a good overall structure for quantifying the costs and benefits associated with pollution prevention projects. Further information on this process can be obtained through the EPA's Pollution Prevention Benefits Manual.

The general procedure outlined in the report is to quantify savings in four different tiers of benefits. These include the following:

Usual costs. These costs include capital expenditures, operating expenses such as raw materials, utilities, labor and disposal, and operating revenues.

Hidden regulatory costs. These costs include monitoring equipment and expenses, preparedness and protective equipment and expenses, reporting costs, special training costs, recordkeeping costs, medical surveillance, and other often neglected costs.

Liability costs. These costs include penalties and fines as well as future liabilities from the releases through storage, transportation of disposal.

Less tangible costs. These costs include items associated with corporate image, employee/union relations, and consumer/customer acceptance.

In addition to identifying the operating costs and savings, the up-front (capital) costs of implementing the change must also be developed.

During the technical review, any equipment or resources needed to implement the change should have been identified. The next step is to develop cost estimates for these items. As with the technical review, cost estimates should be developed using a variety of resources. Equipment vendors will be the best source of cost estimates for equipment purchases, but they may tend to underestimate installation or operational costs. For this reason, other people who have implemented the change are the best resources if they can share their information.

Examples of costs to incorporate in estimates include the following:

Process/product development costs.

Equipment costs.

Start-up chemical/catalyst costs.

Installation costs.

Start-up and debug labor costs.

Plant or line shutdown costs—lost production.

Shipping.

Taxes.

Engineering costs—design and installation.

Permitting costs.

Product and process certification/customer approval costs.

Training costs.

Documentation costs.

While the factual economic evaluation may be discouraging for pollution prevention projects, the economics must be looked at in comparison to a no-action decision. While the return on investment may not be great, it is important to remember that pollution control systems have essentially no return on investment (an exception is heat recovery from air pollution oxidation systems). As companies with wastewater treatment systems have seen, these systems cost money to install, require regular operator oversight and maintenance, require expensive chemical testing, and inevitably the operation still goes out of compliance. The violations then necessitate hiring a consultant and an attorney and eventually spending more money on upgrading the system. Some companies write off these costs as an investment in the environment. The only problem is that there is no real return on that investment.

5.2.3 *Selecting Options for Implementation*

Once all the options identified to achieve the desired result have been reviewed, some should stand out as being ideal for implementation. Besides the direct technical and economic feasibility, there are a number of additional factors to consider in selecting and prioritizing options. These include:

Environmental/economic significance.

Technical requirements/risk.

Staffing abilities.

Anticipated product changes.

Ease of modifications.

Availability of capital.

Each of these factors will have a different importance for different facilities. If capital is not readily available for a plant, then low-cost items will be selected before capital-intensive projects, regardless of the environmental impact. If obtaining immediate results is important, then a project that may require an extensive preconstruction permitting process will not be favored.

Significance.   One of the key factors is the environmental and economic significance of the option. An option that has a high environmental impact, such as eliminating the use of chlorinated solvents, will be preferred over one that would reduce wastewater sludge volume by 10 percent, even though the wastewater project may have a higher return on investment. Likewise, if a facility is having difficulties meeting wastewater permit levels for cadmium, an option to implement a noncadmium alternative coating may be the best project for immediate implementation.

Technical requirements/risk.   Projects that have a higher technical risk or require significant technical support may not be applicable for a facility with limited technical staff. This issue is often a concern for smaller facilities and can make an economically sound option unfeasible until adequate technical support is available.

Staffing abilities.   Some projects require additional attention from manufacturing or maintenance staff for start-up and initial operation. This factor can be a concern and may require initial implementation of projects that do not need staff support.

Anticipated product changes.   If major product or process changes are planned for a facility, it may make sense to delay implementation of certain projects and incorporate them into the planned changes. For example, it would not be wise to invest in a dry powder coating system as a solvent paint replacement if the facility planned to change to plastic parts.

Ease of modifications.   Some customers, such as the military and automotive manufacturers, require that any major product revisions be approved before being implemented. This review process can take many months and much effort. A company may choose to wait on proposing certain modifications until other modifications are also desired to minimize the number of customer approval projects.

Availability of capital.   For plants with lean capital budgets, this factor is extremely important. Often facilities operate with minimal capital expenditures for brief periods. During this period, efforts can be focused on low-cost projects, leaving larger projects for later.

5.3 DOCUMENTING THE PLAN

Once the options for implementation are selected, a written plan can be completed. Obviously, this plan is meant to be a living document, since it is only current until the next major process change is made. The benefits of having a written plan are many. First, it serves as a signal location for all the information developed through the opportunity identification and review process. Second, it can be distributed both internally and externally to communicate the common objectives and plans to meet those objectives. Third, it helps focus future efforts when reevaluating the pollution prevention activities. Last, it is required under a variety of state and federal regulations as well as environmental consent orders and citizen suit settlements. Unfortunately, the last reason is the sole motivating factor in preparing a written plan for many companies.

Many federal and state laws now ''encourage'' businesses to evaluate their operations and prepare plans for improving their environmental status. While these laws have been passed with good intentions, they seem to be largely ineffective for most companies. The environmental compliance mind-set that has been beat into manufacturers by the historical activities of the regulators makes it extremely difficult for them to cope with ''voluntary'' programs. The most successful corporate pollution prevention efforts have been formed and driven as a result of a vision provided and directed by senior management. In this sense, pollution prevention no longer becomes a voluntary issue within the corporate environment.

The pollution prevention plans that are prepared in response to a regulatory requirement are often completed with the thought, ''What is the minimum requirement that needs to be completed in order to produce an acceptable plan?'' It is not surprising that these plans rarely are implemented and only serve to help fill another shelf in the environmental coordinator's office. If a plan is to be effective, it must be communicated to the operators, maintenance personnel, product designers, engineers, and managers.

5.3.1 *Components of the Written Plan*

What should be incorporated into a plan? As with any business plan, it should state the conditions under which the plan was developed, the options evaluated, the results of the evaluation, the plan and schedule for implementation, the expected results, and the method by which those results will be measured. For a plan that will also be used as an internal or external public relations tool, it may be valuable to tout the results of past environmental efforts.

Baseline conditions. This section presents the results of the data gathering conducted in the opportunity assessment. It identifies the products made, the processes used, the key specifications, and the environmental impact of the processes. This section also identifies any relevant policies and objectives for the facility as well as any special environmental concerns.

Past environmental results. This section can be used to tout the gains from past environmental programs. In many states such as California, written pollution prevention plans are public documents. Preparing the plan presents a unique opportunity to discuss the environmental gains that have been achieved through the years.

Options. In this section the pollution prevention options that were identified through the brainstorming process are listed and described. Explanations are given for those options not selected for detailed evaluations.

Evaluation results. In this area the results of the technical and economic evaluations are presented. Options are then prioritized for implementation.

Implementation plan and schedule. This section contains the actual plan. The selected options are listed and the implementation schedule is outlined. This schedule may include a set time for reevaluating and updating the plan.

Expected results. The expected results are presented along with the schedule for obtaining those results. These expectations should be stated in terms that are measurable and meaningful for the facility. Expected results should be compared to the overall goals presented in the beginning of the plan. Since the ECM process entails continuous improvement, it is not likely that all the long-term environmental goals of the facility will be met in a two-year period. Plans should address ways to meet all the immediate environmental goals.

Measurement techniques. One frequent shortcoming of environmental plans is that they do not address how their success will be measured. A good plan should incorporate measurement techniques and assign measurement responsibilities. The importance of measurement has been demonstrated through the many quality management programs that are in existence.

In summary, the plan should be a useful document that can be understood by the people needed to carry out the plan. A sample pollution prevention plan can be found in Appendix A. Note that this is not a comprehensive environmental business plan but instead is a process plan that meets the requirements of most state legislation.

5.4 IMPLEMENTING THE PLAN

As previously mentioned, the plan must be implemented or all the development efforts will be wasted. This is all too often a problem with programs that are developed by outside consultants and turned over to an unenthusiastic crowd for implementation.

Two factors will help ensure implementation: maintaining high-level management pressure on obtaining concrete results and utilizing an energetic supporter within the facility to spearhead implementation efforts. There are similarities to total quality management programs, since failure of a TQM program is often accredited to a lack of sufficient upper management or staff support.

The following discussion presents a number of steps to be followed in implementing a significant process modification within a facility. Obviously, all these steps will not be required for all projects, and some facilities may require additional items to complete.

1. Obtain funding. The funding authorization process is different for each company. The ease of obtaining money seems to vary inversely with the size of the corporation. Typically a financial review form that calculates the return on investment must be completed. Completing this form requires that you develop more detailed cost estimates.

In the development of cost estimates, one should also develop a structure for tracking adherence to these estimates as a tool to control the project costs. Often, in the development of a budget tracking system for a project, you realize you failed to account for some costs. It is best to come to this realization before you have funding rather than after senior management has approved the project.

2. Build a project team. The success of the project lies in the talents of the project team and the ability of the project manager to keep the team working together toward the desired end result. For a large project, the project team could include personnel from the following groups:

Operations

Maintenance

Plant engineering

Product design/research

Process design

Environmental engineering

Electrical engineering

Process controls

Mechanical engineering (HVAC)

Fire protection

Safety

Structural engineering

Accounting

Quality control

Legal

Public relations

Sales and marketing

Obviously, coordinating the efforts of all these groups can require a dedicated project manager. Team members must work together and communicate their concerns and design issues. In smaller companies, specialists in all these disciplines may not be available, but the issues in each field are still important and must be addressed.

3. Outline project tasks and schedule. This step entails listing all the specific tasks that must be implemented to successfully complete the project. A precedence diagram is developed by indicating which items must be completed before others can begin. By adding the expected lengths of time to complete each task, an overall project schedule can be developed. Through this, the critical path items that directly affect the completion date of the project are identified. Many good computer programs are available to assist in developing and tracking project schedules.

4. Make project assignments. Once the project tasks are identified, they must be assigned to members of the project team. Assignments should include the schedule for task completion and the budget for the task. Team members should be informed as to the importance of their tasks to the overall project and the importance of the project to the company. The supervisors of team members should also be informed as to the commitments and workloads assigned to their employees.

5. Track schedules and budgets. At this point, one of the key responsibilities of the project manager is to track the schedules and budgets. The goal is to make sure the project meets the technical requirements and is completed on time and on budget. As mentioned, there are a number of computer programs to assist in project management.

6. Conduct pilot testing. Pilot testing may be required for new processes or significant process changes. This testing should be well thought out to make sure all necessary information is collected and undesirable variables are removed. Pilot test plans should be prepared and distributed to make sure the results will be applicable to full-scale operations.

7. Complete final designs and specifications. Once all the necessary information is collected, construction drawings and specifications are prepared. These documents must be reviewed for accuracy by various members of the project team and typically must be approved by the professional engineer in charge of the design activities.

8. Obtain permits. Permits can often be required before commencing construction, especially on projects that affect the environment. Some permits, such as air emission permits, may be required before ordering equipment.

9. Order equipment. Often the delivery of process equipment directly affects the project schedule. While some preconstruction activities can drag out for an extra month and not affect the final start date, delays of a week in equipment delivery can delay the project and drive up project costs. It is imperative to communicate regularly with key equipment vendors to make sure they are meeting their schedules. These vendors should be able to provide regular project schedules for their equipment fabrication activities.

10. Prepare bid/contract. Final drawings are typically sent out to contractors to obtain bids for construction. Contractors are given an opportunity to walk through the job site and ask any questions before preparing their bid. Bids are all due and opened at one time. Based on these bids and subsequent information, the project is awarded to one or more firms and they are contracted to complete the work.

11. Complete equipment runoffs. For major equipment items it may be beneficial to test the equipment at the vendor's site before it is shipped to the construction site. This runoff process allows any equipment bugs to be detected and fixed at the vendor rather than at the job site. This process can save time and money by avoiding the need to have equipment specialists from the vendor at the job site for extended time periods.

12. Oversee installation. The construction of any major process will never exactly follow every plan and drawing. The experienced contractors working on the project will always have suggestions to save money or make the installation proceed easier. They will also find problems with the drawing and specification package that will require field modifications. For this reason, it is important to have technical field support staff available to respond to these questions and assure contractor adherence to process requirements.

13. Prepare start-up plans. As the system is being constructed, plans must be made for testing and start-up. This is especially important for systems dealing with hazardous chemicals, since leaks and failures are most likely to happen during start-up and maintenance procedures. Piping should be flushed and pressure tested with water or air. Electrical devices with arcing potential should all be tested before the system is exposed to flammable solvents. Safety, fire protection, and spill response equipment and plans should all be in place before operating the system.

14. Check out devices. At this point, individual components are tested and calibrated to make sure they are properly installed. Faulty components are repaired or replaced. Leaking pipes are repaired and vent systems are checked to assure proper air flow.

15. Start system. Rarely does a major process system start on the first attempt. Process control bugs are identified and resolved and eventually the system is made to work.

16. Conduct training. Once the system is ready to be turned over to production, training of operators and maintenance personnel is required. This training should include hands-on instruction of the system and be supplemented by written hand-outs. Testing may be required to certify operator proficiency with the new system. Training should be videotaped to assist future operators in learning the system.

17. Complete manuals and as-built drawings. Finally, the documentation on the project must be completed to assist in the inevitable future process modifications. Drawings should be marked up and revised to reflect the changes made during the construction process. Performance testing results and initial process settings and readings should also be documented. Documenting final costs and schedules will also assist in completing the cost and schedule estimates for future similar projects.

18. Obtain final customer approvals. For some products, final customer approvals are not granted until the production system is up and running. Obviously, if there are problems in meeting customer specifications, then the project is not complete and modifications must be made.

If all these items are completed diligently, the project will progress smoothly and meet customer expectations. While every step may not be required for smaller projects, taking shortcuts can lead to problems in start-up.

The methods described in this chapter set up the supporting structure and identify and implement opportunities for product and process modifications. The next chapter addresses the longer-term methods to reduce the overall environmental impact of the production, use and disposal of the product. This is measured through the life cycle assessment process.

RESOURCES

U.S. Environmental Protection Agency, *Pollution Prevention Benefits Manual* (Washington, D.C.: 1989).

New Product/Process Design

Despite the efforts placed on waste reduction, pollution prevention, and source reduction training at the federal, state, and local levels, industry still relies heavily on end-of-pipe control technology to prevent pollution. According to the U.S. Department of Commerce, U.S. industry spent $80.6 billion on pollution abatement in 1991. Over 90 percent of the dollars spent were for recycling and waste treatment. Only 10 percent was spent on regulation, monitoring, and R&D.

In case after case, it has been demonstrated that environmental protection and bottom-line returns are interrelated. Yet the pollution prevention training that was conducted in the late 1980s and early 1990s failed to yield appreciable results. Companies continue to rely on pollution prevention gains derived from "picking the low hanging fruit," or making the easy changes. The fear of making a mistake prevents companies from taking a leap of faith and investing in new technology that will reduce pollution. These fears have to be salved through process demonstrations, performance and quality testing, economic evaluations, and giving product and process designers the tools, authority, and guidance to develop low-impact manufacturing.

Developing environmentally conscious designs will become more important as consumers and governments demand lower impact products. An example of the demands placed by governments is the use of take-back legislation. Take-back legislation requires that a company take back specified components of a product after the consumer is finished with it. The most famous of the take-back legislation is the German requirement that companies selling products in Germany collect and recycle packaging. Other legislative attempts at take-back legislation include a Yugoslavian proposal for appliance manufacturers to collect and recycle used consumer appliances rather than allowing consumers to dispose of them. The Yugoslavian proposal was aimed at forcing manufacturers to increase the life cycle of consumer products and decrease the volume disposed of in landfills.

The impacts of legislation developed to influence product design need to be thought out thoroughly. Legislation passed in Minnesota and other states requires all packaging and coatings applied to products manufactured in the state be free of heavy metals including lead, mercury, chromium, and cadmium. The legislation was designed to prevent these metals from leaching out of landfills and contaminating groundwater. However, one of the unanticipated effects could be the shortening of product life cycles due to the use of less corrosion-resistant coatings. The increased environmental impacts from additional raw material use, energy use, and disposal of products due to this shortened life cycle may far outweigh the intended benefit.

The benefit to Minnesota's groundwater would also be questionable. The majority of products purchased and disposed of in Minnesota come from out of state. Likewise, the majority of products manufactured in the state are sold and eventually disposed of out of state. As a result the likely benefits to Minnesota are minimal. The impacts on Minnesota businesses are more substantial. The legislation in effect prohibits Minnesota companies from applying some military specification coatings and high temperature resistance coatings, putting these companies at a competitive disadvantage for work requiring these coatings.

This chapter will investigate two different strategies for integrating environmental concerns into new product and process designs. Design for the environment (DfE) is the less rigorous of the two methods and generally can be quickly and easily adopted by a company's design group. Life cycle analysis (LCA) is an extremely in-depth process that is still in its infancy. All the tools and databases required to fully implement LCA are not currently available. The concepts that underpin LCA can be used on a smaller scale by providing the proper design tools and training.

6.1 DESIGN FOR THE ENVIRONMENT

DfE is based on the "design for X" (DfX) concepts. In using DfX, X, a design variable with desirable characteristics, is assigned a high priority in the design process. As the design process unfolds, trade-offs in various characteristics must be made. Under the DfX process, characteristics with a higher priority will dominate, shaping the final product design. DfE uses DfX concepts, assigning environmental considerations a high priority.

A detailed example of how a portion of a product's design might be shaped can be seen in the decision of what coating to apply to a steel panel. The potential design criteria that could be used in making the decision are environmental impact, coating quality, costs, energy consumption, and marketability. The alternative coating options and relative strengths and weaknesses are shown in Exhibit 6–1.

How the design criteria are prioritized will determine the coating selection. Assigning marketability a high priority will shift the design requirements to offering a large number of color choices, at a low cost. This limits the options to high solids and high solvent-based paints. If environmental impact is assigned a secondary priority, then high solids paint will be selected based on the lower amount of hazardous emissions. Reversing the priorities with environmental impact given primary consideration and marketability secondary consideration, then the choice will be between powder coating and high solids paint.

The main consideration in adopting DfX concepts is to determine where the company's priorities are to be placed. These priorities then need to be communicated to the design or R&D department. These departments are then responsible for incorporating the priorities while sorting through the trade-offs presented by the various design options.

Because DfE emphasizes selected areas, companies may develop a false sense of environmental security. The design processes will only reflect the priorities selected and do not necessarily form the basis for sweeping green marketing claims. The basis for valid green marketing claims, which is covered extensively in Part II, is in its infancy and gaining in complexity. Including environmental considerations in the design process does not necessarily eliminate all environmental concerns. Companies cannot relax their environmental management programs using the assumption that all problems will be eliminated or greatly reduced in the design process.

6.1.1 Barriers to DfE

Attitude forms the greatest barrier to adopting alternative design methods. The commonly held belief is that "if it ain't broke, why fix it?" However, the system is "broke." Why should waste be considered a necessary and inevitable by-product of manufacturing? Is not waste a reflection of poor design? What about the increasing cost of managing hazardous and toxic materials? Do these costs represent an outdated approach to product manufacturing? It is indeed broke, and it needs to be fixed.

Other barriers include the equipment and technology available at manufacturing plants. Often product design is based on the production processes available or limited

EXHIBIT 6–1

Design Criteria	Alternative Coating		
	Powder Coating	High Solids Paint	High Solvent-Based Paint
Environmental Impact	No waste or emissions	Low emissions; high waste	High-volume waste and emissions
Coating Quality	Excellent	Average	Average
Costs	Cost-effective in high volume	Cost-effective at high and low volumes	Cost-effective at high and low volumes
Energy Consumption	High temperature curing	Low temperature curing	Air dry
Marketability	Color changes difficult	Color changes easy	Color changes easy

by the capital available for new technology. In some cases, such as the electronics industry where product life cycles can be as short as 18 months, new investment in equipment and technology is necessary for survival. This presents a unique opportunity to develop environmentally conscious designs. However, industries such as electroplating may not experience a complete turnover in new technology for a decade or more. In the job shop, electroplating industry, the cutthroat pricing used to gain market share has dried up the capital needed to invest in new technology. This industry may be at a severe risk if chemical exposure regulations continue to tighten, requiring re-investment in new automatic, fully enclosed plating lines.

6.1.2 *EPA's DfE Program*

The EPA has developed a design for environment (DfE) program with the goal of fostering the design of products, processes, systems, and technologies considering environmental impacts within the design priorities. Under this program, the EPA will work directly with targeted industries to develop alternative design options and general analytical tools that can be used by all industries. Industries involved in the initial cooperative DfE program include computer and electronics, printing, and dry cleaning.

The EPA's program to provide industry with analytical tools for use in DfE applies EPA's strength in chemical risk analysis. It is this strength in developing chemical risk analysis that will provide industry with more quantitative methods of determining and avoiding chemical risks. While the EPA DfE program is still in its infancy, the analytical tools to be developed deserve mention because they will be readily available and widely applicable to

industry. The analytical tools being developed by the EPA include:

- Use clusters.
- Use cluster scoring system.
- Cleaner technology substitutes assessment.
- Cleaner technology inventory.

A use cluster is a set of chemicals, processes, and technologies that can substitute for one another to perform a specific function. The Solvent Alternatives Guidelines (SAGE) software is an example of the use cluster tools that are being developed by the EPA. SAGE allows users to input specific cleaning process data and receive a list of process alternatives to the use of solvent cleaners.

Because alternatives identified in use clusters are rarely direct substitutes, a use cluster scoring system was devised to compare the various considerations. Included in this scoring system are such factors as human and ecological risks, exposure, and pollution prevention opportunities. The use cluster scoring system is to provide a first cut for developing alternatives. This scoring method can also be used to direct research and development toward options with lower environmental impacts.

The Cleaner Technology Substitutes Assessment (CTSA) is an on-site, hands-on EPA program that helps develop the prioritized use clusters. The substitutes assessment compares the trade-offs associated with various options. The factors considered during a substitutes assessment are risk, releases, performance, cost, energy impacts, resource conservation, and pollution prevention opportunities.

The information collected through the use of the above tools will be incorporated into a Cleaner Technology Inventory. This database will be available through the

EPA's Pollution Prevention Information Clearinghouse (PPIC) network.

6.2   LIFE CYCLE ANALYSIS

Life cycle assessment or analysis (LCA) is a process to evaluate the resource consumptions and environmental burdens associated with a product, process, package, or activity. The evaluation is conducted by identifying and quantifying the energy and material usage and environmental releases across all stages of the life cycle of the product. This is done to assess the impact of the energy and material uses and releases on the environment and to evaluate and implement opportunities to affect and improve the impact on the environment.

There are four basic components in a life cycle analysis:

1. *Goal setting and scoping.* Before beginning an assessment, a company must evaluate the purpose and scope of the assessment.

2. *A life cycle inventory.* This inventory involves examining the inputs and outputs of each stage in a product's life cycle.

3. *An impact assessment.* This component of an LCA involves the potential and actual environmental and human health effects that are related to the use of resources and environmental releases.

4. A life cycle improvement analysis. This involves analyzing the changes that are needed to bring about an environmental improvement in a product or process.

These four separate components represent an integrated, yet not necessarily linear, approach that provides the information needed to maximize environmental improvements. For example, the inventory alone may identify opportunities for reducing emissions, energy, and material use. The impact and improvement analyses can be used to help ensure that the potential reduction possibilities are optimized and do not produce unanticipated problems. Furthermore, each component may and should be continually changed and modified as the LCA continues. For example, the goals and the scope may change as the inventory advances and the scope appears either too narrow or too extensive.

6.2.1   History

The basic concept of life cycle analysis is not a new one. As far back as the early 1960s, individuals and companies were developing reports outlining energy requirements for production. In the late 1960s, predictions were made that increases in raw material and energy demand would lead to exhaustion of fossil fuels. Concerns over global warming, melting of polar ice caps, and other climatological changes began to emerge. These alarming predictions resulted in an interest in carrying on more detailed energy calculations on industrial processes.

In the United States, major fuel cycle studies were performed to estimate the costs and environmental implications associated with alternative energy sources. These studies included estimates of gaseous, solid, and liquid emissions. During the early 1970s, when the oil shortage occurred, both the U.S. and British governments began extensive studies of a wide range of industrial systems. This work provided data on raw material requirements and on the mass of solid waste emissions.

As the oil crisis disappeared, interest in energy analysis and environmental concerns faded, although a number of individuals in the United States, the United Kingdom, Scandinavia, Switzerland, and Germany continued their activities. During the late 1980s, the Green Movement, including the formation of Green political parties in Europe and Germany, in particular, rekindled interest in the subject. When problems of solid waste accumulation led to barges of garbage sitting in the New York Harbor, solid waste concerns again came to the forefront in the United States. Green consumerism, among other factors, prompted cities and states to develop recycling projects. At the same time, acid rain and dwindling rain forests increased global attention on waste management strategies. As a result of these pressures, gaseous, solid, and liquid emissions were routinely added to energy, raw material, and solid waste considerations. The continued problems with solid waste disposal, loss of habitat, and prospective oil shortages have emphasized the need for comprehensive environmental improvement programs, including life cycle analysis.

Now, proactive companies, including AT&T, General Electric, IBM, Procter & Gamble, Whirlpool, and Xerox, have begun to use life cycle analysis. These companies have found it is too costly to change products already in existence; therefore, they are beginning to design environmental compatibility into their products. This process has been termed designing *greenness* into a product.

A "green" product can be characterized by a number of factors including: (1) reduced use of raw material; (2) non-polluting manufacturing; (3) no unnecessary animal testing; (4) no impact on protected species (i.e., dolphin free tuna); (5) low energy consumption during production, use, and disposal; (6) minimal or no packaging; (7) reuse or refillability where possible; (8) long useful life and/or updating capacity; (9) postconsumer collection and disassembly system; and (10) remanufacturing capability.

6.2.2   Life Cycle Stages

There are four stages in a product's life cycle: (1) raw material extraction and processing; (2) manufacturing; (3) consumer use, reuse, and maintenance; and (4) product retirement.

Stage One—Raw Material Extraction and Processing. In the first life cycle stage, the environmental impacts of extracting and processing raw materials needed for the product should be considered. Simply using renewable resources and substituting recycled materials in the product for virgin material can decrease the environmental costs of this stage.

For example, Montagne Jeunesse, a British company that makes cosmetics and toiletries, takes great care not to use plant materials that cause destruction of natural habitats such as rain forests and wetlands.

Another example is the use of recycled aluminum. One pound of recycled aluminum saves up to 4 pounds of bauxite and 95 percent of the energy used in making one ton of virgin aluminum while producing 95 percent less air pollution and 97 percent less water pollution. The main challenge of this particular stage is to identify and specify material that can minimize environmental costs.

Stage Two—Manufacturing. Much attention has been concentrated on this particular stage of a product's life cycle. As discussed previously, the difficulty is that most of this attention has been focused on end-of-pipe controls. In other words, focus has been on *treating* wastewater and other industrial waste rather than minimizing the waste at production. In contrast to industry's focus on treatment, the EPA's waste hierarchy places prevention first, then source reduction, then reuse or recycling, and only as a final recourse, treatment and disposal. Despite this fact, 99 percent of federal and state environmental spending goes to end-of-pipe controls.

More recent attention has concentrated on developing methods for reducing the generation of pollutants at their source. For example, products can and should be designed for easy assembly to help reduce defects and rework, products should use molded-on finishes on plastic surfaces to avoid the environmental effects of paint, and materials should be designed to be more compatible with environmentally benign cleaners such as water-based solvents.

Another more long-term approach to improving greenness at this stage involves developing and building "eco-factories." Montagne Jeunesse is currently building one of the United Kingdom's first eco-factories. This factory will use renewable energy and energy efficient systems in the design of the factory itself. It will use a 60-kilowatt wind turbine and advanced vacuum solar collectors to provide hot water and solar panels to provide lighting. Special glazing on the windows will increase lighting and, finally, the design will actually use recycled and toxin-free materials such as insulation made from recycled newspaper.

Stage Three—Consumer Use, Reuse, and Maintenance. This stage involves, for example, federally mandated fuel efficiencies and exhaust emission standards for automobiles and design-based initiatives that reduce the energy use required for products such as computers, refrigerators, and dishwashers.

Stage Four—Product Retirement. Until recently, this stage has been given the least attention. With the recent focus on the dangers of excessive solid waste, this has become a more important element in the life cycle analysis. Companies are now encouraged to design their products with reuse and recycling in mind.

Products are often designed to encompass both recyclable and nonrecyclable materials. Therefore, the design of these products must take this into consideration. Designing for product retirement must involve the consideration of the following three criteria.

1. Design for recyclability. Design for Recyclability (DfR) focuses on reducing the product's impact on the environment when it reaches the end of its useful life. This involves an attempt to facilitate the complete separation of product components so that recyclable components can be separated and recycled and nonrecyclable components can be disposed of in another manner. This makes the recyclable material more valuable; therefore, people will be more apt to recycle. Also, separation must be made easier and therefore cheaper. This is also referred to as "design for disassembly," or DfD.

2. Design for remanufacture. Parts in a product usually wear out at different rates. Designing for remanufacture allows for replacement components to be created and thus for overall waste to be diminished. An example of this is Xerox Corporation's development of customer replacement units designed to replace the toner in the copy cartridge without replacing the entire copy cartridge. Another example is Germany's Volkswagen, which has announced that it will take back its new Golf model free. This is a step toward the company's goal of 100 percent recycling.

3. Design for disposal. Design considerations in this stage should focus on ways to ensure that the product can be disposed of safely.

6.2.2.1 Closed loop analysis.

Another way to look at life cycle or cradle-to-cradle analysis is to term it as a value chain being remodeled to a circular remanufacturing reconsumption process. At one time, the value chain was a linear pattern starting with design and ending with service. This chain is now being remodeled into a circular pattern, and leaders are striving to close this loop entirely to minimize waste and emissions, thus, the phrase cradle to cradle as opposed to cradle to grave.

An example of the most successful effort in this close-looped process is Anheuser Busch. Anheuser Busch is the world's largest brewer but also the world's largest recycler of aluminum cans. In 1991, the company reached a milestone by recycling one used can for every one of the 17 billion cans it produced. Not only did this cut down on waste, but it also cut costs since production from recycled materials saves 95 percent of the energy needed to make aluminum from raw ore.

6.2.2.2 Limitations of life cycle analysis. There are limitations of a comprehensive LCA. For example, to determine whether one product is truly greener than another, it is necessary to know how the inputs and outputs should be weighed to reflect their relative environmental risks. To determine this balance, additional environmental data, currently in short supply, are needed.

Another limitation is that the data needed for a complete evaluation can be immense. It is not yet clear where boundaries of the analysis should be drawn; this, combined with the uncertainties and complex interactions among design choices, requires an amount of data that can make LCAs hopelessly complicated.

LCAs will be very useful tools in designing products for a better environment, but further development and research must be done before they become an effective and accurate method for every industry. Still, many companies are developing design projects using information gleaned from generic LCAs.

6.2.3 Implementing Life Cycle Analysis

As discussed above, there are four basic components in life cycle analysis: (1) goal setting and scoping; (2) inventory; (3) impact assessment; and (4) improvement assessment. The following discussion on performing an LCA will center around these four components.

6.2.3.1 Goal definition and scoping. There are many purposes and uses for life cycle analysis. Before beginning an assessment, a company must determine which of these goals and objectives are behind the assessment. A company must also determine whether the results will be for internal use only or whether they will be released to the public or used in a public policy context. These issues are important because of possible legal ramifications of releasing the information to the public or to governmental agencies. For example, by releasing this information to the public, confidentiality and privilege may be lost. Also, proprietary information may be released.

There are several helpful questions for developing and defining the goals and scope of an LCA and of the impact assessment component of the LCA:

- Does the goal of the impact assessment match the overriding goal of the life cycle analysis?
- Have limits been imposed allowing concentration on certain impacts and processes due to

  a. Time.
  b. Finances.
  c. Requests of the client.
  d. Existing facility constraints.
  e. Technical limitations?

- Is the aim to optimize a product or develop a new one?
- How comprehensive is the approach intended to be?
- What limits have been placed on the extent of the analysis of the impacts?
- Has consideration been given to the audiences and their interest, priorities, and level of scientific expertise?

By answering these questions, a company can determine what the level and scope of the life cycle analysis should be and if the limits have been adequately defined.

One overriding concept to keep in mind is that there is an interrelationship between all the steps and components of an LCA. Therefore, the initial goals and scope are only a starting point and will need to be reexamined and reformulated throughout the process.

6.2.3.2 Inventory. After defining the goals and scope of LCA, an inventory must be completed (see Exhibit 6–2). An inventory includes an examination of the entire life cycle of the product, including: the extracting and processing of the raw materials; manufacturing; transportation and distribution of the product; use, reuse, and maintenance of the product; and the recycling and final disposal. Often, more specific scoping and goal definition will be needed as the inventory begins and expands.

The inventory is a fairly complex and in-depth process. It can be completed by consultants engaged to conduct such a process or by several internal teams with knowledge and experience in each stage of the life cycle. In some cases, the information may already be available in various formats and can simply be compiled to complete the inventory. For example, information regarding output levels such as air emission, water pollutants, and even habitat destruction may have already been gathered in order to apply for government permits and comply with regulations.

Raw materials and energy. The considerations included in this category involve all the activities needed for and resulting from acquisition of a raw material. This particular component of the life cycle begins with acquisition of the raw materials and ends with the first manufacturing or processing stage.

When conducting the inventory, inputs and outputs to every stage of the life cycle must be examined. One input to the raw material and energy stage involves the energy utilization and the type and mix of energy sources that are utilized to obtain the raw materials; for example, fossil fuels used to operate mining machinery or agricultural equipment.

Materials should also be an input at this stage. The materials to be reported should either be consumed or used in maintaining the raw material source; for example,

EXHIBIT 6–2
Life Cycle Inventory

Life Cycle Stages	Input Examples	Output Examples
Raw materials and energy	• Energy used to obtain raw materials • Pesticides • Infrastructure and capital equipment	• Air emissions • Water pollutants • Solid wastes • Habitat changes
Manufacturing processing and formulation	• Raw materials (maybe recycled) • Energy sources • Grease • Cleaning products	• Product itself • Co-products • Waste, including emissions, water pollutants and solid waste
Distribution and transportation	• Product itself • Fuel • Motor oil • Spare parts	• Noise • Air pollutants • Product itself
Use/reuse and maintenance	• Product itself • Maintenance facility • Retail outlet • Motor oil	• Air emissions • Noise • Product itself • Water pollutants • Used motor oil • Detergent
Recycling	• Product itself • Receiving facility • Processing facility	• Air emissions • Raw materials
Waste management	• Product itself • Waste	• Raw materials • Air emissions • Water pollutants

pesticides on crops and chemicals to control dust emission in mining activities. Another input in this area is the infrastructure and capital equipment needed; for example, roads, buildings, and equipment used to extract or obtain the raw materials.

The outputs in the acquisition of raw materials are the air emissions, the water pollutants, the solid wastes and other releases, and the habitat changes or destruction that result from the gathering of raw materials. Examples of other outputs that must be considered are oil spills, agricultural run-off, and aesthetic degradation. Some of these things are more difficult to quantify but must at least be considered and listed as potential outputs in the life cycle inventory.

There are two basic categories of raw materials: energy and nonenergy. Examples of materials in the energy category are natural gas, petroleum, coal, nuclear, and hydroelectric. Examples of nonenergy raw materials include, among others, bauxite, cotton, corn, brine, chalk, gas, lead, limestone, coal, oil, sand, water, wood, and zinc. When taking inventory of these materials, distinctions must be made between renewable and nonrenewable resources in both the energy and nonenergy categories. Credit must be given for use of renewable resources. A

resource is considered renewable if it is replenished at a rate equal to or greater than its rate of depletion.

Credit also must be given for use of recycled raw materials. Companies must note that there is a debate over what is waste and what is a co-product and as such, what should be given credit for being a recycled raw material and what is simply a co-product. Co-products are products, created as a result of producing the main product, that can be used for another purpose. For example, when processing beef, the beef is the product and the leather and fat (used for soap) are the co-products. Waste has no market, intrinsic value, or alternative function.

When examining renewable resources, one must also consider the quality of the resource when it is renewed or whether habitat is destroyed even if the resource is renewable. For example, wood is a renewable resource, but habitat for some creatures, for example, the spotted owl, is destroyed; and although water is also considered a renewable resource, the quality of the water may be lower after being renewed.

The inventory of this area involves listing all the inputs and outputs and giving consideration to use of renewable and recycled raw materials.

Manufacturing, processing, and formulation. This component of the life cycle involves the conversion of raw materials to final products. It begins with the initial receipt of raw materials and includes on-site storage and handling. The process is complete when the product is in its final manufactured form and transferred to distribution. The primary packaging and any packing required to transport the product to the first distribution site are included in this component of the life cycle.

The inputs in this category are the raw materials and energy sources and various ancillary materials, such as grease and cleaning products, that are used to produce the products but do not become part of the products. Recycled materials can also be an input in this stage.

The outputs include products, co-products, and waste. In this stage, when doing the inventory, it is important to assign a proportion of the waste created and the energy needed to both the co-products and the products. Waste needs to be evaluated in its post-treatment stages to reflect the emissions that are actually discharged to the environment.

Distribution and transportation. Distribution and transportation are not separate components of the life cycle itself. Instead, they involve every component of the life cycle and every transportation involved within the entire life cycle.

Transportation is defined as the movement between operations at different locations. Distribution is defined as all non-transportation activities that are carried out to facilitate the transfer of a manufactured product from the final manufacturer to the ultimate end user. This includes movement of goods within a warehouse or within a retail establishment. Also included in both transfer and distribution are any environmental controls such as temperature and humidity. Again, this is not an independent portion of the product life cycle, but instead can be involved at various stages.

Inputs at this stage could possibly include the product itself, the fuel, motor oil, coolant, spare parts, and other products used by a refrigerated truck. Outputs might include noise, refrigerant loss, air pollutants, and the products. The data required to do an inventory of this area are: the energy consumption and form required for the transportation, the supplies required, the energy conversion and pollutant emission factors, the product and materials loss rate, the discarded and added packing materials, and the results of accidents and spills.

It is important to set boundaries at this stage. In other words, will the inventory be more or less detailed? For example, if the method of transportation is a refrigerator truck, do you consider the inputs and outputs of just the truck, the truck and the roadways, or the truck, the roadways, and the truck stops? Also, do you consider just the truck and the outputs its use generates, or do you also consider maintenance on the truck such as the inputs and outputs involved in the production of spare parts, or do you go all the way back and consider the actual production of the truck such as truck factories? Answering these questions will help you define what is involved in your inputs and outputs.

Use/reuse/maintenance. This stage begins after distribution of the finished product to the end user and ends when the product is discarded and enters a waste management process. Use includes such activities as consumption of a product, operation of equipment, storage of a product for later use such as in a refrigerator, and preparation of a product for use such as cooking.

Inputs and outputs of maintenance include on-site and off-site repair and preventive maintenance. It also includes the transportation needed to take the product to an off-site repair or maintenance facility. Furthermore, it includes the need for traveling to retail outlets to purchase supplies for on-site repair and maintenance. Reuse includes both on-site and off-site reuse. It can be intentional or incidental and includes things such as donations to charities, rental of equipment, and return of a product to the manufacturer to be reused for the original purpose.

Data for this component of the life cycle can be gained from three sources: (1) asking consumers how they use a product, (2) published studies and articles, and (3) assumptions. The information that is needed includes, for example, (1) how long a product is used before it is disposed of; (2) what is used with the product for maintenance (such as oil for a car and detergent and bleach to wash a washcloth); (3) the frequency of the product's repair or maintenance; (4) other uses of the product beyond the original purpose (such as using a mayonnaise jar to hold paintbrushes); and (5) disposal of a product.

Recycling. Recycling is another stage that is not distinct in and of itself but does apply to some products. Recycling begins when a discarded material or product is delivered to a collection system for recycling. It includes the collection and handling of the materials plus any processing that is needed to prepare and deliver the materials for reentering a manufacturing system as a raw material. There are two types of recycling—closed loop and an open loop. The closed loop process involves a product that is recycled to use for the same purpose it was originally designed for. An open loop product is recycled to use for a different purpose.

In a closed loop, one must examine:

A.—All inputs and outputs associated with production of primary materials.

B.—All inputs and outputs associated with disposal of the product.

C.—All inputs and outputs associated with recycling.

D.—All inputs and outputs associated with a nonrecycling system.

E.—Any inputs and outputs from converting a product with recycled material instead of raw materials.

To determine the net inputs and outputs of a closed loop system as opposed to disposal, the following formula is used:

Net inputs and outputs $= D - B(a) - A(b) + C(a) + E,$

where

a = recycling rate

and

b = recycled contents of the product.

In an open loop system, the following must be considered:

A.—All inputs and outputs associated with production of the virgin material for product two.

B.—All inputs and outputs associated with disposal of product one.

C.—All inputs and outputs associated with recycling of product one.

D.—All inputs and outputs associated with no recycling system for product one.

E.—All inputs and outputs associated with a virgin system for product two.

F.—Any conversion inputs and outputs incurred as a result of product two using recycled matter.

To determine the net inputs and outputs of an open loop, the following formula can be used:

Net inputs and outputs $= D + E - B(a) - A(b) + C(a) + F,$

where

a = recycling weight of product one

and

b = recycled content level of product two.

Waste management.   Within waste management, the EPA developed a hierarchy for which method of waste management is the best. The hierarchy is: (1) waste minimization and waste reduction; (2) reuse; (3) recycling/material recovery; (4) composting; (5) thermal physical, chemical, or biological treatment; and (6) disposal on land, ocean, and in groundwater.

The waste management component of the life cycle begins where waste is generated. Each stage of the life cycle will generate waste; therefore, this is not a component in and of itself but must be considered at each stage of the life cycle. Airborne, waterborne, and solid waste must each be analyzed separately, and each stage of the life cycle where waste is generated must be analyzed separately. By separating the types of waste and the life cycle

stage in which the waste is generated, it becomes easier to identify the phases where waste is generated and can be reduced.

6.2.3.3   Impact Assessment.

Impact assessment involves the evaluation of impacts on any system as a result of some action. There are four impact categories: (1) ecological health, (2) human health, (3) resource depletion, and (4) social welfare.

Social welfare is not yet considered to be an important or necessary impact category, but companies may chose to consider the impacts on social welfare depnding on the purpose and scope of the LCA. To determine the impacts of these categories, one must determine and evaluate the stressors. Stressors are conditions that may lead to human health or ecological health impairment or to resource depletion. The stressors are discovered in the life cycle inventory stage discussed previously and are linked to the impact stage.

One stressor may have many different impacts. Therefore, the first step in an impact assessment is to organize the life cycle inventory data into stressor categories that are linked to ecological and human health and welfare impacts. An impact assessment does not involve how much actual quantitative impact an individual stressor has. Instead, it is important to know the relative magnitude and contribution a stressor has to an impact category. In other words, you want to link the inventory data with the impacts the stressors cause and assess the potential magnitude of the impact by the quantity of the release, the potency of the release, the expected environmental concentrations, and the probable exposure.

There are three steps to an impact assessment. The first is the classification of the data into the stressor categories within larger impact categories. The second is characterization that involves the analysis and estimation of the potential impact on the ecological health, human health, or resource depletion for each stressor category. The third is the evaluation, or the relative values or weights given to each of the different impacts. The classification into individual stressor and impact categories can be done by using the life cycle inventory and the impact categories.

Determination of potential impacts.   The classification into individual stressor and impact categories can be done using the life cycle inventory and the impact categories. Several levels can be used in this step. The levels chosen will depend on the purpose and scope of the LCA. As an example, the impacts of chemical stressors can be examined (see Exhibit 6–3).

Level one is loading assessments, which involves little more than simply using the life cycle inventory. The weight and volume of each chemical contaminant released to the environment either by life cycle stage or in total for the product is examined and placed into impact categories. This particular level of examining a potential impact

EXHIBIT 6–3
Potential Impacts

	Advantages	*Disadvantages*
Level 1—loading assessments	• Easy • Less expensive • Can ID stages of life cycle where loading could be reduced	• Little improvement in environmental impact • Wasted resources not IDed • Too simplistic
Level 2—impact equivalency/ assessment	• Comparisons between outputs can be made	• Difficult to create equivalency units
Level 3—toxicity persistence and bio-accumulation profile	• Areas for improvement can be identified more easily	• More expensive
Level 4—generic exposure	• Impacts can be more easily valued	• Models detailing effect of stressors are limited • Expensive
Level 5—site-specific exposure	• Impacts on site more easily valued	• Expensive • Models almost nonexistent

can identify stages of the life cycle where loading could be reduced, and it can compare loading between products. It is an easy and less expensive method. The disadvantages are that this analysis often results in little improvement in the environmental impact of the product and wasted resources may not be identified and opportunities for improvement may be missed.

Level two is an impact equivalency assessment. In level two, chemicals are assigned to chemical stressor groups, and for some groups, it is feasible to develop impact equivalency units that can be summed up and used to assess the collective contribution of emissions to environmental problems. The advantages to this level are that different inventory outputs can be compared within the exposure and toxicity framework. The disadvantages are it is difficult to create an understanding of an equivalency unit approach and there is not a direct relevancy to environmental improvements.

Level three involves a toxicity persistence and bio-accumulation profile. This includes an examination of the physical and chemical properties of substances to assess their fate and potential environmental effects.

Level four is generic exposure. This links environmental loading to individual populations and ecosystems by using models to examine the potential impacts on an organism's population and an ecosystem's characteristics. This creates a database of potential impacts that can be examined and assigned relative values. The disadvantage is that the models detailing the effects of stressors are limited.

Level five involves site-specific exposure and effect assessment. This involves actual examination of a site-specific field study.

Besides examining only chemical stressors as the levels above discuss, the same levels can be used to examine the effects of ionizing radiation, heat, noise, environmental disturbance such as habitat alteration, and physical change to water. One must also examine the physical change to soil such as compaction, including paving and other impermeable soil coverage, erosion, and soil loss. Loss of wetlands, regional climate change, and species change must also be considered.

Decision analysis. After the impact assessment, products or processes may be compared by assigning weights to the various impacts. It is important to weigh the relative importance of each impact and compare the relative impacts of different processes. When doing this, uniqueness of the resource or area being impacted, the size of the area being impacted, the reversibility of the impact, and the magnitude of the impact can be used to compare the relative importance of the impacts (see Exhibit 6–4).

Uniqueness involves the examination of the impact on ample resources, moderately available resources, or scarce resources. The area of the impact involves an examination of the spacial extensive impacts—local, regional, continental, or global. The reversibility of the impact involves an examination of the short-term, moderate, or long-term effects of the impact. Finally, the magnitude involves the size of the impact.

After doing an impact assessment, how do you make

EXHIBIT 6–4
Impact Analysis Criteria

Impact Analysis Criteria	Definitions
Uniqueness	Is the resource ample, moderately available, or scarce?
Area	Is the impact local, regional, continental, or global?
Reversibility	Is the impact short term, moderate or long term?
Magnitude	How big is the impact?

decisions about what to do with the product? It may be that the product has a small enough impact that the manufacture and use of the product should continue with no or only slight modifications. It may be that certain impact categories are determined to be too high and some processing changes need to be made. Finally, there may be too much total impact, resulting in a decision to drop the product from production.

Evaluation.   As discussed above, values and relative weights are assigned to various categories of impact. This is not an entirely objective process and involves the subjective beliefs and values of the individual or company making the evaluations. In fact, every step of LCA involves the subjective beliefs and values of the individual making the decisions. This fact should be recognized and accounted for when using the results of the LCA.

For example, when originally developing goals, a company may decide it is concerned only with resource depletion. This is a value judgment that could be based on many factors, including fear of imminent future government regulation in this area, fear of consumer problems related to resource depletion, or desire to develop alternative sources of raw materials in order to continue production of a product. In other words, the evaluation results will be affected by the values and beliefs of those making the decisions, and this fact should be recognized.

6.2.3.4 Improvement assessment. After relative values and weights are assigned to the impact categories, a plan for improvement must be developed. Currently, there is little research or writing regarding this particular component of the LCA process.

Opportunities for impact reduction include: minimization of energy and raw material consumption, closed-loop systems for chemicals, minimization of activities that result in habitat destruction, and minimization of waste releases. To achieve these goals, companies can design products for minimal environmental impact, use good manufacturing practices, improve energy and material efficiency, and improve recycling systems. Finally, companies can keep ownership of products. For example, some companies rent chemical solvents, thus preventing others from inadequately disposing of the dangerous products. As mentioned above, some manufacturers take back the product for disassembly and recycling.

LCAs are new processes and are still very much in the development stage. The methodologies discussed here are examples of possible means for conducting effective LCAs. As industries begin conducting LCAs, the processes and methodologies will evolve to fit the circumstances and the industry, but the advantages and benefits of completing the studies will only increase.

Case Study

Ingersoll-Rand's Experience with Waste Reduction

In 1987, Ingersoll-Rand completed a study to determine the feasibility of developing an in-house hazardous waste incineration capacity. Since 1986, Ingersoll-Rand had been tracking the volume of hazardous and nonhazardous wastes generated at its manufacturing and service facilities. Even at this early junction, the waste survey and projections were showing an alarming increase in the volume of waste generated and in the cost to dispose of these wastes. In addition, the company was listed as a potentially responsible party (PRP) at several hazardous waste landfills. Even though the hazardous waste was disposed of in accordance with the laws at the time and in "state-of-the-art" landfills, the company was still faced with millions of dollars in cleanup costs.

The company believed the solution to this problem was to purchase, install, and operate its own incinerator, which would provide control over disposal methods and costs. The results of the incineration feasibility study set in motion an environmental management program that quietly became one of the most effective and little noticed in the United States. Following are some highlights of the feasibility study:

- The return on investment from an in-house incinerator was estimated at between seven and eight years. Due to the potential for drastic changes in environmental regulations, this was an unacceptable risk.

- To limit transportation costs, siting and permitting the incinerator in the southeastern United States was the most cost-effective option. At the time of the feasibility study, the states of North Carolina and South Carolina were embroiled in bitter debates over the siting of commercial incinerators.

- And finally, an in-house incineration capacity would actually discourage waste reduction. The lower in-house costs for incineration would lower the economic incentive to reduce waste at the source.

Based on the negative economic, public relations, and environment aspects of the proposal, it was decided that the money and effort would be better spent on a waste reduction program.

Program Goals.   Based on the data developed from the waste survey, five areas were identified as posing environmental risks and problems, but also having potential solutions. The areas identified were:

  Chlorinated solvents

  Plating wastes

  Oils and coolants

  Paints and coating

  Grinding swarf

These areas made up over 90 percent of the waste generated throughout the company.

A five-year program was laid out to address each of these areas. The goals of the program were aggressive—a 50 percent reduction in hazardous waste generation by the end of the third year and a 90 percent reduction by the end of the five-year program.

Program Implementation.   The five-year waste reduction program was designed to provide a large amount of technical assistance. The specific components of the project included:

- Waste reduction assessments.
- Research and development with an emphasis on waste reduction.
- Waste-specific workshops.
- Engineering assistance.
- Demonstration projects.

Waste reduction assessments were added to the corporate regulatory compliance and risk reduction assessment program. During the compliance assessment, a waste engineer was assigned to complete a tour of each facility, assemble process- and product-specific data, and identify waste reduction opportunities. The results of the waste reduction assessment were included in the assessment report that was delivered to the facility manager and corporate officers. The intent of the waste reduction assessment was not to provide definitive answers to waste generation, but rather to present the various options and demonstrate the potential environmental effects and any side benefits.

Several areas of waste generation had limited options, or the options available also created their own waste disposal problems. An example was the use of copper-cyanide plate as a heat treat stop-off. Another option included the use of a heat treat stop-off paint. However, to remove the paint after the heat treatment operation required the use of a chlorinated solvent. Ingersoll-Rand's product development group developed a proprietary method that eliminated the use of copper plate, improved product throughput, and reduced operating costs.

Continued on pg. 60

Continued from pg. 59

Waste-specific workshops were held to provide information on waste reduction alternatives to facilities with specific operations. These workshops targeted plating operations, chlorinated solvents, painting operations, and coolant and oil recycling. One of the unanticipated results of the workshops was the information exchange between facilities. During one discussion, a facility was describing the steps to be taken to convert from a zinc cyanide plating process to an alkaline zinc process. The response from one of the other workshop attendees was that they had accomplished that very same project several years ago and were able to provide information that accelerated the conversion.

Engineering assistance was contracted through an environmental consulting and engineering firm. When facilities had a need for information, an alternative opinion, or additional manpower, they were able to contact this firm for the necessary support.

In cases where one waste stream was identical in a number of plants, a demonstration project was arranged at one site. Environmental coordinators and engineers from other sites could then view the completed project and use the information to implement similar projects at their facilities.

Program Results. The waste reduction program started in 1988 and was to end in 1992, although the successes achieved and yet to come have kept the program alive. By the end of 1993, Ingersoll-Rand's generation of hazardous waste had been reduced by 67 percent, short of its goal of 90 percent but still a considerable achievement. The waste reduction process has slowed considerably as product- and process-specific issues had to be resolved through research and development efforts. Many of these R&D efforts are still being implemented with the benefits to be realized over several years.

The side benefits of the waste reduction program have also been considerable. If no waste reduction gains had been made over the period, the cost to dispose of the additional waste would have been $9 million. Also, improvements in product quality, increased production throughput, and reduced production costs were also realized throughout the company.

In 1991, Ingersoll-Rand was contacted by the EPA about participating in the industrial toxics project (ITP), also known as the 33/50 program. Based on the data gathered in the waste survey, Ingersoll-Rand decided to participate by setting its goals for a 33 percent reduction by 1992 and a 60 percent reduction by 1995. In 1991, Ingersoll-Rand was recognized for achieving a 57 percent reduction in emissions of the 17 targeted ITP chemicals. By 1994 the company forecasted a 90 percent reduction in ITP chemical emissions. Of particular note was the fact that Ingersoll-Rand did not need to implement a new program to achieve the ITP emission reductions.

In describing the Ingersoll-Rand waste reduction program, Jim O'Dell, vice president of operations, had this to say:

> We recognized that from a pure cost standpoint and later on from a pure responsibility standpoint that we had to eliminate the hazardous materials we were generating and then having to dispose of. We dealt with it as we deal with any other management issue. Basically if you want to move the organization in some direction, you have to tell them where we're trying to get to, why we're trying to get there, and set up the management systems that enable you to move forward. Our management systems include a very good measurement system and a very good feedback system. Then you apply all the resources available to achieve the objectives. We succeeded because we did these things and provided all the resources we felt were needed, capital and information.

RESOURCES

Designing for the Environment, p. 55.

Garcia, Shelly. "Getting Labeled: Green Seal's Sticky Situation." *Adweek*, eastern edition, April 15, 1991.

Lasota, Jolanta. "Environment Friendly from Cradle to Grave." *Cosmetics and Toiletries Manufacturers and Suppliers*, January 1992, p. 1.

Oakley, Brian T. "Total Quality Product Design—How to Integrate Environmental Criteria into the Production Realization Process," *Total Quality Environmental Management*, Spring 1933.

Simon, Francoise L. "Marketing Green Products in the Triad." *Columbia Journal of World Business*, Fall/Winter 1992, p. 13.

Society of Environmental Toxicology and Chemistry, *A Conceptual Framework for Life-Cycle Impact Assessment* (1993).

Society of Environmental Toxicology and Chemistry, *A Technical Framework for Life-Cycle Impact Assessment* (1993).

Personal conversation with James O'Dell, senior vice-president of technology, Ingersoll-Rand, 1993.

II

MARKETING, LABELING, AND PACKAGING

Chapter Seven

Understanding Environmentally Conscious Marketing: The Conceptual Framework

Although environmentally conscious marketing preceded the early 1980s and enjoyed a surge in the mid- to late-1980s, it remains a new and rapidly changing area of business. At the same time, environmentally conscious manufacturing is simply a modern facet of public relations and advertising, and it is becoming as essential to successful competition as other forms of public relations and advertising. Yet, if environmentally conscious marketing is just another kind of marketing, why does it deserve special attention? The answer is twofold.

First, consumers have a very specialized set of expectations with respect to environmentally conscious marketing claims, and these expectations may change rapidly. The risk that an environmental claim may backfire is substantial if claims are made without careful attention to the specific concerns of the environmental marketplace.

Second, there is a highly specialized regulatory focus on environmentally conscious marketing claims that involves a more complicated set of rules for environmentally conscious marketers than traditional marketers. Currently, there is also a high level of regulatory zeal that subjects an environmentally conscious marketer to a greater risk of regulatory attention, and associated liability or penalties, than other aspects of marketing.

Despite fits and starts in early attempts at environmentally conscious marketing, there is no doubt that environmentally conscious marketing is here to stay, and the opportunity is there for those willing to commit the resources to become effective in the green marketplace. The best news of all for those following a corporate philosophy of environmentally conscious manufacturing is that environmentally conscious marketing easily follows a solid foundation of sound environmental engineering and management in the design and manufacturing of your products. Indeed, there is no substitute for a good product. Environmentally conscious marketing is therefore most effective as one piece of an integrated, environmentally conscious operation. Environmentally conscious marketing issues should be a consideration during the entire design and manufacturing process; conversely, design and manufacturing considerations should drive environmentally conscious marketing efforts by building a solid foundation that will support environmentally conscious marketing claims.

7.1 THE GREENING OF THE MARKET

It is no secret that the market is greening. There are at least three factors at work to define environmentally conscious marketing and to push it forward: (1) regulatory pressures from the local level to the national level and even the international level are forcing firms to become green; (2) consumers, inspired by public education, are becoming increasingly green in their attitudes, creating a significant demand for green products; and (3) the competition is engaging in aggressive environmentally conscious marketing campaigns, in response to consumer demand. Accordingly, environmentally conscious marketing is attractive both because of the carrot of new markets or increased market share and because of the stick of stigmatization, liability, and loss of market share that may occur absent corporate greening.

7.1.1 Consumer Demand

Market research on consumer attitudes and consumer behavior regarding environmentally conscious marketing claims varies widely, in part because of different definitions of "green" products or claims. However, market research consistently finds that environmental concerns make a difference in purchasing decisions: Consumers do choose one product over another based on claims that a product has positive environmental attributes. For example, in a poll conducted in May 1990, 96 percent of all respondents said environmental concerns are important to their purchasing decisions. In one study of actual consumer behavior, 51 percent of the consumers surveyed said they had made an environmental purchase decision within the past six months, either to buy or to avoid a product based on its environmental attributes. A Nielsen survey of 1,000 people in January 1993 concluded over 80 percent of consumers are buying more green products than they were two years ago. A survey by the Roper organization for S.C. Johnson found that 29 percent of U.S. consumers say they have chosen one product over another based on an advertisement or label claiming environmental benefits.

Another study suggests that 87 percent of consumers would pay more for an environmentally safe product or package, which is corroborated by a July 1990 poll of the Roper organization that found consumers would pay an average of 6.6 percent more for environmentally sound products. An often-quoted report published by FIND/SVP found that $1.8 billion was displaced from product sales into the green market in 1990; the report predicted this figure would increase to $8.8 billion by 1995. A more speculative estimate of the U.S. market in environmental goods claims total sales were worth $25 billion to $50 billion annually in 1991 and climbing.

Market research in Europe also concludes that there is consumer demand for green products. A 1989 survey of 12,000 adults throughout the European Community revealed 78 percent of the respondents considered environmental protection and pollution prevention "very important," topping the list even ahead of concerns about unemployment and poverty. A survey of 3,015 consumers in Germany commissioned by the German government found that 52 percent of the respondents said they would be willing to spend more for environmentally superior products, with 36 percent claiming they would be willing to spend 5 percent extra for such products. A British study in 1989 concluded 42 percent of the British public responding to the survey stated they had made a product purchasing decision based on environmental claims about packaging or formulation. A 1993 study by Gallup that compared consumer attitudes in Europe, America, and Japan concluded that 81 percent of Germans, 50 percent of Americans, and 40 percent of Japanese have refused to purchase products they considered environmentally stigmatized.

The demographics of the environmental consumer suggest the green market is fertile. The largest group of self-identified environmentalists who have taken action on an environmental issue are families with a household income over $50,000. In the United States, market research has segmented the green market in a number of ways, with general agreement that (1) 30 to 50 percent of consumers are very concerned about environmental issues, and (2) this 30 to 50 percent are educated and relatively affluent. The accompanying table, "Environmental Consumer Segmentation," summarizes the findings of market segmentation efforts.

7.1.2 The Competition

Competition in the green market is highly visible, from compostable diapers to "safer" cleaning products to reduced packaging to green public relations campaigns. According to one survey, introductions of new "green" products between 1985 and 1990 grew as much as 100 percent or more each year. The same survey found that 57,000 new green packaged goods would be introduced in 1989 alone. Green campaigns such as those of Wal-Mart and Target stores have been high profile and successful. Moreover, the green market is not for large companies alone. Small companies account for as much as two-thirds of all new green product introductions.

Environmental Consumer Segmentation

Survey Organization	Degree of Awareness and Activity			
	Highest	*Strong*	*Moderate*	*Not Active*
Environmental Research Associates	Very concerned (50%)*	Somewhat concerned (37%)	Not very concerned (3%)	Not at all concerned (1%)
Green Market Alert	Visionary Greens: committed greens (5–15%)	Maybe-Greens: swing group (55–80%)		Hard-Core Browns: adamant non-environmentalists (15–30%)
JWT Greenwatch (J. Walter Thompson)	Greener-than-Greens: make many sacrifices for the environment (24%)	Greens: concerned about the environment but make only some sacrifices (59%)	Light Greens: concerned but not willing to make any personal sacrifices (15%)	Ungreens: don't care about the environment (3%)
Kaagan Research Associates	Young White Collar: most environmentally conscious, affinity for environmental groups, at odds with corporate America (22%)	Substantial Means: strong believers in (and practitioners of) individual environmental responsibility, more vocal on abstract and global environmental debates (15%)	Older White Collar: self-satisfied with personal environmental efforts and optimistic about the future (14%). Blue Collar: lack the belief that individual effort can make a difference, believe that industry and government will pick up the slack (24%)	Limited Means: lacking the educational background to grasp the complexity of some environmental issues, or the incomes to make discretionary pro-environmental purchases; environmentalism not a high priority (18%)
The Roper Organization	True-Blue Greens and Greenback Greens: earn more, have more education, politically liberal, and tend to be female (22%)	Sprouts: well-educated, wealthy, "swing" group (26%)	Grousers: high school education or less, income below $25,000, rationalized indifference (24%)	Basic Browns: most socially and economically disadvantaged, virtual absence of environmental consciousness or activity (28%)

*Percentage of survey sample.

Source: U.S. Environmental Protection Agency, *Evaluation of Environmental Marketing Terms in the United States* (1993), Pub. No. EPA F41-R-92-003 (prepared by ABT Assoc.).

The total quality movement has also had an impact on the greening of business. Standards like ISO 9000 and BSI 7750 including environmental components are becoming a requisite for any participant in markets in Europe and even in parts of the United States.

7.1.3 *Regulatory Pressures*

Anyone operating in the United States, Europe, or Japan is aware of increasing regulatory pressures to develop cleaner processes. For many, becoming green is not merely a matter of staying competitive, but a matter of being allowed to enter or stay in the race at all. The much-publicized German packaging law, for example, simply excludes from the German market virtually any product unless its packaging can be eliminated, reused, or recycled. In the United States, pollution prevention plans are a requirement for many medium-sized and large companies, and this trend will certainly focus on smaller and smaller companies and greater and greater reductions in the future. Numerous regulatory programs create incentives to reduce or eliminate waste.

7.1.4 *Overcoming Stigma*

Consumer demand, competition, and regulatory pressures create obvious incentives to conduct environmentally conscious marketing. However, for environmentally stigmatized consumer products, environmentally conscious marketing may be a necessity for survival. Consumer fears threaten to eliminate smaller businesses producing products such as pesticides, household cleaners, other products formulated with toxic chemicals, and any other product suffering from an environmental stigma. For these products, communicating the sound environmental practices of the producer are essential.

7.1.5 Leadership

Finally, environmental marketing is an opportunity to demonstrate leadership. The dramatic impact of environmental issues far beyond the marketplace makes them a vital connection between business and customers' everyday lives. Environmentally conscious manufacturing produces a unique opportunity for businesses to make a difference in their communities in ways that matter to individuals on a very personal level.

7.2 METHODS OF ENVIRONMENTALLY CONSCIOUS MARKETING

Because consumer attitudes regarding the environment continue to develop along with scientific research about the environment, the green market itself remains somewhat undeveloped and undefined. Broadly speaking, environmentally conscious marketing means any communication that fosters a positive public attitude about the environmental attributes of a firm or its products. The challenge for business lies in the fact that what is and what is not a "positive environmental attribute" is a moving target.

Despite the potential for debate over what is "green," environmentally conscious marketing generally can be classified into two types of marketing. First, there is marketing that attempts to create a positive public attitude about the environmental attributes of your business. This type of marketing concentrates on communicating your firm's positive environmental philosophy and commitment to environmental improvement. Second, there is product-specific environmentally conscious marketing, which focuses on communicating the positive environmental attributes of a specific product. In public relations marketing, you might, for example, attempt to communicate through various media that your business has adopted environmentally conscious policies, that your firm is a member of one or more green industry groups, that your firm sponsors or contributes to a green organization, and so on. In product-specific marketing, by contrast, you might communicate that your product creates less pro-

cess waste than it once did, uses less packaging than competitors' products, or is reusable or recyclable.

7.2.1 Public Relations: Communicating Your Firm's Environmental Commitment

Public relations campaigns are a fine tradition of business in the United States. Corporate sponsorship of scientific, cultural, and charitable organizations—from small-business participation in local activities to Fortune 500 sponsorship of major cultural events—is a cornerstone of many public relationship efforts. A number of vehicles exist to translate this concept into environmentally conscious marketing.

Green public relations is no different from traditional public relations except that it involves communicating positive environmental attributes of your business. In the United States, Wal-Mart is a good and easy example of a successful green public relations campaign. In a December 1990 survey of 1,514 consumers by Advertising Age and the Gallup organization, 19 percent of respondents ranked Wal-Mart as "very concerned about the environment." Wal-Mart's environmentally conscious marketing has involved a public call to its suppliers for more green products. Wal-Mart has launched an innovative program of marking its shelves to call attention to green products. In addition, Wal-Mart has promoted a program of recycling in its parking lots. The key to Wal-Mart's success has been extensive television publicity of its environmental commitment. This has included a TV spot supporting recycling, a number of TV spots publicizing Wal-Mart's support for a children's environmental group, and Wal-Mart's announcement of its intention to print all of its ads on recycled paper.

Corporate sponsorship of green organizations is common. Beginning with Mutual of Omaha's sponsorship of "Wild Kingdom" through Target Stores' sponsorship of Kids for Saving Earth, firms have attempted to communicate their commitment to the environment by association with environmental organizations. On a smaller scale, every business can publicize its own efforts to produce, reuse, and recycle. This may be as simple as communicating to customers that you have taken steps to reduce your office or process waste and, where reduction has proved impossible, to reuse or recycle. Communications can be made by posting signs in areas where customers are present, through newsletters, or in any other regular communication with customers.

Wal-Mart has been successful in communicating its environmental commitment to customers using methods that can easily be adapted to any size business. Any retailer can duplicate Wal-Mart's use of special shelf labels indicating green products. Such a measure should probably be tailored carefully to simply point out claims made by the manufacturer or supplier, or efforts should be made to ensure that your claim complies with any applicable regulations on environmentally conscious marketing.

Wal-Mart has also been successful in publicizing its efforts to use recycled paper in all its ads. Virtually all businesses communicate with customers on paper and also use a substantial volume of paper internally. Once a switch is made to recycled paper, it takes very little additional effort to communicate your commitment to customers by indicating on each sheet of paper its recycled content, provided such an indication complies with applicable laws and regulations.

Public relations green marketers must exercise great caution, however, to avoid the backfire of consumer skepticism. The rush to demonstrate environmental commitment in the 1980s has created a significant legacy of "green hype." Consumer skepticism has in part been fueled by the aggressive efforts of environmental groups, regulators, and scientific researchers to examine and debunk not only specific product claims that do not hold up but also public relations efforts that are not founded on a genuine companywide environmental commitment.

Public relations campaigns can backfire for any number of reasons. The biggest risk is probably to those businesses that publicize a positive environmental effort before they have integrated environmental sensitivity into their entire design, manufacturing, production, and distribution process. To engage in a public relations campaign is to invite public scrutiny, and all of the business's operations must be able to withstand this scrutiny before any of them are offered as examples of the company's environmental commitment.

An important lesson from failed environmental public efforts is that a firm should be able to point to concrete steps it has taken that demonstrate its environmental commitment. It is not enough to adopt a policy. Instead, a firm should delay launching a public relations campaign until actual steps implementing the policy have been taken. Moreover, a public relations campaign is much more likely to succeed if the concrete actions on which it is based are relevant to the firm's business. For example, in an article titled "Beware: Green Overkill," one commentator lambasted General Motors for print ads saluting Earth Day 1990. In the same article, the author also criticized the makers of WD-40 for its print ads publicizing its contributions to the National Park Service, terming the ad "transparently opportunistic." Donations to unrelated environmental organizations or other actions that are not changes to your way of doing business are likely to meet a high level of skepticism. Sponsorship of environmental groups or projects, tree planting campaigns, or other pro bono efforts can be important, but should accompany steps directly relevant to your business or your customers.

7.2.2 *Specific Product Attributes: Communicating the Positive Aspects of Your Product*

Environmentally conscious marketing of specific product attributes can be made through the packaging and labeling of the product, at the point of sale, or in advertisements or other communications to the public. Early environmentally conscious marketing efforts tended to make vague and generalized claims such as "environment friendly" or "safe for the environment." These generalized claims were quickly challenged in the United States by the Federal Trade Commission and attorneys general of various states under truth-in-advertising laws—often successfully because they are very difficult to substantiate. Under general truth-in-advertising principles that are being adapted to environmentally conscious marketing claims, claims must be factual, true, and documented or substantiated. This has led to attempts to define specific green terms by statute, regulation, or interpretive guidance; such a move has the disadvantage of limiting the availability of these claims for certain products but the advantage of providing the green marketer with a certain level of confidence that a claim that complies with the relevant statute or guidance will not be challenged successfully.

A May 1990 poll attempted to measure and rank consumers' most widely desired environmental attributes for products. The five most widely desired or recognized attributes were (1) energy efficient (rated "important" by 79 percent of consumers), (2) non-polluting (77 percent), (3) recyclable (75 percent), (4) ozone safe (72 percent), and (5) pesticide free (69 percent). The five least widely desired or recognized attributes were (1) plant based (28 percent), (2) unbleached (33 percent), (3) petroleum free (35 percent), (4) body friendly (37 percent), and (5) organic (42 percent).

The exact wording of a specific product claim is critical. It must hit home with the consumer and stay within the guidelines for environmentally conscious marketing claims in each market where the claim is made. Choosing a claim that consumers desire involves careful and up-to-the-minute market research, since consumer demand for specific attributes is evolving rapidly. Staying within regulatory limits requires a careful analysis of which markets the product will be sold in and of the requirements applicable to the product in those markets.

The Federal Trade Commission and the Environmental Protection Agency have focused on a fairly narrow set of factual claims, including that a product is degradable, compostable, recyclable, made of recycled content, reusable, refillable, safe for the ozone, or a reduced source of pollution. The FTC has issued nonbinding guidance on when these terms will be considered true and fair as applied to a product (see Chapter Eight).

Various states have enacted specific definitions of these terms, which must be complied with before they can be used.

European countries and Japan have truth-in-advertising laws that also require a claim to be factual, true, and substantiated. Europe and Japan have not yet, however, moved to specific definitions for specific terms. In Europe, this is partly because regulatory efforts have focused on product certification standards. That is, the Europeans have chosen to endorse products based on government-selected green attributes, rather than regulating the truth of claims about the attributes and allowing the market to determine which attributes will be in demand. In Japan, regulation is somewhat haphazard.

In Europe, product-specific claims tend to use an emblem approved either by a government or a private certification program. Private certification and emblem programs are being created in the United States as well. These types of claims are made by applying for permission to use a particular emblem or logo that is widely recognized by consumers as a certification that the product has green attributes. In the United States, Scientific Certification Systems and Green Seal are emerging as private party emblems. In addition, the states of New York and New Jersey have developed state emblems. In Canada, Europe, and Scandinavia, government emblems are much more advanced than in the United States, and the use of emblems is more widely recognized by consumers. Certifications may have two particular advantages over individualized product claims: (1) their sponsors invest substantial resources in creating wide market recognition of the emblems, and (2) the certification by a third party may carry greater credibility in the mind of the consumer than does a direct claim by a green marketer.

Product-specific claims have been the subject of the most intense scrutiny by regulators and by the public and therefore represent the most developed area of environmentally conscious marketing. By the same token, this scrutiny subjects specific product claims to the greatest level of regulation.

Developing a specific product claim should begin with the design of the product. A product that will best withstand the scrutiny of its green claims is one that has its environmental attributes incorporated from the very beginning. Accordingly, members of the environmentally conscious marketing team should be included from the very first steps in the design process.

The input of the ECM team in the design of the product is valuable because design changes may affect dramatically whether and how claims can be made. For example, many products must be disassembled before the individual parts can be recycled. If you wish to make a claim that your product is recyclable, the designers need to be aware that they should design into the product a process for easy separation of parts of the product made from different materials. To maximize environmentally con-

scious marketing opportunities and minimize expensive and time-consuming changes in basic design, designers must be aware of marketing issues.

Another aspect of environmentally conscious marketing that should be included in the design and manufacturing process from the beginning is documentation. The key to successfully maintaining environmentally conscious marketing claims is substantiation that the claim is true. For many environmental claims, the best substantiation of a claim is careful documentation of the design and manufacturing process.

For example, it is increasingly common to label a product with its recycled content. Emerging regulation requires very specific disclosure of the exact percentage of such recycled content, including the percentage of "preconsumer" waste and "postconsumer" waste. It may be difficult and expensive to document the amount of recycled content after the fact. If, on the other hand, the design and production teams are aware that recycled content is an important factor to track and document, they may be able to do so much more accurately and cheaply and to have the figures available when they are needed.

The following sections provide an overview of different types of specific product claims.

7.2.2.1 Traditional advertising claims.
Product-specific environmentally conscious marketing may take the form of traditional claims on the package or label that identify specific green attributes of a product. Packaging and labeling may identify a product, for example, as "recyclable" or "degradable" or "reusable."

Traditional ECM claims are subject to the traditional regulation of advertising. That is, the claims must be true, they must not be misleading, and they must be substantiated. These aspects of truth-in-advertising regulations are discussed in detail in Chapters Eight and Nine. In addition, traditional environmentally conscious marketing claims may be subject to highly specialized regulation applicable to specific terms such as recyclable, degradable, or a host of similar claims. These specific regulations are also discussed in detail in Chapters Eight and Nine.

7.2.2.2 Government certifications.
Many government agencies have recognized that environmentally conscious marketing may provide a powerful market tool to help governments deal with waste management problems. If the market can create demand for products that produce less waste, or for reusable products, or for recycling, a significant portion of the waste stream may be diverted from public management. The market may be able to solve other problems currently managed by governmental agencies as well. Accordingly, many government agencies support environmentally conscious marketing efforts as a private-sector solution to public waste problems.

In an effort to promote the creation of green markets, a number of government agencies have created official emblems of government approval for green products. Typically, these programs involve application to the government body for permission to use the government emblem. A fee is usually associated with the application, and the applicant must demonstrate that the product is in fact environmentally beneficial. Government certification programs are discussed in detail in Chapters Eight and Nine.

7.2.2.3 Private certifications.

Private certifications are similar to government certifications discussed in the preceding section. They differ from government certifications only in that they are administered by private organizations that may be for profit or not for profit. The most visible examples to date are the Green Seal and Scientific Certification Systems programs in the United States.

Private seals of approval have the potential to exert much influence on consumers. A survey conduct by *Advertising Age* and the Gallup organization found that 34 percent of 1,514 consumers questioned indicated a certification program such as Green Seal or Scientific Certification Systems would have a great impact on their purchasing decision, while an additional 40 percent said a certification program would have some impact. As a result, some have expressed concern over the power of the private certification program absent government standards. These programs are discussed in detail in Chapter Eight.

International Standards Organizations also may provide a form of environmental certification using standards such as ISO 9000 or BSI 7750. These certification programs certify that the manufacturing process, rather than the product itself, is environmentally sound (some of the European environmental emblem programs examine process as well). Such international standards certifications have not proved highly visible to consumers, but they may take on a status similar to the UL listing in the United States in the future.

7.2.2.4 Take-back programs.

The concept of the mandatory take-back program, under which producers are required to provide recycling opportunities for the packaging associated with their products or take the packaging back themselves, originated in Germany. The "green dot" was developed in Germany to identify packaging that is eligible for collection in the recycling/take-back system. Although the emblem was not originally conceived as a marketing measure, the green dot has taken on symbolic value as a positive environmental attribute of the products that bear it. France has followed the German example with the Eco-Emballage symbol. These take-back symbols are discussed in detail in Chapter Nine.

7.3 CHALLENGES FACING GREEN MARKETERS

The concept of environmentally conscious marketing evokes a mixed reaction. Everyone recognizes the tremendous potential for green products and the disadvantage that "nongreen" products will suffer in the coming years. At the same time, substantial fear and confusion exist regarding developing regulations and changing consumer expectations. Like the oil fields in the early part of this century, the green market appears to present boundless opportunities for those with courage, innovation, and determination and to present serious challenges and risks for those that are uninformed, hesitant, or uncommitted. The following sections describe several of these major challenges.

7.3.1 *Developing Market Data*

Developing information on the green market presents a chicken and the egg problem. One of the barriers to entry into the green market is consumers' lack of clearly defined environmental preferences—what has been described as consumers' "inadequacies in basic 'green knowledge.' " Accordingly, a significant body of market research today has identified only a vague but substantial body of undifferentiated "green" demand. In the current market, research may not reveal any concrete preferences for specific green attributes of specific products, particularly when green choices involve trade-offs in price or performance.

Market research in the context of environmentally conscious marketing therefore requires creativity and a willingness to engage in dialogue with consumers. The ideal situation is to have sufficient resources to engage in frequent market research and to adapt and respond as the information develops. A number of advertising consultants are emerging with expertise in environmentally conscious marketing. (See Appendix M for a list of contacts.)

A specific field of ECM research is emerging that is accessible to medium-sized or smaller businesses without the resources to perform original research. One useful source of research is specific environmental marketing research publications. Such research services include *Green Action Trends* (Yankelovich, Clancy, Shulman), *Green Gauge* (The Roper Organization), *The Kaagan Environmental Monitor* (Kaagan Research Associates), *The Environmental Report* (Environmental Research Associates), and *EnvironMonitor* (Burke Marketing Research) (see Appendix M for addresses). The EPA has published a 196-page research study entitled *Evaluation of Environmental Marketing Terms in the United States* with a significant body of research that may be obtained free from the EPA Office of Pollution Prevention and Toxics. Mainstream advertising publications, such as *Ad-*

vertising Age, also provide significant information on green markets on an ongoing basis.

7.3.2 *Substantiating Your Claim*

Consumers and regulators alike are demanding greater substantiation of environmental claims. Substantiation has always been a requirement of truth-in-advertising laws. As a result of "green hype"—the inflated claims and empty window dressing of the late 1980s discussed below—consumer skepticism toward environmentally conscious marketing is very high.

Fundamentally, substantiation means documenting that your marketing claim is true. For businesses using a total quality management or similar approach, documentation is already a fundamental part of the process. Creating the documentation after the fact is much more difficult.

Exactly what facts must be documented is a function of the specific claim made. For the claims that are specifically defined in regulations on environmentally conscious marketing, the regulations provide substantial guidance on the exact facts that must be proved. For example, use of the term *recycled* in California requires that the product contain at least 10 percent postconsumer material by weight. Substantiating a recycled claim in California means documenting the total weight of the product, the raw material inputs into the product, the source of any postconsumer material used as a component of the product, and the weight of the postconsumer material input. Strategies for identifying facts that must be substantiated and developing adequate substantiation are discussed in Section 10.3.1.

Although conceptually simple, documentation can become a very difficult process, particularly when it requires substantiating factors outside the production process. For example, *recyclability* is often defined in terms of whether actual recycling opportunities exist in the geographic areas where the product is distributed. Substantiation of a recyclability claim under these regulations requires the green marketer to document where the product is distributed and what recycling opportunities exist in those areas. Developing this information requires you to go outside your processes and chain of supply to identify sources of the required information, if it exists. As discussed in greater detail in Chapter Eight, in some cases this requirement can be met in part by industry associations that track recycling opportunities throughout the country. In other cases the requirement may be near impossible, or at least economically impracticable, to meet.

One of the highest forms of substantiation can be found in the field of life cycle analysis (see Chapter Six). Some forms of environmental product certification already require at least a limited form of life cycle analysis. Even when life cycle analysis is not applied directly to a particular product to be certified, life cycle analysis may be the basis for establishing specific criteria for certification for a particular product type. For example, the criteria for certification of washing machines under the European Community's Eco-label program, including limits on the amount of energy and water that may be consumed per kilo of washing and parameters on the wastage of detergent, were developed using life cycle analysis.

For large companies or large industries, life cycle analysis of specific products is becoming increasingly common as a management tool. In the United States, for example, life cycle analysis of disposable diapers, plastic trash bags, and fast-food packaging has received widespread attention. If smaller companies cannot afford to commission life cycle studies, other resources may be available to take advantage of life cycle analysis. Life cycle analysis of generic materials can, in the right circumstances, be applied to final products using those materials. It must be understood, however, that the actual use and disposal of a specific product by consumers is a critical aspect of life cycle analysis and must be considered for any life cycle analysis to be meaningful. For example, many plastics, particularly those formulated with certain additives, have been shown to degrade rapidly in air, light, and water. When buried in a landfill, however, these products tend to degrade much more slowly because landfills are designed to prevent air, light, and water from reaching their contents. A simple life cycle analysis of the degradable plastics materials could be used to suggest that products formulated from the materials are "degradable." When life cycle analysis is applied to the actual use of plastic products that end up in landfills, however, the claim could not be supported.

Another option for smaller businesses that cannot afford to commission full-blown life cycle assessments is private-party certification by a program such as Green Seal or Scientific Certification Systems, which is discussed in greater detail in Chapter Eight. Because full-blown life cycle assessment is too costly for many producers, these programs have developed abbreviated versions of life cycle assessment, which they offer at a cost in the range of $2,500 and $10,000 per product. Green Seal uses Underwriters Laboratories to conduct its testing. These compressed forms of LCA, while cost-effective, are not without criticism. Scientific Certification Systems, for example, has had some difficulty with its abbreviated approval process, which has been criticized by the Environmental Defense Fund.

7.3.3 *Dealing with Inconclusive Science*

Environmental marketing would be much simpler if there were scientific consensus on which substances are harmful, how they are harmful, and in what amounts they are harmful. In fact, however, scientific debate over the real impact on the environment of human activity is often inconclusive. Scientists often disagree on environmental issues, and even where there is agreement, conclusions may change and evolve relatively rapidly. This creates two problems for the green marketer: (1) consumer demand may shift rapidly with each change in the sway of one scientific theory or another regarding the environmental attributes of their product; and (2) it may be very difficult to substantiate an environmental claim when there is fundamental disagreement about what the facts really are.

The debates over diapers and fast-food packaging have illustrated both of these problems. Much has been written about the relative environmental impact of disposable diapers and cloth diapers, but no conclusion has been reached as to which is environmentally preferable. For example, in 1989, a study commissioned by a trade group of the cloth diaper industry concluded the disposal of paper diapers in landfills constitutes a significant environmental problem that should be solved by the use of cloth diapers. In response, the disposable diaper industry commissioned studies showing that laundering of cloth diapers has a significant impact on the air and water because of the chemicals, energy, and water consumed in the laundering process as well as the impact of cloth diaper delivery trucks. To date, there is no definitive answer on whether cloth diapers or disposable diapers are measurably greener than one another, in large measure because it is difficult to rank or compare impacts on landfills against impacts on water and air, and in part because of the problem of expanding life cycle analysis to take even more remote impacts into account.

The debate over fast-food packaging by McDonald's has run a similar course. The McDonald's story illustrates how science can change its conclusions rapidly. As early as 1976, McDonald's commissioned a study to determine how to make its packaging more environmentally sensitive and concluded polystyrene was an environmentally superior alternative to paper. In the 1980s, research on the use of ozone-depleting chemicals in the formulation of polystyrene resulted in a large body of research concluding polystyrene has a substantial negative environmental impact. While it is fair to say the debate between paper and polystyrene packaging was, and remains, as inconclusive as the debate between cloth diapers and disposable diapers, the new research on polystyrene struck a responsive chord with consumers, who developed fears about dangers to the ozone and the nondegradability of plastics. Consumer groups and public interest groups concerned about CFCs put significant pressure on McDonald's to stop using polystyrene, despite McDonald's research showing polystyrene is as good or better than composite paper for the environment. Reacting to these new consumer perceptions, fueled by new (but inconclusive) scientific research, McDonald's has switched back to paper packaging.

In a climate of changing or disputed factual predicates, such as the diaper and packaging debates, it may be difficult to defend any claim as true. The challenge of inconclusive science for green marketers raises the same problem as any rapidly changing market factor. The most important tools to deal with changing information are staying informed, staying responsive, and being candid about the limitations of scientific knowledge on the environment. McDonald's has succeeded in maintaining the perception that it is green in the minds of its customers by keeping aware of consumer preferences and adapting its process to customer demands: switching packaging materials as demand required.

Procter & Gamble, as a player in the diaper debate, has adapted to the uncertainty of scientific information with a policy of candor, conservatism, and education. Rather than attempting to win a debate on which product is greener—a debate that probably cannot be won under the current state of knowledge—Procter & Gamble has launched an education campaign to advise consumers about the difficult environmental issues raised in the diaper debate and Procter & Gamble's commitment to finding better alternatives. The environmental marketing approach these companies have taken has paid off: In a survey of 1,514 consumers in December 1990 conducted by *Advertising Age* and the Gallup organization, both McDonald's and Procter & Gamble significantly outperformed all other companies in creating a public perception of environmental consciousness.

7.3.4 *Educating the Consumer*

Because of the complexity of environmental issues as well as the controversy over what is truly green, many consumers are confused or ignorant about environmental issues. It could be argued that this very consumer ignorance contributed to the overreaching claims of early environmentally conscious marketing efforts and the resulting disillusionment discussed later in this chapter. In any event, the burden is on the environmental marketer to provide a contextual background to consumers in which to make sense of environmental claims. In essence, this is no different from the familiar phenomenon of food marketers providing nutrition information to give consumers a better background understanding to make nutrition-based claims more effective.

In the United States, there is a range of depth to consumers' green knowledge. The most commonly used marketing terms, such as *recyclable, recycled, degradable,* and *no CFCs*, are relatively well understood. The term *source reduction*, however, was understood by only 16 percent of the respondents in a survey of 1,000 adults in 1990. Even among the relatively well-understood terms, a full 20 percent of consumers did not understand the term *recyclable*, and in another study as many as 25 percent did not understand the term *recycled*.

Yet, these standardized terms are relatively simple to understand. Much more complex claims involving trade-offs may be even more difficult for consumers to understand. McDonald's switch to polystyrene packaging and its reversal back to composite paper packaging is an example. McDonald's concluded polystyrene packaging is a comparatively greener alternative than coated paper because coated paper is very difficult to recycle and consumes significant energy in production as well as significant volume in landfills relative to polystyrene. McDonald's was unable to communicate these advantages to its customers, however, who simply viewed plastic as an environmentally unsound choice. In response to customer demand, McDonald's switched back to coated paper. While McDonald's continued to recognize that some studies indicate foam packaging is environmentally sound, the company was simply unable to communicate this message to its customers.

Consumer education is in the forefront of several highly successful national ECM campaigns. Procter & Gamble, for example, has done direct mailings of a pamphlet outlining its efforts to address its contributions to the solid waste problem. Procter & Gamble intends the pamphlet to be informational and educational, preparing the market with sufficient information to make future, targeted ECM efforts effective.

The lesson for environmental marketers is that environmentally conscious marketing messages must be accompanied by sufficient context so the market will develop an appreciation for the environmental attributes of the product that will withstand scrutiny. The risk of advancing claims in a market that lacks the context to understand them is substantial. As discussed in the following chapters, a claim that is technically true may still be improper if consumers don't understand it.

7.3.5   Remaining Competitive in the Eyes of the Consumers

Persuading the market that your green product is competitive in price and performance is an important responsibility for a green marketer. Market research suggests consumers say they are willing to pay 5 to 15 percent more for green products. Actual consumer behavior suggests this self-proclaimed commitment may be exaggerated, but a respectable 50 percent of consumers actually make purchasing decisions based on a product's environmental attributes. Thirty-seven percent who have actually purchased environmentally oriented products said they have paid more, while the remaining 63 percent said they have paid less or the same for a green product over an alternative that lacks the environmental benefit.

Unless you can determine that your market falls within one of the segments more willing to pay a green premium, it may be wiser to base environmentally conscious mar-keting decisions on competitive pricing with an increased market share, rather than on hopes for increased margin on individual sales. The green market is too immature, however, for marketers to conclude consumers will never be willing to pay increased costs for green attributes, particularly if a launch is accompanied by significant public education about the benefits of the green product.

Producers are familiar with the performance trade-off in formulating a green product. In many cases, environmentally harmful components of a product are linked closely to the product's effectiveness or convenience. Eliminating those attributes may diminish the performance of the product.

Market research specifically on consumers' performance concerns about green products is limited. Nonetheless, there is a consensus among marketing researchers that performance of green products is a significant concern among consumers. The lesson for green marketers is that reassurance about product performance must accompany marketing of green attributes.

7.3.6   Overcoming Public Skepticism

Green consumers tend to be affluent, well educated, and skeptical. They have experienced a market that was saturated with efforts to cash in on green claims, whether or not a given marketer's claims were substantiated or even justified. Consumers also experienced the collapse of that wave of green hype under intense pressure by environmental regulators and public interest groups. The legacy of environmentally conscious marketing out of control is substantial enough to create strong consumer skepticism about ECM claims.

Market research suggests consumers continue to have a high level of skepticism about environmentally conscious marketing. In 1990, only 14 percent of survey respondents stated they would be "very likely" to believe an advertisement by a small company concerning the effect of a product on the environment. In an *Advertising Age*/Gallup study, only 8 percent of respondents said they were "very confident" that product-specific green advertising is "accurate."

In another study in 1990, 47 percent of consumers said they generally dismiss environmental claims as "mere gimmickry." A study by the Angus Reid Group, Golin/Harris Communications, and Environomics in the spring of 1991 developed a sliding scale of consumer skepticism depending on the familiarity and the specificity of environmental claims. The results are summarized in Exhibit 7–1. The familiar and specific claims of *recyclable, recycled,* and *biodegradable* were met with confidence by 65 to 80 percent of consumers. Vague terms such as *environmentally friendly, ozone friendly,* and *cruelty free* received much more skeptical ratings at or below 50 percent confidence.

EXHIBIT 7–1
Consumer Confidence/Skepticism

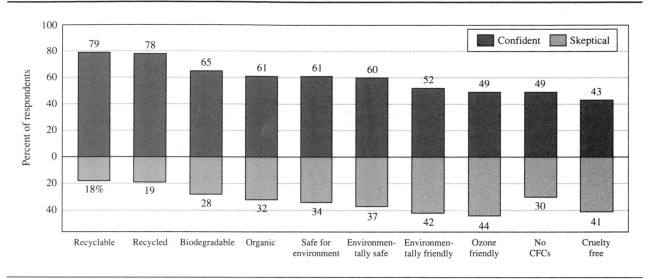

Note: Respondents total less than 100% due to omission of "unsure" responses.
Source: EPA, *Evaluation of Environmental Marketing Terms in the United States* (1993), Pub. No. EPA 741-R-92-003 (prepared by Abt Assoc.) From Environmental Today, Angus Reld Group et al.

Green skepticism means several things for green marketers. First and foremost, you must be prepared to face skepticism and to substantiate your claims. Second, you should be conservative in your claim and avoid "green hype" or "green fluff." Third, you should make sure your market is educated enough to understand your claim, and if it is not, you must be prepared to educate your market by presenting contextual background for your claim and an explanation of your claim at or before the time you make it.

7.3.7 *Deciphering the Regulatory Maze*

Environmentally conscious marketing claims have been the subject of increasing regulation during the past decade, especially since the early 90s. ECM regulations pose at least three challenges for the green marketer. First, different jurisdictions often have slightly different regulations. It may be difficult to identify every jurisdiction in which your product will be distributed in connection with the environmentally conscious marketing claim and to identify the specific environmental marketing regulations applicable in that jurisdiction. Many states have their own regulations, and the national governments of the United States, Canada, many European countries, and Japan all have environmentally conscious marketing regulations.

A second challenge for green marketers is complying with the different regulations in different jurisdictions. You may be able to make a claim that your packaging is "degradable" in 49 states, but not in California simply because your product does not meet the statutory defini-

tion of degradability in California. Chapter Ten suggests strategies for dealing with this problem.

Finally, a third challenge faced by green marketers is the rapid development of regulations governing environmentally conscious marketing claims. Along with rapid developments in the green market itself, regulators are hurrying to keep ahead (or catch up), resulting in new regulations all the time. The best strategy for dealing with this problem is to stay informed. It is important to review the regulations applicable to your product in every jurisdiction where it is sold as often as once a year to determine whether any new regulations have developed. Such review can be conducted by legal counsel. Staff devoted to environmentally conscious marketing can stay advised of developments through the environmentally conscious marketing press. A useful tool for tracking regulation is the loose-leaf service offered by the Thompson Publishing Group for about $400, which is updated on a regular basis and tracks the development of new regulations around the world (see Appendix M for an address).

7.4 CONCLUSION

Despite some false starts and significant challenges, environmentally conscious marketing is here to stay. It seems likely that environmentally conscious marketing will eventually enter the mainstream of advertising and marketing practices. In another 10 to 20 years, environmentally conscious marketing claims will be as ordinary (and no less vital) than claims about price, performance, and convenience. In the meantime, green marketers must adapt to the rapidly changing rules of environmentally conscious marketing and exercise creativity and persistence in developing the rules where they don't exist already.

Although environmentally conscious marketing is not yet a necessity for the survival of most products, it is already making a difference in market share and leadership for a substantial portion of the market. For products perceived by consumers as hazardous or environmentally harmful, environmentally conscious marketing is fast becoming a necessity. Other products may have a larger window before environmentally conscious marketing is a simple requirement for competitiveness, but that development is probably coming within the next 10 to 20 years.

Environmentally conscious marketing requires commitment, research, strategic planning, and a coherent philosophy of environmental responsibility. While it is a mixture of art and science, it is not voodoo, and it can be approached methodically. Indeed, when environmentally conscious marketing is approached haphazardly, it is in danger of becoming green hype or green fluff and backfiring dramatically.

Following the principles listed below is critical to the success of an environmentally conscious marketing campaign:

- Start by being green—develop an environmentally committed organization that can honestly represent to the public that it and its product are green.

- Document and measure your process.

- Leverage your environmental commitment by communicating to the public.

- Figure out who your market is—where is your product sold, who buys it, and what is the buyers' level of environmental understanding?

- Be prepared to educate your market so your buyer understands and appreciates your claim.

- Be honest.

- Be factual.

- Be conservative in your claims—stay a little behind the curve of public opinion in your claims.

- Be proactive in your own environmental awareness—stay ahead of the curve of public opinion in your knowledge of environmental issues and your preparedness to adapt to change.

- Update your environmentally conscious marketing program regularly to account for changes in the state of knowledge and in the regulations that apply to your claims.

The remainder of Part II will build on these essential principles.

RESOURCES

Advertising Age, January 29, 1991, pp. 10, 26.

Billig, Alex & Partner. "Ermittlung des Oekologischen Problembewusstseins der Bervoelkerung," March, 1994.

"Case Study: McDonald's," *Advertising Age*, January 29, 1991, p. 32.

"Case Study: Procter & Gamble," *Advertising Age*, January 29, 1991, pp. 16, 34.

"Coming Clean on Goods: Ecology Claims Faulted," *The New York Times*, March 12, 1991, p. D1.

"Consumers Go Green," *Advertising Age*, September 25, 1989, p. 3.

Environment Reporter, August 3, 1990, p. 691.

"Green Concerns Grow," *Advertising Age*, September 4, 1990, p. 27.

"Green Issues Research," *ENDS*, January 1993, p. 24.

Howett, Ciannat M. "The 'Green Labeling' Phenomenon," *Environmental Consumerism*, pp. 451-52.

McIntosh, Andrew. "The Impact of Environmental Issues on Marketing and Politics," *1990s Journal of the Market Research Society*, July 1991, p. 207.

J. Ottman Consulting, Inc. *Environmental Consumerism: What Every Marketer Needs to Know*, 1991.

J. Ottman Consulting, Inc. "Environmental Consumerism in the U.S.," *Environmental Consumerism: What Every Marketer Needs to Know*, 1991, pp. 36, 53.

"P&G Gets Top Marks in AA Survey," *Advertising Age*, January 29, 1991, p. 8.

Simon, Francoise. "Marketing Green Products in the Triad," *Columbia Journal of World Business*, Fall/Winter 1992.

U.S. Environmental Protection Agency. *Evaluation of the Environmental Marketing Terms in the United States*, pp. 5, 9-20, 22.

"Watchdogs Seeing Over 'Green' Labeling: Environmental Certification Blasted," *Chicago Tribune*, October 29, 1991, p. 1.

Chapter Eight

Environmental Marketing Regulations and Guidelines: The American Model

The United States, Canada, and Australia regulate environmentally conscious marketing claims as part of general consumer protection, primarily using truth-in-advertising laws. Accordingly, in all three countries, environmentally conscious marketing claims are scrutinized by trade regulators (such as the Federal Trade Commission in the United States) rather than environmental regulators. This focus on consumer protection results in similar laws in the United States, Canada, and Australia—the "American model"—which differs from the focus of the European model on protection of the environment. The European model is discussed in Chapter Nine.

The basic rules under truth-in-advertising are that claims must be (1) true, (2) not misleading, and (3) substantiated. Application of each of these simple concepts to environmentally conscious marketing claims is made more difficult, respectively, by (1) the lack of consensus in scientific research on environmental issues, which results in disputes over what is "true"; (2) the complexity of environmental issues and the lack of green knowledge among consumers, which makes consumers easily confused or "misled"; and (3) the complexity of measuring environmental impacts using models such as life cycle analysis, which makes substantiation potentially complicated and expensive.

Further complicating the compliance issue in the United States is the problem of state regulation of environmentally conscious marketing claims. Virtually all states have their own truth-in-advertising laws, which fortunately are generally the same or similar to the federal truth-in-advertising laws. A few states, however, have developed their own laws that specifically regulate environmental claims, and these laws are not necessarily consistent with federal law or the laws of other states. State laws generally define common environmental marketing terms such as *recycled* and *degradable* and set forth the exact requirements for using such terms within the state. The result is that different requirements may apply in different states. New York, California, Wisconsin, Indiana, and the northeastern states have been particularly active in developing their own environmental marketing regulations.

In response to the challenges to environmental marketers under the American model, environmental truth-in-advertising guidelines have been established by a coalition of state attorneys general ("The Green Report II") and by the U.S. Federal Trade Commission (FTC). These voluntary, truth-in-advertising guidelines have been somewhat helpful in harmonizing the interpretation of state and federal truth-in-advertising laws. Canada, Australia, and states enforcing their local truth-in-advertising laws generally employ the same or similar guidelines as the FTC. Specific local environmental marketing regulations, however, continue to contain material differences and remain a challenge to the environmental marketer.

8.1   FEDERAL TRUTH-IN-ADVERTISING LAWS

8.1.1   *General Overview*

Since at least the 1970s, the FTC has applied its general rules requiring truth-in-advertising to regulate green advertising and labeling. The Federal Trade Commission Act (FTCA) declares that "unfair methods of competition in or affecting commerce, and unfair or deceptive acts or practices in or affecting commerce are . . . unlawful." The act grants the FTC authority:

1. To prevent persons, partnerships, or corporations from engaging in unfair or deceptive activity, and
2. To ensure that advertising claims are adequately substantiated.

The FTCA does not preempt state and local regulation of unfair trade practices, truth in advertising, or green labeling, so states are free to develop their own regulations. While the FTC has no authority to establish environmental policy, it has implemented the authority granted by the FTCA to regulate environmentally conscious marketing claims in conjunction with the EPA. The FTC's general approach to applying truth-in-advertising rules to environmentally conscious marketing claims is enunciated in FTC statements on (1) deception (see Appendix G), (2) advertising substantiation (see Appendix I), and (3) unfairness (see Appendix H).

If the FTC finds that a manufacturer/advertiser lacks a reasonable basis for a claim, it can issue a cease-and-desist order. Violation of a cease-and-desist order may result in imposition of penalties as high as $10,000 per violation per day. Additionally, the $10,000 penalty can be imposed by the FTC on a manufacturer/advertiser that commits a knowing violation of the FTCA, regardless of whether the FTC had previously issued a cease-and-desist order. Finally, the FTC has authority to prohibit the sale of existing inventory, in effect requiring the company to remove products with deceptive labels from store shelves.

8.1.1.1   Deception.   Under the Federal Truth in Advertising Law, deceptive practices are illegal. A detailed and complicated body of law attempts to define exactly when a practice is deceptive, but deceptiveness can be an elusive concept. The FTC has attempted to summarize the basic principles of deception in its statement on deception.

1. Deception is a representation, omission, or practice that is likely to mislead the consumer.
2. The representation must be likely to mislead from the perspective of a consumer acting reasonably in the circumstances.
3. The representation, omission, or practice must be "material."

Under FTC guidance, a misrepresentation is "an express or implied statement contrary to fact." The first two factors of the deception test mean you must not make claims that would lead a reasonable consumer to believe something that is not true about your product. If you do, the claim is illegal.

The third part of the test related to materiality is better defined than the first parts of the test, although it may be harder to prove. Express claims are always deemed material; in other words, if you make an affirmative representation about your product ("does not pollute"), that representation is automatically deemed to be material. Implied claims or omissions of facts about your product are deemed material under any of the following circumstances:

1. Information that significantly involves health or safety.
2. Information pertaining to the "central characteristics

of the product or service," such as the purpose, safety, efficacy, or cost of the product or service.

3. As a catch-all, information is material if the consumer would change the decision to purchase the product based on the information.

8.1.1.2 Advertising substantiation.

The basic rule about ad substantiation is that an advertiser must have a reasonable basis for marketing claims before they are disseminated. The kind of ad substantiation required depends on the specific claim. First, when the ad expressly or implicitly states a level of substantiation (for example, "tests prove" or "proved by"), you must have the substantiation you claim. When there is no express level of support claimed, you must have a "reasonable basis" for claims. The FTC considers a number of factors in determining how much substantiation is reasonable, including the cost of developing substantiation and the amount of substantiation experts in the field believe is reasonable.

8.1.1.3 Unfairness.

Unfairness is perhaps the vaguest prohibition in the truth-in-advertising law. Generally, a marketing claim can be considered unfair if it substantially injures consumers. It is difficult to give this concept of consumer injury any concrete meaning. Examples of unfairness are coercion, unfair credit practices, or substantial safety risks. Unfairness is generally found only when a seller takes advantage of the buyer in a substantial way that is recognizable as a matter of common sense. For example, in one case of unfairness a seller's servicemen dismantled a home furnace and then refused to reassemble the furnace until the consumer agreed to buy services or replacement parts.

8.1.2 FTC 1992 Environmental Claims Guidelines

In addition to the generalized guidance discussed in the preceding section, the Federal Trade Commission in 1992 issued guidelines on the application of the federal truth-in-advertising laws to environmentally conscious marketing claims. The guides are reprinted in Appendix D. Although the EPA does not currently regulate environmental claims, the FTC formed a joint task force with the EPA and the U.S. Office of Consumer Affairs to develop the guidelines. The guidelines are nonbinding and indicate how the FTC probably would apply truth-in-advertising laws to various specific or general environmental claims. Although the guidelines leave individual states free to impose their own interpretation on their truth-in-advertising laws, they should generally be regarded as a persuasive guideline on how the truth-in-advertising laws will be enforced both by the federal government and by individual states.

Part F of the FTC guidelines contains a discussion of how the general principles of truth in advertising should be applied to environmental claims. Part G discusses the

requirements for eight specific types of environmental claims, with examples of acceptable and unacceptable uses of the claims. These eight examples are:

1. General environmental benefit claims.
2. Degradable/biodegradable/photodegradable.
3. Compostable.
4. Recyclable.
5. Recycled content.
6. Source reduction.
7. Refillable.
8. Ozone safe or ozone friendly.

The FTC's interpretation of the specific terms, which are also the subject of many specific state regulations, is discussed later. The remainder of this section discusses the general guidance contained in the FTC 1992 guidelines.

First and foremost, the federal truth-in-advertising law requires that environmental claims must not be deceptive. The FTC has summarized these requirements as applied to environmental attributes in Section E of the 1992 guidelines:

[A]ny party making an express or implied claim that presents an objective assertion about the environmental attribute of a product or package must, at the time the claim is made, possess and rely upon a reasonable basis substantiating the claim. A reasonable basis consists of competent and reliable evidence. In the context of environmental marketing claims, such substantiation will often require competent and reliable scientific evidence. For any test, analysis, research, study or other evidence to be "competent and reliable" for purposes of these guides, it must conducted and evaluated in an objective manner by persons qualified to do so, using procedures generally accepted in the profession to yield accurate and reliable results.

This statement can be broken down into a number of distinct elements, each of which must be complied with under the federal truth-in-advertising law.

1. *Are you making an "express or implied claim"?* Generally, an express or implied claim means any objective assertion about the environmental attribute of a product or package. An express or implied claim is any statement or set of statements that would lead a reasonable consumer to form a factual belief about the product. A "claim" may be made not only by an affirmative representation but also by an omission in a context in which a reasonable consumer would expect disclosure.

2. *Are you relying on a reasonable basis substantiating the claim at the time it is made?* The concept of reasonableness is one that is time honored in the law but escapes further definition. The 1992 guidelines specify that a reasonable basis consists of "competent and reliable evidence." At a minimum, you must have formulated the question whether your claim is true or not and undertaken some process to reach a conclusion that it is.

In general, you will be more likely to persuade a regulator that you have a reasonable basis for your claim if most or all of the following factors are present:

- You have documented an investigation of whether your claim is true.

- Your documented investigation includes reasonable and competent evidence.

- Your investigation was performed by a person who is an expert in the field of the investigation.

- Your expert was objective—a third-party consultant is best.

- Your investigator used procedures generally accepted in his or her profession to yield accurate and reliable results.

- Your investigation is persuasive to others with expertise in the field.

- You have quantified as much of the data as possible in your investigation.

General principles.   Section F of the 1992 guidelines restates some general principles of truth in advertising as they may be applied in the environmental context. This section covers (1) qualifications and disclosures, (2) distinctions between product and package, (3) overstatement of environmental attribute, and (4) comparative claims.

Qualifications and disclosures are an acceptable method of hedging to ensure that your claim will not be misunderstood. A qualification and disclosure simply provides information on limitations to your claim. For example, if a claim is made that a product is recycled, a qualification and disclosure might state that the product contains 30 percent postconsumer waste to ensure that consumers do not draw the conclusion that the product is made from 100 percent recycled materials.

Environmental claims should be as conservative as possible and should always disclose qualifications or limitations. Under the guidelines, qualifications and disclosures should be "sufficiently clear and prominent to prevent deception." Specifically, qualifications and disclosures should be made with clear language and should be displayed in such a way that they will be read and understood by a consumer reading the environmental claim. Generally, type size should be comparable, and the qualification and disclosure should be displayed on the same part of the packaging as the claim.

You should clearly state what part of the product or package your environmental claim applies to. The 1992 guidelines state that an environmental marketing claim "should be presented in a way that makes clear whether the environmental attribute or benefit being asserted refers to the product, the product's packaging or to a portion or component of the product or packaging." For example, you may indicate that a cardboard outer package is recycled (assuming the claim is true), but you must be careful to avoid leaving a consumer with the impression that the product inside the package is also recycled. In such a case, the claim should include a qualification such

as "outer package contains 50 percent postconsumer recycled content." It is not necessary to specify the part of your product or package intended by the claim if it is obvious or if the claim applies to "all but minor, incidental components of a product or package."

Do not puff.   The Federal Trade Commission reacted to green hype as strongly as the consumer market. In the guidelines, the FTC specifically prohibits "implications of significant environmental benefits if the benefit is in fact negligible." The guidelines give two examples. In the first a package is labeled "50 percent more recycled content than before." The real increase is from 2 percent recycled material to 3 percent recycled material. In the second, a trash bag labeled "recyclable" is theoretically recyclable but will in fact probably not be separated out from other trash at a landfill or incinerator for recycling. The FTC states that each of these claims is technically true but is overstated and therefore misleading.

"Overstatement" is a vague test that depends largely on consumer expectations and consumer knowledge. The test of overstatement is whether consumers will be likely to misunderstand the claim and believe the environmental benefit is more significant than it is. It is not deceptive, therefore, to make a claim of a relatively insignificant environmental benefit if the limitation is obvious. The FTC gives the specific example of a paper grocery sack labeled "reusable," which can be reused but will fall apart after two or three reuses. According to the FTC, "because reasonable consumers are unlikely to assume that a paper grocery sack is durable, the unqualified claim does not overstate the environmental benefit."

If there is any question that consumers might be confused by your claim, however, the best policy is to disclose the limitations of the claim. At the very least, you should document a reasonable basis for believing that consumers will properly understand the claim and its context. Such documentation might include a survey of how consumers actually understand your claim. Such a survey would have to comply with the guidelines on investigation by persons qualified to do so using procedures generally accepted in the profession.

Comparative claims.   There are two basic principles for comparative claims. First, the basis for comparison must be sufficiently clear to avoid consumer deception. Second, you must be able to substantiate the comparison. Comparative claims are particularly likely to be subject to scrutiny because competitors may be highly motivated to challenge your comparative claim.

The basic rule for disclosing the basis for the comparison is to disclose what you are comparing and how you are comparing it. If you are comparing an improved product to a previous design, your claim should state for example "20 percent more recycled content than our previous package." If you are less specific than this, your claim may be construed as a claim that your product is better than all competing products. In addition, you should avoid vague comparisons. For example, the claim that your product is "20 percent less toxic" than before is ambiguous. A more specific statement, such as "contains 20 percent less mercury" is preferable.

General environmental benefit claims.   General environmental benefit claims are vague statements that fail to disclose specifically what attribute of a product is claimed to be green. For example, claims that a product is "safe," "environmentally friendly," "won't harm the environment," "environmentally improved," "better for the environment," or similar vague claims are vulnerable to challenge because they are virtually impossible to substantiate.

The problem with generalized claims is that the FTC looks to consumers' understanding of what objective feature you are claiming by making a generalized statement. You must be able to substantiate each specific, objective quality of the product that a reasonable consumer understands you to be claiming by a generalized statement. When you make a generalized claim that a product is environmentally friendly or safe, for example, regulators would likely take the position that a reasonable consumer would understand that you are claiming your product causes no harm to the environment whatsoever. Such a claim is virtually impossible to substantiate and rarely, if ever, true.

The best policy is to avoid generalized claims. If a generalized claim is made, it should be accompanied by qualifications and disclosures that explain very specifically what environmental benefits you are really claiming. As discussed above, these qualifications and disclosures must be "sufficiently clear and prominent to prevent deception." In other words, they must be presented in such a way that a consumer reading the generalized claim will also be likely to read and understand the qualifications and disclosures.

8.1.3. *Examples of the FTC Enforcement of Truth-in-Advertising Laws vis-à-vis Environmental Claims*

According to a February 1993 EPA report on enforcement of environmentally conscious marketing claims, there were 48 actions against environmental claims through June 1992. The report stated that 17 of those claims concerned the degradability of plastic products and 10 actions related to claims about propellants, including "ozone friendly" or "no CFCs." Most related to consumer products, with the exception of an action against a lawn care service and another against a manufacturer of home water treatment systems. Exhibit 8–1 is a summary of the actions from the EPA's report (see pp. 80–95).

FTC and individual states have continued to prosecute claims against environmental advertisers since EPA's February 1993 report. Examples of claims that have been challenged or prohibited since then are included in the table below.

Company	Product	Claim	Action
Amoco Foam Products	PS cups and plates	Recyclable	May, 1994, consent agreement with FTC
Keyes Fibre Co.	Chinet tableware	Degradable	May, 1994, FTC consent order to qualify and disclose that product is compostable only where municipal solid waste composting facilities exist.
Oak Hill Industries	Plastic tableware	Recyclable	May, 1994, FTC consent order requires qualifications and disclosures
LePage's Inc.	Transparent tape	Degradable	May, 1994, FTC consent order requires qualification and disclosure
America's Favorite Chicken	Food packaging	Recyclable	April, 1994, FTC consent order
Redmond Products, Inc.	Hair care products	"Environmentally formulated" and "contains natural propellants"	April, 1994, FTC consent order—product contains VOCs
OSRAM	Energy saver light bulbs	Less cost, less energy use	Proposes consent order with FTC requiring disclosure that bulbs provide less light
DeMert & Dougherty, Inc.	All Set hair spray	"Environmentally safe"	August 17, 1993, FTC cease and desist order—product contains VOCs.
Nationwide Industries	Automotive cleaning products	Will not damage ozone	August 26, 1993, Order to stop—product contains class I and class II ozone depleters
White Castle	Food packaging	Recyclable	January 6, 1994, FTC consent order

EXHIBIT 8-1
Legal Challenges to Environmental Marketing Terms (As of June 1992)

First Announced	Company	Product Type	Type of Claim	Language	Regulatory Body[3]
October 17, 1990	American Enviro Products	Disposable diapers	Degradable	• "Biodegradable" • "Will biodegrade before your child grows up"	State AsG
December 9, 1990	Combibloc, Tetra-Pak	Drink boxes	Recyclable	• "As easy to recycle as this page"	NYC DCA
1990	Vons	Vegetables	Toxicity related	• "Pesticide free"	FTC
June 1990	Chemlawn	Lawn pesticides	Toxicity related	• "Safe" • "Non-toxic" • "Fully tested for health and environmental effects"	NY AG
1991	Craftmatic/Contour Industries	Home water treatment systems	Unsubstantiated claim	• Home tap water was "polluted and harmful" to human health	FTC
1991	Advanced Automotive Technology	PetroMiser gasoline additive	Pollution	• "Reduces automotive emissions"	OK AG
1991	Johnson Controls	Plastic bottles	Recycled	• "Environmental packaging . . . the 100 percent recycled container"	NAD–ad modified or ended
March 21, 1991	Webster Industries	Plastic waste bags	Degradable	• "Degrade into harmless organic powder" • "Contains photodegradable additive"	NYC DCA
March 21, 1991	Key Food	Plastic shopping bags	Degradable	• "Degrades in sunlight"	NYC DCA
March 21, 1991	RKO Warner Video	Plastic shopping bags	Degradable	• "This bag is photodegradable"	NYC DCA
March 21, 1991	Daffy's Stores	Plastic shopping bags	Degradable	• "This bag is recycled plastic and is degradable"	NYC DCA
March 21, 1991	Procter & Gamble	Disposable diapers	Compostable	• "Ninety days ago this was a disposable diaper"	NYC DCA
March 21, 1991	Icelandic Spring Water	Drink boxes	Degradable	• "Biodegradable packaging"	NYC DCA

EXHIBIT 8-1

Legal Challenges to Environmental Marketing Terms (As of June 1992) (continued)

First Announced	Company	Product Type	Type of Claim	Language	Regulatory Body[3]
April 22, 1991	Zipatone, Inc.	Spray cement	General environmental	• "Ecologically Safe Propellant . . . you get the job done quickly without damaging the environment"	FTC
May 9, 1991	Sloan's Supermarkets	Plastic shopping bags	Degradable	• "Degradable" • "Will begin degrading within three days of exposure to ultraviolet light . . ."	NYC DCA
May 9, 1991	Pathmark Supermarkets	Plastic shopping bags	Degradable	• "Degradable" • "Non-toxic when incinerated"	NYC DCA
May 9, 1991	Love Pharmacy	Plastic shopping bags	Degradable	• "Degradable bag" • "Will begin degrading within three days of exposure to ultraviolet light . . ."	NYC DCA
May 9, 1991	Down to Earth Stores	Plastic shopping bags	Degradable, recyclable	• "This bag is biodegradable and recyclable"	NYC DCA
June 5, 1991	Jerome Russell Cosmetics	Hair care and other beauty aid products	Ozone-related	• "Ozone-safe" • "Ozone-friendly"	FTC
June 27, 1991	Mobil Chemical Company	Plastic waste bags	Degradable	• "Degradable"	State AsG
July 2, 1991	Webster Industries	Plastic waste bags	General environmental, degradable	• "Environmentally safe" • "Photodegradable"	State AsG
August 5, 1991	Alberto-Culver Company	Hair spray products	General environmental, ozone-related	• "Environmentally safe" • "Ozone friendly"	State AsG
August 28, 1991	Tetra Pak; Combibloc; Lintas, Inc.	Drink boxes	Recyclable	• "As easy to recycle as your daily newspaper"	State AsG
August 30, 1991	American Enviro Products	Disposable diapers	Degradable	• "Will decompose in landfill within 3–5 years or before your child grows up"	FTC
October 9, 1991	First Brands	Plastic waste bags	General environmental, degradable	• "Safe for the environment" • "Degradable"	FTC
October 10, 1991	Clairol, Inc.	Hair sprays	General environmental	• "Environmentally safe"	State AsG

81

EXHIBIT 8–1

Legal Challenges to Environmental Marketing Terms (As of June 1992) (continued)

First Announced	Company	Product Type	Type of Claim	Language	Regulatory Body[3]
October 10, 1991	The Drackett Co.	Household cleaners	Ozone-related	• *"Use with confidence . . . contains no fluorocarbons alleged to damage ozone"*	State AsG
October 10, 1991	Westwood Pharmaceuticals, Inc.	Sunscreen	Ozone-related	• *"Ozone safe"*	State AsG
October 15, 1991	Rockline, Inc.	Coffee filters	General environmental	• *"Environmentally friendly product and packaging"*	NAD–ad modified or ended
October 16, 1991	Colgate-Palmolive	Dishwashing liquid	Vage use of *new*; unsubstantiated exclusivity claims; recycled	• *"New bottle! With 20 percent recycled plastic . . . the only dishwashing liquid made with 20 percent previously used plastic . . ."*	NAD–ad modified or ended
November 14, 1991	Procter & Gamble	Disposable diapers	Compostable	• *"Ninety days ago this was a disposable diaper"* • *"Environmental Information . . . This product is compostable in municipal composting units"*	State AsG
July 17, 1991	Revlon, Inc.	Hair sprays	General environmental	• *"Environmentally safe"*	NYC DCA
July 17, 1991	Procter & Gamble	Antiperspirants	Ozone-related	• *"Contains no CFCs which harm the ozone layer"*	NYC DCA–no action
July 17, 1991	S.C. Johnson & Son	Furniture polish, air fresheners, shaving gel	Ozone-related	• *"Contains no propellant alleged to damage ozone"*	NYC DCA
July 17, 1991	Gillette	Shave cream, hair sprays, antiperspirants	Ozone-related	• *"Ozone-friendly"* • *"No CFCs"*	NYC DCA
1991	Chemlawn	Lawn pesticides	Toxicity related		PA AG
January 1992[1]	Tech Spray, Inc.	Electronic equipment-cleaning products	Ozone-related	• *"Ozone friendly"* • *"Ozone friendly formula"* • *"CFC Free"*	FTC
February 25, 1992[5]	Carlisle Plastics, Inc.	Plastic bags	Degradable		State AsG

EXHIBIT 8-1

Legal Challenges to Environmental Marketing Terms (As of June 1992) (continued)

First Announced	Company	Product Type	Type of Claim	Language	Regulatory Body[3]
March 1992[6]	Sunshine Makers	Household cleaners	Toxicity related	• *"Simple Green is completely non-toxic, so it's safe, even for kids"* • *"Without the chemical pollutants others contain"* • *"biodegradable"* • *"environmentally safe"*	NAD—recommended claims discontinued or modified, advertiser refused, NAD referred case to government agency
March 12, 1992[2]	Statler Industries, Inc.	Bathroom tissue	Recycled, pollution	• *"Made from 100 percent recycled paper"* • *"The Tree Free company has been recognized by Earth Day-NYC as having the lowest toxic emissions of any integrated tissue paper mill in the country"* • *"No elemental chlorine is added in the manufacturing process to whiten our products"*	NAD—claim substantiated
March 13, 1992[2]	Fort Howard Corp.	Paper towels	Recycled, recyclable	• *"Green Forest Paper Towels are made to our highest standards from 100 percent recycled paper fibers, including a minimum of 10 percent postconsumer content."* • *"Green Forest products help the environment in two ways: precious natural resources are saved and paper is recycled instead of entering landfills."* • *"This wrapper may be recycled where plastic film recycling facilities exist."*	NAD—the first two claims were substantiated, the last was considered misleading and was discontinued

EXHIBIT 8–1
Legal Challenges to Environmental Marketing Terms (As of June 1992) (continued)

First Announced	Company	Product Type	Type of Claim	Language	Regulatory Body[3]
March 31, 1992[2]	Celestial Seasonings, Inc.	Herbal teas	Oxygen bleached, exclusivity, recycled, source reduced	• "... now the first tea company in America to use only oxygen bleached tea bags ...", • "We print our boxes on 100% recycled paperboard." • "Our tradition of using English Pillow Style tea bags ... saves 2 million pounds of packaging annually."	NAD–first claim substantiated, recommended modifying second two. Celestial Seasonings agreed to consider recommendations
April 1, 1992[2]	A.V. Olsson Trading Company, Inc.	Paper products	100% unbleached, general environmental	• "100% unbleached" • "Environmentally friendly products" • "... using them will have a positive effect on the environment"	NAD–first claim substantiated, recommended modifying second. Advertiser appealing decision.
April 8, 1992[2]	3M	Dishwashing pads	Recycled, Biodegradable, Absence of phosphorus	• Plastic fibers made from recycled PET • "All detergents ... are biodegradable" • "0.0% phosphorus" • "Packaging made with Recycled Paper and Paperboard"	NAD–claim substantiated
April 17, 1992[2]	ICD Products/Confab Corporation	Paper towels	Recycled	• "This product is made from 100% recycled fiber" • "And with every purchase, 1% will be donated to the National Audubon Society, to help protect our natural resources" • "No new trees were destroyed to create this product. We add no inks, dyes, perfumes, or chlorine bleaching agents that can pollute the atmosphere"	NAD–claims substantiated

EXHIBIT 8–1
Legal Challenges to Environmental Marketing Terms (As of June 1992) (concluded)

First Announced	Company	Product Type	Type of Claim	Language	Regulatory Body[3]
May 14, 1992[1]	RMED International, Inc.	Disposable diapers	Degradable, unsubstantiated claim	• TenderCare disposable diapers "biodegrade in landfills"	FTC
June 12, 1992[2]	Stone Container	Paper lawn and refuse bags	General environmental, biodegradable, recyclable	• "Environmentally safe" • "Biodegradable and recyclable paper, lawn, and refuse bags" • "Paper, the natural package, recyclable, biodegradable, reusable, renewable" • "Compost—the 'Yard Master Refuse Bag' will biodegrade along with its green contents"	NAD—case appealed by Stone, Review Board recommended claims be qualified
June 25, 1992[2]	Melitta USA, Inc.	Bleached and unbleached coffee filters	General environmental, Recycled	• "Oxygen cleansed filters are environmentally safer . . ." • "Made from . . . pulp produced with no chemical bleaches . . . thereby minimizing the release of unwanted by-products" • ". . . there is no better white filter made today—safer for the environment . . ." • "Recycled Paperboard. Environmentally Sound." • ". . . environmentally friendly . . ."	NAD—unbleached claim substantiated, others discontinued or modified
July 27, 1992	Mobil Oil Corporation	Hefty trash bags	Degradable	• "Degradable" • "These bags will continue to break down into harmless particles even after they are buried in a landfill"	FTC

[1]Personal communication with FTC spokesperson.
[2]NAD press release.
[3]Unless otherwise noted, all cases resulted in consent agreements.

[4]FTC press release.
[5]*Environment Reporter*, April 3, 1992.
[6]*Advertising Age*, March 23, 1992, p. 21.

Source: Except where noted, *Green Market Alert*, January 1992, February 1992.

85

8.2   OTHER GENERAL GUIDELINES—UNITED STATES

In response to the federal government's inability to formulate binding uniform national environmental advertising standards, several states, and regional organizations comprised of multiple states, have taken the initiative to adopt laws and standards to address environmental and consumer concerns arising from advertisers' environmental claims. Almost every state regulates environmental marketing claims either through truth-in-advertising regulations or through environmental claim-specific statutes and regulations. The regional organizations that have addressed this issue include the Environmental Marketing Task Force and the Northeast Recycling Council. The problem with this state-by-state/region-by-region approach is that it worsens the lack of uniformity of regulation. The lack of uniformity may require manufacturers that market nationally to forgo the benefits of environmental claims because it may be too expensive to change labeling to comply with each jurisdiction's law.

8.2.1   *Environmental Marketing Task Force*

The attorneys general of 11 states formed the Environmental Marketing Task Force to create standards for the use of environmental claims by advertisers. The task force is comprised of the attorneys general from California, Florida, Massachusetts, Minnesota, Missouri, New York, Tennessee, Texas, Utah, Washington, and Wisconsin. Compiling the information obtained from research and public hearings, the task force published its guidelines in a document titled "Green Report II: Recommendations for Responsible Environmental Advertising" ("Green Report II"). (See Appendix E for a copy). Although "Green Report II" does not function as law, it contains guidance on how regulators are likely to apply state laws.

Essentially, "Green Report II" contains two sections: (1) recommendations for federal action and (2) recommendations to industry. In the recommendations for federal action, the task force calls on the federal government to: (1) adopt national standards for environmental marketing claims used in the labeling, packaging, and promotion of consumer productions; (2) create uniform definitions for terms such as *degradable, compostable, recycled,* and *recyclable* to ensure that marketers know what properties their products must possess before they can make those claims and to ensure that consumers get the accurate information they need to make informed purchasing decisions based on environmental considerations; and (3) address the product life or life cycle issue.

In making its recommendations to industry, the task force acknowledged the power of the "free market economic system" to achieve the goal of protecting the environment and the need for consumers to possess "accurate, specific and complete information" to generate the demand necessary to drive the free market's pursuit of that goal. The task force issued four main recommendations with many subrecommendations in each of the four categories:

1. Environmental claims should be as specific as possible and not general, vague, incomplete, or overly broad.
2. Environmental claims relating to the disposability or potential for recovery of a particular product (e.g., "compostable" or "recyclable") should be made in a manner that clearly discloses the general availability of the advertised option where the product is sold (in other words, claims should reflect current solid waste management options).
3. Environmental claims should be substantive.
4. Environmental claims should be supported by competent and reliable scientific evidence.

8.2.2   *Northeast Recycling Council*

The Northeast Recycling Council (NERC) was formed to accomplish goals similar to those of the task force. NERC is comprised of the following states: Connecticut, Delaware, Maine, Massachusetts, New Jersey, New Hampshire, New York, Pennsylvania, Rhode Island, and Vermont. In 1990, NERC adopted a labeling resolution setting forth three standards "intended to be used by states in establishing a regionally consistent program guiding the labeling of packages and products for such attributes as their reusability, recyclability, and recycled content." The essential elements of the guidelines are as follows:

Reusable standard.   NERC will not grant the use of a regional reusable emblem or term unless the original package or material is used or refilled a minimum of five times in a program established by a manufacturer, distributor, or retailer.

Recyclability standard.   NERC calls for states in the region to work with producers of packages and products to create successful recycling programs. Additionally, NERC's resolution calls for the creation of a technical advisory team to work with states to initiate and develop recycling programs.

Advertisers may qualify their products to use NERC's recyclability emblem or term in two ways. First, an emblem may be attached to a shelf of products for which the advertisers can demonstrate that an approved recycling program exists in the community. Second, an emblem or term may be granted if a manufacturer meets at least one of the following criteria in at least five NERC states that represent at least 75 percent of the region's population:

1. 75 percent of communities or population have an approved recycling program for the material category.
2. The material category has achieved a 50 percent recycling rate (by weight) statewide.

3. The manufacturer, retailer, or distributor of the product can demonstrate that it has achieved a statewide recycling rate of more than 50 percent (by weight) for that particular product or package.

Recycled content standard. NERC requires that "use of a regional 'recycled' or 'recycled content' emblem or term must include an identification of the pre- and postconsumer composition (by weight) of the product or package as demonstrated by an annual mass balance of all feedstocks and outputs of the manufacturing process."

NERC defines the terms *preconsumer material, postconsumer material,* and *end product* in the following manner:

Preconsumer material means any material generated during any step in the production of an end product, but does not include any waste material or by-product that can be reused or has been normally reused within the same plant or another plant of the same parent company. Preconsumer material for the paper industry does not include mill broke, wet or dry, rejected unused stock, obsolete inventories, butt rolls, or other paper waste generated by paper or paper product mills. Waste generated by converting operations that are used by the same parent company whether for the same or different product are also not included within this definition.

Postconsumer material means only those products generated by a business or a consumer which have served their intended end uses, and which have been separated or diverted from solid waste; wastes generated during production of an end product are excluded. Examples of exclusions include: printers' waste, undistributed finished products, or lathe wastes.

End product means only those items that are designed to be used until disposal; items designed to be used in production of a subsequent item are excluded.

8.3 REGULATION OF SPECIFIC TERMS AND CONCEPTS—UNITED STATES

In addition to the federal and state guidelines under truth-in-advertising laws, a number of states have adopted specific regulation of environmentally conscience marketing claims. While each state may have slight variations in its approach, the rules are based on a limited number of concepts. The following discussion treats each of the commonly regulated terms or concepts in turn and describes the conceptual models that have been applied to each particular term or concept. You should always evaluate specific claims you wish to make by reading current state laws in detail and evaluating your specific claim. See Appendix M for a list of contacts to obtain current state regulations.

8.3.1 CHASING ARROWS

The symbol of three chasing arrows has taken on the commonly understood meaning that a product is recycled or recyclable. As such, the symbol is regulated by a number of states in a number of different ways.

8.3.1.1 New York state official emblem. The State of New York has established an official recycling emblem based on the three chasing arrows or use of the terms *recycled, recyclable,* or *reusable.* Any person who uses the three chasing arrows emblem or one of these terms on or in the promotion or advertisement of a package or product offered for sale in the State of New York must apply to the New York Department of Environmental Conservation (DEC) and demonstrate that the product or package meets the definition of the terms under New York law. (See Appendix L for a copy of the applicable rules and a application for permission to use a term or emblem.)

Recyclable. New York defines recyclability generally using the Coneg definitions. Specifically, *recyclable* means a material that meets one of four alternative tests: (1) community recycling programs for that material are available to at least 75 percent of the population of New York, or (2) the material is recycled at a rate of at least 50 percent in New York, or (3) the manufacturer's specific product is recycled at a rate of at least 50 percent of the product sold within New York, or (4) the recyclability claim will be made only in municipalities that have a source separation and recycling program that provides an opportunity for recycling the product or package.

Recycled. To represent that a package or product is recycled, the package or product must meet a somewhat complicated formula for minimum secondary material and postconsumer material content by weight as determined by an annual mass balance of all feedstocks and outputs of the manufacturing process. For packaging, each material component of the package must meet the recycled content standards for that material. That is, if any one component of a composite package is not "recycled" within the meaning of the statute, it is improper to claim that the package is recycled. The claim that the product is recycled must include a disclosure of the percentage of secondary material content in the center of the arrows or immediately adjacent thereto.

Reusable. To carry a claim that a product or package is reusable, the original package or product must be returnable for refilling or reuse a minimum of five times as demonstrated by an annual accounting in a program established by the manufacturer, distributor, or retailer.

Generally the DEC is required to act on a request for certification within 75 days. Authorization is granted for a maximum period of five years and can be renewed by request at least 120 days before expiration. A claim must be displayed in a manner that clearly indicates exactly what part of the package or product to which it applies.

Although the rules could be interpreted to cover any use of the three chasing arrows or the words *recyclable, recycled,* and *reusable,* the DEC has limited the scope of the rule by interpretation. Specifically, the DEC believes use of the words within a textual claim are not regulated by the program. It is proper, for example, to include a textual claim that a particular communication is "printed on recycled paper" without obtaining permission from the DEC. Such textual claims are, however, subject to truth-in-advertising requirements and therefore must be fair and accurate.

The DEC has also determined by interpretation that the program does not apply to use of the three chasing arrows by themselves without one of the words. Similarly, the program does not apply to use of the SPI voluntary material coding for plastic containers.

As of November 1993, the DEC has received 140 applications, 90 of which have been granted, 14 of which have been denied, and the remainder of which are pending.

The DEC also maintains a registry of companies that have been granted certification under the program. This registry is generally available to the public and is being promoted as a source for environmentally conscious consumers to identify environmentally conscious products. The registry is also circulated to state and local governmental purchasing agents throughout New York. It may be useful, therefore, to apply for certification merely to obtain a listing in the registry even if you do not choose to display the New York emblem on your product or package.

Some changes to the program are under consideration, including changes to the minimum contents standards.

The DEC does respond to informal inquiries by telephone regarding interpretations under the program. In addition, the DEC issues letter rulings on interpretive issues.

8.3.1.2   American Forest & Paper Association Guidelines.
The American Forest & Paper Association has issued nonbinding guidance based on the FTC guidelines for the use of the three chasing arrows symbol. For products made with less than 100 percent recycled fiber, the guidelines require the use of black arrows on a white background with a specific disclosure of the total percentage of recycled content, stating "XX percent total recovered [or recycled] fiber." Calculation of the percentage must be made on a fiber-to-fiber basis. The textual disclosure should appear in type no less than six points in type size. The claim should make clear whether it applies to the product or the package. For products made from 100 percent recycled fiber, the arrows may be printed in white on a black background.

8.3.1.3   SPI Resin Coding.
Many states require the use of a resin-coding symbol on plastic products. For a discussion of these requirements, see Section 8.3.7.

8.3.2   *Compostable*

Claims of compostability are similar to claims of degradability and generally raise the same concerns. The focus of regulations on compostability claims are whether the material will actually decompose in the types of composting environments in which it will be disposed of by consumers, whether it will decompose in a reasonably prompt amount of time, and whether composting opportunities are convenient to consumers. Under truth-in-advertising laws, for example, when a product is compostable only in commercial composting facilities, such facilities must be reasonably available to a majority of consumers of the product.

Examples of regulatory definitions of compostability include the following:

1. *Compostable* means the material will decompose into soil-like material in less than one year under controlled biological circumstances (Indiana).

2. *Compostable* means "all the materials in the product or package will break down into, or otherwise become part of, usable compost (e.g., soil-conditioning material, mulch) in a safe and timely manner in an appropriate composting program or facility, or in a home compost pile or device. This definition under the FTC guides would probably apply to most truth-in-advertising laws.

The FTC guides caution against making claims of compostability when (1) the product contains toxic ingredients that would be released into compost material, (2) when applicable, the claim fails to disclose that the product or package must be composted in a municipal composting facility and not a home composting pile, or (3) the product contains noncompostable components.

8.3.3   *Degradable (Bio-, Photo-)*

Specific regulation of degradable claims generally focuses on whether a product will actually degrade in the environment into which it is commonly disposed. Regulation of degradability claims is based largely on the highly publicized case of Mobil's "degradable" trash bags. Under criticism by public interest groups and intense regulatory pressure, Mobil was forced to abandon its claim that its plastic trash bags were degradable because of allegations that the bags did not degrade when they ended up in landfills, where they were deprived of air, light, and water. Although the bags were in fact degradable in the presence of air, light, or water, consumers typically disposed of the bags into landfills and accordingly, under conditions of actual use, the bags tended not to degrade.

The other issues targeted by regulation of degradability claims are the period of time required for degradation and the products of degradation. Regulations require that the product degrade rapidly, sometimes within a specified time period. In addition, regulations require that the product degrade into natural elements, which are defined in different ways.

Following are examples of degradability definitions in different regulatory programs:

1. *Biodegradable* means a material has the proven capability to decompose in the most common environment in which it is disposed within one year through natural biological processes into nontoxic carbonaceous soil, water, or carbon dioxide (California, Indiana).

2. *Photodegradable* means a material has the proven capability to decompose in the most common environment in which it is disposed within one year through physical processes, such as exposure to heat and light, into nontoxic carbonaceous soil, water or carbon dioxide (California, Indiana).

3. *Degradable/biodegradable/photodegradable* means "the entire product or package will completely break down and return to nature, i.e., decompose into elements found in nature within a reasonably short period of time after customary disposal." This definition is given by way of example in the FTC guides for the use of environmental marketing claims. A number of states with local truth-in-advertising laws will probably find this example a persuasive guide to how their state truth-in-advertising laws should be interpreted.

4. Products that are currently disposed of primarily in landfills or through incineration—whether paper or plastic—should not be promoted as *degradable, biodegradable* or *photodegradable*. This recommendation in "Green Report II" will probably be persuasive to most states with local truth-in-advertising laws and will almost certainly be enforced by states whose attorneys general are members of the "Green Report II" task force.

5. *Degradability* means (1) the product decomposes completely to the point where it is entirely broken down into soluble nutrients, minerals, inorganic compounds, water, carbon dioxide, methane, or any combination of those breakdown products; (2) the product decomposes completely within a reasonably short time period; and (3) the product's decomposition does not release any contaminant at a level that violates state or federal environmental quality standards or that is toxic to humans or the environment (Wisconsin).

8.3.4 *Documentation*

Truth-in-advertising laws require extensive written substantiation or documentation. In addition, some states have imposed specific documentation requirements under their environmental marketing regulatory programs.

These are modeled on the truth-in-advertising requirements for substantiation, but they include some additional requirements. The requirements of the California, New York, and Wisconsin statutes are representative.

California. The California program requires any person "who represents in advertising or on a label or container of a consumer good that the good is not harmful to or is beneficial to the natural environment" using terms defined by the California statute or makes any other environmental claim to document:

1. The reason the persons believe the assertions are true.

2. Any significant adverse environmental impacts directly associated with the production, distribution, use, and disposal of the consumer good.

3. Any measures taken to reduce the environmental impact directly associated with the production, distribution, use, and disposal of the consumer good.

4. Violations of any permits (federal, state, or local) directly associated with the production, distribution, use, and disposal of the consumer good.

5. Whether or not (if applicable) the consumer good meets the definition of *recycled, recyclable, biodegradable, photodegradable*, or *ozone friendly*, as defined in the statute.

The information should be available to the public upon request.

New York. New York requires the following documentation for participation in its environmental emblem program:

1. Upon what package or product and how the emblem will be displayed.

2. Whether the label covers all sides of the container.

3. For packages or products with different material components, a description of each component.

4. The name, title, address and telephone number of the responsible party making the certification of the information.

5. The outlet to be used for marketing the package or product displaying the emblem.

6. A statement that a commitment is made to supply corroborative data at the request of the Department of Environmental Conservation to confirm certification.

7. The waste implications of the package or product after use.

8. The extent to which secondary material and postconsumer material contents standards are limited by technology.

9. Availability of secondary material and postconsumer material feedstocks.

Wisconsin. In Wisconsin, documentation for any environmentally conscious marketing claim must have existed at the time the claim was made. The documentation must be provided to the Department upon request within 30 days. The documentation must include the following:

1. Adequate support for every direct and implied representation made in a claim.

2. Substantiation based on competent and reliable evidence consisting of tests, analyses, research studies, or other evidence developed and evaluated in an objective manner by persons qualified to do so, using procedures that are generally accepted in the profession as yielding accurate and reliable results.

3. Adequate support to prove that the representation is not false, deceptive, or misleading.

4. Adequate documentation to support the claim under all reasonably foreseeable conditions specified or implied by the claim.

Many states impose a requirement that documentation of environmental marketing claims must preexist the claim and must be immediately available for inspection upon request of regulatory authorities. Typically, documentation may become a public record once provided to a regulatory agency, unless specific procedures are followed to designate such information as confidential trade secrets.

8.3.5 *Minimum Content Requirements*

Minimum content requirements are regulations that require all products made out of a given material to contain a minimum amount of recycled content for that material. For example, many states and the federal government have regulations encouraging or requiring their purchasing departments to purchase paper products only if the paper contains certain minimum amounts of recycled content. In addition, certain programs require manufacturers of significant quantities of paper or paper products to include minimum recycled content in their product.

8.3.6 *Ozone Friendly or Ozone Safe*

Restrictions on ozone-related claims typically are aimed at ensuring that a product does not use or give off any product that is considered an ozone-depleting chemical. Currently, the principal chemicals that have been identified as ozone-depleting substances are:

1. Chlorofluorocarbons (CFCs).
2. Halons.
3. Carbontetrachloride.
4. 1,1, -trichloroethane.
5. Hydrochlorofluorocarbons (HCFCs).

Any product or package that gives off any of these substances, or that is formulated using one of these substances, probably cannot be labeled ozone safe or ozone friendly. In addition, it would probably be considered deceptive to label a product that contains one of the non-CFC ozone depleters with the claim "no CFCs" without the explanation that the product contains other ozone-depleting ingredients.

Specific regulation of claims regarding ozone-depleting chemicals simply focus on whether ozone depleters are used in the formulation of the product or are given off as a result of the use or production of the product. Typical definitions are as follows:

1. "*Ozone friendly*, or any similar term, means that any chemical or material released into the environment as a result of the use or production of a product will not migrate to the stratosphere and cause unnatural and accelerated deterioration of ozone" (California).

2. The claim that a product is "ozone friendly," "ozone safe," or "contains no CFCs" is deceptive if the product contains any ozone-depleting substance, including those substances listed as Class 1 or Class 2 chemicals in Title VI of the Clean Air Act Amendments of 1990 or other substance subsequently designated by EPA as ozone-depleting. This definition, paraphrased from the FTC guides for the use of environmental marketing claims, would probably be persuasive in interpreting state or federal truth-in-advertising laws.

Some states, such as Virginia, ban the sale of products that contain fully halogenated chlorofluorocarbons as a blocking or expansion agent. Some states, for example, Iowa, also limit or regulate the sales of polystyrene products, in particular in connection with packaging of food service items.

8.3.7 *Plastic Resin Coating Requirements*

The Society of the Plastics Industry, Inc., (SPI) developed voluntary guidelines for the marking and coding of disposable plastic containers to identify their material types. The purpose of the coding system was to help create and facilitate recycling markets by making it easy for recyclers to identify different plastic resin types so they could be separated and recycled appropriately. Many states have now adopted resin container coding requirements similar to the SPI voluntary guidelines and made them mandatory by statute.

The SPI guidelines are reprinted in this text as Appendix K. The SPI guidelines divide materials into seven categories:

1. PETE (polyethylene terephthalate).
2. HDPE (high density polyethylene).
3. V (vinyl/polyvinyl chloride).
4. LDPE (low density polyethylene).

5. PP (polypropylene).

6. PS (polystyrene).

7. Other.

Under the guidelines, plastic containers must be coded by molding a symbol based on the three chasing arrows into the center of the bottom of the container. Efforts are currently underway to change the symbol to avoid confusion with the chasing arrows. The resin categories are indicated by molding the number into the center of the three arrows and molding the appropriate acronym underneath the three chasing arrows. A vast majority of states have adopted laws requiring the use of coding. The states usually require the markings for all persons "manufacturing, selling, or offering for sale plastic bottles and rigid plastic containers." Generally, the following definitions determine whether you are required to use the resin coating symbols:

Plastic means a material made of polymeric organic compounds and additives that can be shaped by flow.

Rigid plastic container generally means a formed or molded container intended for single use, composed predominantly of plastic resin, and having a relatively inflexible finite shape or form with a capacity of not less than 8 ounces or more than 5 gallons.

Mandatory resin coating statutes usually include civil penalties for violations that can be as much as $1,000 per violation. Some regulators may take the position that every single container distributed without the symbol constitutes a single violation. Many mandatory coding statutes also can be enforced by court order requiring use of the coding symbol.

It is generally not cost-effective to differentiate between containers distributed in states requiring use of the SPI symbols and states that don't. Accordingly, manufacturers of plastic containers commonly mark all their containers with the SPI symbol regardless of where they are distributed. This could raise a problem for users of the SPI symbol that distribute their product in states that otherwise regulate the use of the three chasing arrows symbol or of the terms *recycled,* or *recyclable*. However, states that restrict the use of the three chasing arrows exempt use of the SPI symbol from their regulations to avoid any conflict. The Federal Trade Commission has also noted in its guideline that use of the SPI symbol on the bottom of a plastic package will not be considered to constitute a claim of recyclability subject to truth-in-advertising requirements. If efforts to alter the symbol are successful, the issue will be avoided.

8.3.8 *Recyclable or Recyclable Material*

Regulations of recyclable materials generally focus not just on whether a material is theoretically capable of being recycled, but whether meaningful opportunities exist for

consumers to conveniently dispose of the materials through recycling. Accordingly, regulations typically require, for example, that convenient collection facilities exist for the material, and that the collection facilities are capable of performing any necessary source separation. Recycling typically means reprocessing of the material into a useful product, and not merely incineration for energy recovery. Typical definitions of recyclable include the following:

1. *Recyclable* means a material for which any of the following standards are met:

   a. Access to community recyclable recovery programs for that material is available to no less than 75 percent of the population of the state; or

   b. A statewide recycling rate of 50 percent has been achieved within the material category; or

   c. A manufacturer, distributor, or retailer achieves a statewide recycling rate of 50 percent of the product or package sold within the state; or

   d. A product or package may be recyclable within the jurisdiction of a municipality where an ongoing source separation and recycling program provides the opportunity for recycling of the product or package.

A more simplified version of this type of "opportunity" regulation was enacted by California, which states that a product is "recyclable" if it can be recycled conveniently in every county in California with a population of more than 300,000. At the writing of this book, this statutory definition has been struck down by a court in California because the requirement of "convenience" is too vague. The statute has been generally upheld, however, and should be enforceable with some adjustments and refinements to the concept.

Other programs list specific materials or product categories that may be labeled as recyclable. For example, some programs employ a definition such as the following:

Recyclable materials include but are not limited to glass food and beverage containers, plastic milk containers, plastic soft drink containers, newspaper, tin-coated cans and steel cans, aluminum, corrugated cardboard and mixed office paper (New Hampshire).

Under truth-in-advertising laws, recyclable generally means the product or package exclusive of "minor, incidental components" can be deposited at collection sites that are available to a substantial majority of consumers or communities where the product is distributed. An example of a "minor, incidental component" under the FTC guidelines is the cap of a bottle.

8.3.9 *Recycled/Recycled Content*

Regulations governing the use of the terms *recycled* or *recycled content* generally regulate (1) the minimum content of "preconsumer" and "postconsumer" waste used to formulate a package or product and (2) specific disclosure of the percentage by weight of recycled content used in the package or product. Minimum content standards specify, for example, that a product must contain at least X percent postconsumer waste to be labeled recycled. Disclosure standards specify that a claim of recycled content must include a disclosure of percentage by weight, for example, "contains 10 percent postconsumer recycled content by weight."

Thresholds for minimum recycled content vary dramatically. In some states, regulations may impose a simple and single threshold for all products, such as California's requirement that products labeled "recycled" must contain at least 10 percent postconsumer material by weight. Minimum content regulations can become much more sophisticated, such as the New York model. The minimum standards for recycled content in New York vary by product category and include standards for both minimum preconsumer content and minimum postconsumer content. So for example, a claim of recycled content in New York when applied to tissue products is proper only when the product contains at least 80 percent "secondary material" of which at least 40 percent must be postconsumer waste.

When applying recycled content standards, it is important to evaluate the definitions of *preconsumer waste* and *postconsumer waste* carefully. Many states base their definitions on the NERC model. Generally, *postconsumer material* means products, packages, or materials that have "served their intended end uses" and have been "diverted from the waste stream." By contrast, *preconsumer waste* is generally material that never reached a business or consumer for an intended end use but has still been recovered or diverted from the waste stream. Typical examples include process wastes (when they are recycled into the same process by which they were generated), scrap material, overstock, or obsolete inventory.

While it is often obvious whether a material is preconsumer or postconsumer, it is not always obvious whether a material is "waste." Generally, the intention of regulatory programs is to divert materials from the solid waste stream. The key question, therefore, is whether the material would have gone to a landfill or other disposal facility if it had not been recycled. Accordingly, by-products of the manufacturing process that are recovered and reintroduced into the same process generally do not qualify as recycled materials because they never were intended for disposal. To illustrate the distinction, consider a manufacturer that collects spilled raw material and scraps from trimming finished products and combines those spills and scraps with virgin material for use in further production of the same product. The spills and scraps do not qualify as waste, and cannot be counted in a measure of recycled content. By contrast, if the spills and scraps would typically be discarded and cannot be reused by the manufacturer, they probably are waste. If the spills and scraps

are then sold or otherwise given to a manufacturer of a different product instead of sent for disposal, these spills and scraps can be counted toward the recycled content of the second product and would be preconsumer waste.

There are several methods for determining the recycled content of a product. One method is to measure the amount of virgin material and recycled material used as inputs in each production run. The average or minimum percentage of recycled content by weight of all production runs is a reasonable measure of recycled content.

Some states require that recycled content be measured by an annual mass balance of all feedstocks and outputs of the manufacturing process. In addition to this annual measure, it may be necessary to demonstrate that smaller units of production, such as production runs or monthly production, contain a minimum recycled content. In New York, for example, it is necessary to perform annual mass balancing and to demonstrate that secondary material used in any month is no less than 80 percent of the average annual use.

Some regulations of recycled content for paper products or packages require disclosure of the recycled content by fiber content rather than by weight.

8.3.10 *Responsible Party*

Generally, any person manufacturing or distributing a product that displays an environmentally conscious marketing claim is considered to be responsible for the claim and can be charged with a violation of an applicable law. Manufacturers are responsible for claims they associate with the product. A manufacturer typically would not be responsible for a claim affixed to the product by a distributor or wholesaler without knowledge or control of the manufacturer.

Wholesalers and distributors typically are responsible for claims affixed by a manufacturer. There are at least two exceptions to this rule, however. First, some programs recognize that a wholesaler or retailer is not responsible for representation if the wholesaler or retailer did not initiate the representation by placing it on a package or product. Indiana, for example, has written this defense into its definitional law. Second, a wholesaler or distributor may be able to rely on the defense that it reasonably relied on the manufacturer's representation that claims were true and substantiated. There is a proposal in Wisconsin at the time of this writing that claims in that state must clearly identify the product manufacturer, distributor, lessor, or seller who is responsible for an environmentally conscious marketing claim.

8.3.11 *Reusable or Refillable*

Regulation of claims that a product is refillable or reusable focuses on requirements that (1) the product actually be constructed to be durable enough to be reused a minimum number of times and (2) actual opportunities exist for refilling the product. Initial concerns focused on eliminating claims that were based on the theoretical possibility that a product could be refilled or reused when in actual practice the product simply ended up in a landfill.

The principle regulation of refillable/reusable claims comes through truth-in-advertising regulations. Under the FTC guides, for example, refillable/reusable means "a system is provided for (1) the collection and return of the package for refill, or (2) the later refill of the package by consumers with products subsequently sold in another package."

There is some specific regulation of refillable/reusable claims through statutory definition. In New York, a claim that a product is reusable means that "the original package or product can be returned for refilling or reuse a minimum of five times as demonstrated by an annual accounting in a program established by a manufacturer, distributor, or retailer for refilling or reuse of the manufactured product or package."

8.3.12 *Toxics in Packaging*

A number of states prohibit or limit the heavy metal content allowable in packaging or packaging waste. The principal toxics regulated are:

1. Lead
2. Mercury
3. Cadmium
4. Hexavalent chromium

Regulation of toxics in packaging generally specifies a maximum concentration of the amount of these heavy metals allowed to be "incidentally present" in packaging. The regulations generally prohibit the intentional introduction of the heavy metals into packaging through the use of inks, dyes, pigments, adhesives, stabilizers, or any other additives that contain the heavy metals.

For purposes of most of the regulatory packages, the following definitions apply:

1. *Distributor* means a person who imports any container used to package products into the state whether filled or unfilled.
2. *Manufacturer* means any person who makes containers to be used to package products.
3. *Package* means a container providing a means of marketing, protecting or handling a product and shall include a unit package, an intermediate package, and a shipping container. *Package* shall also mean and include but not be limited to such unsealed receptacles as carrying cases, crates, cups, pails, rigid foil and other trays, wrappers and wrapping films, bags, and tubs. Some states refer to the American Society for Testing and Materials definition ASTM D996 for packages.

The statutes generally prohibit any manufacturer or distributor from offering packages for sale or for promotional purposes if the package includes any of the listed heavy metals that have been intentionally introduced or exceed the concentration levels.

The sum of the concentration levels of lead, cadmium, mercury, or hexavalent chromium present in any package shall not exceed the following limits, which are typically phased in from the time of enactment of the statute:

1. 600 parts per million by weight in year 1 following passage of the statute.
2. 250 parts per million by weight in year 2.
3. 100 parts per million by weight (.01 percent) in year 3 and afterwards.

Certain limited exemptions may be available as follows:

1. Packages that contain the heavy metals in order to comply with health or safety requirements of federal law or for which there is no practical alternative as determined by an applicable regulatory body, provided authorization or permission is obtained from the applicable regulatory body to manufacture and distribute the package.
2. Packages that exceed the standard because of the addition of postconsumer materials. This exemption may be subject to a phaseout.
3. Glass, ceramic, or metal receptacles that are intended to be reused or refilled. This exemption may be subject to a phaseout.
4. Bottles containing liquor that have lead foil tops and baskets as seals.
5. Packages manufactured before enactment of the regulatory program.

Most statutes do not specify how the heavy metal's content should be measured. The Iowa program requires testing using ASTM materials test standards or U.S. Environmental Protection Agency test methods for evaluating solid waste, S-W 846.

Some toxics in packaging laws provide that it is a valid defense to a claim of a violation under the Act that a manufacturer or distributor relied on good faith on the written assurance of the packaging manufacturer that the packaging complies with packaging regulations. The written assurance must be a certificate of compliance stating the package is in compliance with the requirements of the program; and if the package has received an exemption, the certification must state the specific basis on which the exemption is claimed.

8.3.13 *Truth in Advertising—State Laws*

Most states have not adopted legislation specifically addressing environmental marketing claims, indicating they will follow the FTC guidelines for the use of such claims.

However, environmental claims in the following states are regulated by truth-in-advertising laws that are enforced by the states' attorney general's offices: Alabama, Alaska, Arkansas, Colorado, Delaware, District of Columbia, Georgia, Idaho, Kansas, Kentucky, Louisiana, Maryland, Michigan, Mississippi, Montana, Nebraska, Nevada, New Mexico, North Carolina, North Dakota, Ohio, Oklahoma, Oregon, Pennsylvania, South Carolina, South Dakota, Vermont, Virginia, West Virginia, and Wyoming.

A task force of 11 attorneys general has produced "The Green Report II: Recommendations for Responsible Environmental Advertising." The Green Report generally discusses requirements for fair and accurate environmental marketing claims with examples.

8.3.14 *Warnings—California Cancer and Reproductive Toxicity Exposure Labeling*

The California Safe Drinking Water and Toxic Enforcement Act could be characterized as an antienvironmental marketing requirement. The act, commonly known as Proposition 65, requires anyone doing business in California who may expose anyone to one or more of certain hazardous chemicals to display a warning in connection with the business. The scope of the statute is incredibly broad. Proposition 65 requires products containing a listed chemical to display a warning containing the following information:

> WARNING: This product contains a chemical known to the State of California to cause cancer or reproductive harm.

Proposition 65 essentially applies to any product that is sold within the state. The only exemption is for products that must carry a warning governed by federal law in a manner that preempts Proposition 65. Categories of such exempt products may include medical products, pesticides, cigarettes, and alcohol.

The list of hazardous chemicals includes literally hundreds of chemicals and compounds. The state maintains separate lists for chemicals known to the state to cause cancer and chemicals known to the state to cause reproductive harm. A manufacturer must warn of cancer risk, reproductive risk, or both depending on how the chemical is listed.

For the cancer warning, the manufacturer must display the warning unless the manufacturer can demonstrate that exposure to the product poses no significant risk assuming lifetime exposure. For the reproductive health warning, the manufacturer must display the warning unless the manufacturer can demonstrate that exposure will have no observable effect assuming exposure at 1,000 times the level in question for the substance. A showing of no signif-

icant risk or no observable effect must be based on competent scientific research. The act also exempts businesses employing fewer than 10 employees.

The state has established "no significant risk levels" and "no observable effect levels" ("NOELS") for certain of the listed chemicals. In such a case, the manufacturer may rely on the level established by the state rather than conducting individualized research on the warning threshold. For most chemicals, no threshold has been established, and the expense of establishing a threshold level is significant. Demonstrating the no significant risk level for cancer means demonstrating exposure would result in less than one excess case of cancer per 100,000 individuals exposed over a 70-year lifetime.

Warnings must be communicated in such a way that they are likely to reach every potentially exposed individual. Directly labeling the product is a "safe harbor" method of meeting this requirement—that is, it will automatically be deemed to be in compliance. Other methods, such as public advertising or the display of warnings at the point of sale, may also be acceptable.

Green marketers have an opportunity to work with the requirements of Proposition 65, because the statute does not exclude the provision of additional information with the warning, provided the warning remains "clear and reasonable." It remains to be seen exactly what would be considered to deprive a warning of its clearness or its reasonableness. It would probably be wise to consult with the California Department of Toxic Substances Control for a formal or informal opinion on whether a given statement would satisfy the clear and reasonable warning test.

Examples of environmentally conscious manufacturing principles that apply to Proposition 65 include providing context and providing further information on the benefits or positive environmental attributes of a product. For example, a distributor of lead crystal might modify the warning as follows:

> WARNING: Like all fine crystal, this product contains lead, a chemical known to the State of California to cause cancer and reproductive harm. For more information call [phone number] or see the fact sheet enclosed with this package.

Additional factual information could be provided such as the necessity for use of lead in crystal, the rate of exposure resulting from normal use of the product, the company's efforts to reduce the lead content of its crystal without compromising quality, steps consumers can take to minimize the risk of lead exposure in using the product, and comparison of the lead content of the product to other popular brands. Obviously, principles of truth-in-advertising, including truthfulness and documentation, would apply to any such claims.

8.4   CANADA

Recognizing that Canadian consumers are concerned about the environmental effects of the products and materials they use, and that the proliferation of a broad range of descriptions, logos, vignettes, and other representations used to describe or imply environmental features of consumer products "requires action to ensure responsible labeling and advertising," the Canadian Department of Consumer and Corporate Affairs (CDCCA) published a document titled "Guiding Principles for Environmental Labelling and Advertising." CDCCA, which administers the relevant statutes controlling misrepresentation and misleading labeling and advertising, created the document within the context of the Consumer Packaging and Labelling Act (CPLA) and the Competition Act (CA).

The guiding principles of the document are:

1. Industry is responsible for ensuring that any claims and/or representations are accurate and in compliance with the relevant legislation.

2. Consumers are responsible, to the extent possible, for appropriately using the information made available to them in labeling and advertising, thereby enhancing their role in the marketplace.

3. Environmental claims and/or representations that are ambiguous, vague, incomplete, misleading, or irrelevant, and that cannot be substantiated through credible information and/or test methods should not be used.

4. Claims and/or representations should indicate whether they are related to the product or the packaging materials.

The guiding principles are not binding law but, like the "Green Report II," provide guidelines to business for complying with existing laws.

8.5   AUSTRALIA

Environmental claims in Australia are governed by the Commonwealth Trade Practices Act (CTPA) administered by the Trade Practices Commission (TPC). The CTPA is very similar to the FTCA in the United States; both are aimed at protecting consumers from misleading advertising. The TPC, like its American counterpart the FTC, has created guidelines that firms should follow when making environmental claims. The TPC developed the guidelines using a life cycle approach that focuses on the principal environmental problem area of a product. The guidelines have two objectives:

1. To ensure that consumers and ethical traders are not disadvantaged by false or misleading claims about the environmental impact of particular products on their method of manufacture.

2. To encourage product innovation and marketing competition based on sustainable environmental claims.

The guidelines make several other points about environmental claims:

- All claims must be sustainable.
- There must not be other factors that negate the claims being made.
- The claims must be achievable and truly new or different.
- An environmental endorsement scheme must not be misinterpreted as being approved by a group unless it actually is.

A professor of marketing at the University of Newcastle, Australia, has opined that the guidelines "would follow very closely the overall objectives of similar consumer protection activities in the U.S."

8.6   EMBLEMS AND CERTIFICATIONS

Due to the lack of uniform federal regulation of environmental marketing claims, other than through truth-in-advertising enforcement, two U.S. nonprofit organizations, Green Seal and Scientific Certification Systems, have begun issuing "environmental seals of approval" to a variety of consumer products. Manufacturers/advertisers participate in these programs voluntarily, paying a fee to have their products evaluated and to use the seal of approval. Tension exists between the two organizations due to their respective focuses and methodologies of testing.

8.6.1   *Green Seal*

Green Seal does not simply verify a manufacturer's/advertiser's claim; instead it grants its seal of approval to products that meet the organization's standards as environmentally preferable. The group had planned to use the life cycle assessment method to analyze the products' impact on the environment, but technical and cost issues forced the group to base its standards on a modified, shorter version of life cycle assessment the organization calls "environmental impact evaluation." Green Seal intends to update the standards every three years and to educate consumers about the standards and their significance. Underwriters Laboratories, Inc., tests the products for Green Seal at a cost between $1,000 and $10,000.

8.6.2   *Scientific Certification Systems*

The Scientific Certification Systems (SCS) program—formerly known as Green Cross—simply verifies manufacturer/advertiser claims about the environmental impact/benefits of products. SCS has also decided to issue general seals of approval based on life cycle inventory of the products. The cost of SCS testing generally falls between $3,000 and $5,000. SCS has been under attack by the Environmental Defense Fund (EDF). EDF claims that SCS fails to apply "state-of-the-art standards when certifying products, provides very limited public access to its standards and procedures and almost no opportunity for public comment, fails to monitor how manufacturers use the seal and, because SCS will award its logo based on certification of one claim, has allowed its logo to appear on products in a manner that suggests that additional product claims have been verified."

8.6.3   *Canadian National Emblem*

Canada's "Environmental Choice" certification program has existed since 1988. The federal agency Environmental Canada created the program to help consumers identify products that are more environmentally sound. Products that meet certain specified criteria are awarded the Environmental Choice "EcoLogo."

Manufacturers apply to receive the EcoLogo. Criteria for evaluation are developed using the life cycle assessment. As of 1991, criteria had been established in 15 categories of consumer products. If the manufacturer's product is approved, a license fee of $1,500 to $5,000 must be paid.

RESOURCES

FTCA, 15 USCA Section 45 (Supp. 1993).

FTCA, Section 5, 15 USCA Section 45 (Supp. 1993).

"Green Report II: Recommendations for Responsible Environmental Advertising," May 1991.

Holland Furnace Company v. FTC, 295 F. 2d 302 (7th Cir. 1961).

Howett, Ciannat M. "The 'Green Labeling' Phenomenon: Problems and Trends in the Regulation of Environmental Product Claims," *Virginia Environmental Law Journal*, 1992, pp. 431, 448-50.

Ministry of Consumer and Corporate Affairs. "Guiding Principles for Environmental Labelling and Advertising," May 1991.

Northeast Recycling Council. "Labeling Resolution," November 27, 1990.

Polonsky, Michael Jay. "Australia Sets Guidelines for 'Green Marketing'," *Marketing News*, October 14, 1991, pp. 6, 18.

U.S. Environmental Protection Agency. *Evaluation of Environmental Marketing Terms in the United States*, February 1993, p. 77.

International Regulations and Guidelines

Regulation of environmentally conscious marketing outside the United States (with the exception of Canada and Australia, which follow the U.S. model) tends to be as much an issue of protection of the environment as it is an issue of consumer protection. Accordingly, foreign regulations focus not just on the truth of an environmentally conscious marketing claim, but also tend to set specific environmental performance standards for different categories of products that must be met in order to make an environmental claim. The most common model for such regulation is certification of product for the display of a nationally promoted emblem.

Regulation of environmental claims outside the United States has been most prominent in the European Union, Scandinavia, and Japan. Germany, in particular, has been a leader in the regulation of ECM and tends to be much further along in implementation of its regulations than many other countries. The European Union as a whole is pressing forward with regulation of environmental claims, particularly in the areas of certifications and packaging reduction laws. Regulation in Scandinavia and in some EU member states is still in development. Regulation in Japan is more haphazard, although a national certification program in Japan has provided some guiding principles.

Many nations around the world regulate advertising, and many countries have self-regulation of advertising. The International Advertising Association (IAA) in New York tracks advertising regulation throughout the world and is an important source of information about international advertising regulations. The IAA has identified 14 countries as being the most highly restrictive of advertising:

- Germany
- United Kingdom
- France
- United States
- Canada
- Australia
- Sweden
- Austria
- Belgium
- Argentina
- Mexico
- Italy
- Finland
- Denmark

Accordingly, analysis of environmentally conscious marketing claims should include an analysis of generally applicable advertising laws or self-regulation in each country in which the claim will be made.

Although laws regulating truth in advertising exist in many countries, the rapid development of environmental performance criteria for classes of products, using life cycle analysis, has essentially overtaken truth-in-advertising laws in the environmental marketing arena internationally. To some extent, the certification approach limits entrepreneurial freedom and market mechanisms for the development of environmentally conscious products and claims. Instead of leaving the market open for the development of those attributes and corresponding claims that succeed on their own merit in the market, certification programs mandate particular attributes deemed environmentally preferable. The intended trade off for the loss in flexibility and innovation is greater certainty in the exact meaning of environmental claims and a higher level of confidence that a product meets the standards implied by a particular claim.

9.1 EUROPEAN UNION (EU)

The Maastricht Treaty, which took effect November 1, 1993, established the European Union (EU) as the successor to the European Community. The European Union is a group of European states that have joined in an effort to form a common market through which goods and services may be traded freely. There are currently 12 members in the European Union: Belgium, Denmark, Greece, France, Germany, Ireland, Italy, Luxembourg, the Netherlands, Portugal, Spain, and the United Kingdom. The European Union has legislative, executive, and judicial bodies in which each of the member states are represented. Through these bodies, the EU has developed environmental legislation that is intended to apply communitywide.

To date, environmental legislation in the EU has been implemented with greater enthusiasm in some member states than others. Accordingly, even though there already exists a framework for use of an EU-wide environmental emblem known as the Eco-Label, not all European member states have developed local programs for certifying products as eligible to display the emblem. It is therefore important to evaluate both communitywide programs and local laws and regulations when considering environmental marketing claims in Europe.

Two major EU programs most directly affect green marketing in Europe: (1) the EU Eco-Label and (2) EU initiatives on packaging, recycling, and waste reduction. As discussed below, the Eco-Label may be applied by manufacturers that can demonstrate their products meet certification standards based on performance criteria developed through life cycle analysis. Details of the program are discussed in Section 9.1.1. Packaging and general waste reduction efforts in the community are discussed in Section 9.1.2.

Like the United States, the individual European countries often impose a patchwork of differing laws and standards, making compliance for products distributed throughout the community potentially difficult. Communitywide programs adopted by the EU are intended to eliminate or at least ease this problem by creating uniform, communitywide laws or standards. This process, known as *harmonization*, can be slow and difficult as individual member states are often reluctant to subordinate national interests and national regulatory programs to communitywide programs.

In the case of environmentally conscious marketing, Germany, in particular, and also France and the United Kingdom have developed and implemented national programs directed at environmentally conscious labeling and packaging issues. Existing and proposed national programs may conflict with, or at least exceed the minimum, communitywide rules and regulations. Over time, national differences may give way to uniform standards throughout the community.

9.1.1 European Union "Eco-Label"

The Eco-Label is a government-sponsored emblem in the European Union, which may be awarded by application to any individual member state within the EU. The Eco-Label was established by a regulation of the European Union in 1992 and is intended to influence consumers' purchasing decisions by identifying products with a lesser impact on the environment. The EU regulation provides an overall framework for an "award" of the Eco-Label to a particular product.

The Eco-Label is available only to products for which standards have been established by the European Commission, an executive body of the EU. Standards that have been developed apply to product groups such as washing machines, dishwashers, light bulbs, and so on. Standards for product groups are developed on a communitywide basis, so that the same standard applies in all 12 member states. (There may be some variations in implementation by individual member states, as discussed below.)

It appears that standards for an award of the Eco-Label, which have so far been developed only for dishwashers and washing machines, will set specific performance standards the product must meet to qualify for the Eco-Label. For example, the following standards apply to washing machines:

1. A maximum energy consumption of 0.23 kilowatt hours (KWH) per kilo of cotton products at 60 degrees Celsius without prewashing.

2. A maximum water consumption of 17 liters per kilo of cotton products washed at 60°C.

3. A maximum "wastage" of 5 percent of the detergent placed in the detergent drawer for a given wash.

4. A clear warning on any machine that is formulated with polymers weighing more than 50 grams.

5. User instructions for the machine must include appropriate settings for optimal ecological use in different situations.

6. The machine must meet or exceed European performance and efficiency standards.

Hoover is the first company to have been awarded the right to use the EC Eco-Label, which was awarded to three of Hoover's washing machines in November, 1993.

Similar standards will apply to additional product groups. The initial product categories slated for implementation are light bulbs, paper towels, toilet paper, hair sprays, detergents, paints and varnishes, soil improvers, T-shirts, bed linens, and refrigerators.

Product award criteria are developed by individual member states, that each propose product categories and then develop proposed standards based on life-cycle studies for those categories. The following product categories have been proposed and are being worked on by the member state indicated:

Product Category	State
Antiperspirant	United Kingdom
Batteries	France
Building materials	Italy
Cat litter	Netherlands
Ceramic crockery	Portugal
Ceramic tiles	Italy
Dishwasher detergents	Germany
Feminine sanitary products	United Kingdom
Glassware	Portugal
Hair spray	United Kingdom
Household cleaning products	Germany
Insulation materials	Denmark
Laundry detergents	Germany
Lightbulbs	United Kingdom
Paints and varnishes	France
Paper products	Denmark
Packaging materials	Italy
Photo-voltaic cells	Germany
Refrigerators	Italy
Shampoos	France
Shoes	Netherlands
Soil improvers	United Kingdom
Solar heating systems	Germany
Textiles/T-shirts/bed linen	Denmark

Criteria that are in the proposal stage for product categories that are further along include the following:

Lightbulbs. The United Kingdom has retained Environmental Resources Limited to perform life-cycle research on lightbulbs, and is considering energy use as the primary criterion for certification. A requirement of energy efficiency of 40 lumens/watt is under consideration, which would probably exclude most bulbs except compact fluorescent lights from qualification for use of the Eco-Label. A secondary criterion requiring packaging to be reusable or recyclable is also under consideration.

Paints and coatings. France has retained Ecobilan to study and propose criteria for paints and varnishes. Ecobilan intends to study inputs and outputs as a basis for proposing criteria, and will probably focus on emissions to CO_2, NO_x, and other substances that may harm the atmosphere.

Detergents. Germany has formed a working group to study detergent impacts, and has proposed focusing on reduction of the amount of detergents entering the liquid waste stream. Criteria under consideration are: (1) limits on the amount of components in detergents; (2) a preference for ingredients that biodegrade quickly; (3) exclusion of phosphates, APEOs, EDTA, and possibly NTA; and (4) recycled or reusable packaging. A European industry group known as AIS has proposed criteria focusing on waste per wash, with a requirement of a performance standard as well. Similar criteria for dishwasher detergents may follow.

Hair spray. The United Kingdom has retained Chem Systems to study hair sprays, and has proposed VOC emissions during production and use as a primary criterion for study, with impacts of the packaging and dispenser as a secondary criterion. Chem Systems is already working on or will probably work on standards for other hair care products, as well as deodorants and antiperspirants.

The criteria for product groups are developed according to the product life cycle matrix.

It appears in the initial implementation that a relatively discreet number of performance criteria will be selected from the chart for particular products based on their greatest areas of environmental impact. According to the regulation, the standards must be set using the following method:

1. The criteria must be set using a "cradle-to-grave" approach (i.e., life cycle analysis).

2. The criteria must be precise, clear, and objective so as to ensure uniformity of application by the competent bodies.

3. The criteria must ensure a high level of environmental protection, be based as far as possible on the use of clean technology, and, where appropriate, maximize product life.

4. The criteria should be reviewed about every three years or more frequently if necessary.

Although each product category is subject to a single, communitywide standard for the Eco-Label, each member state has its own implementing authority. Accordingly, a manufacturer or distributor must apply for an award of the Eco-Label to an individual member state, usually the state

Environmental Fields	Product Life Cycle				
	Pre-production	Production	Distribution (including packaging)	Utilization	Disposal
Waste relevance					
Soil pollution and degradation					
Water contamination					
Air contamination					
Noise					
Consumption of energy					
Consumption of natural resources					
Effects on eco-systems					

in which the product is manufactured, first offered for sale, or imported. In requiring standards to be "precise, clear, and objective," the EU intended to ensure uniform application by different member states. It is more likely, however, that member states will legitimately interpret standards differently or exercise a bias in favor of local interests in deciding whether to award the Eco-Label to a particular applicant. It may, therefore, be possible for an applicant to shop around to different member states for permission to use the Eco-Label, although the program is intended to restrict applications to the country of origin or first importation. In an effort to minimize preferential treatment, any state that awards permission to use the Eco-Label must report the decision to the European Commission, which then distributes the recommendation to the other member states. Each state has an opportunity to object, and conflicts are resolved by the Commission.

An application for permission to use the Eco-Label is made to the member state in which the product is manufactured or first marketed or into which the product is imported. The application must include the results of independent testing demonstrating compliance with the applicable standard. Individual member states may impose reasonable requirements for the form of the application or additional documentation. There is an application fee of 500 European currency units, or about $625. If the product is approved, there is a fee of .15 percent of annual sales. According to the EU Commission, a certification obtained in one member state is valid in all other EU states, although there is some confusion on this point.

Permission to use the EU Eco-Label is granted for a definite period, which cannot exceed the expiration of the applicable standard. As noted, standards are generally expected to be reviewed at no less than three-year intervals. If a standard is unchanged after review, permission to use the Eco-Label may be extended without further application.

Environmentally conscious marketers who are considering applying for the Eco-Label for products manufactured outside the European Union must consider whether their products meet European health, safety, and performance standards. The overall Eco-Label program generally specifies that imported products must meet European Safety and Health Standards to qualify for the emblem. In addition, individual product group standards may incorporate performance criteria.

The Eco-Label program is intended to set strict standards. In the case of the first product groups to be evaluated, the standards were set with the expectation that about 10 to 20 percent of products on the European market would be capable of complying with the standards. It appears that standards will generally be set with the intention of qualifying only about 10 percent of the existing product in the category under consideration.

In its first year of operation, consumer interest in the Eco-Label was moderate. One survey indicated 25 percent of consumers in the United Kingdom were aware of the Eco-Label with no less than 76 percent expressing an interest.

Some have questioned whether products should be disqualified from the Eco-Label program based on negative environmental attributes. One suggestion, for example, has been that products with excessive or environmentally burdensome packaging should not be allowed to carry the Eco-Label. Another consideration, which has been raised less openly, is whether local waste handling and waste disposal practices in foreign manufacturing locations should be considered. For example, manufacturing operations in developing countries may handle process wastes in ways that would be unacceptable within the European Union. It could be argued under the Eco-Label regulation that such practices would effectively disqualify a product from consideration for the Eco-Label.

9.1.2   *European Union Packaging Legislation*

The European Union is considering, but has not yet enacted, legislation that would require member states to recover between 50-65 percent and recycle 25-45 percent by weight of total packaging waste over a 5-year phase-in period.

The European waste proposal faces three obstacles: One, European industry has urged that recovery and recycling goals are unattainable. Second, the German experience with this type of waste reduction (see Section 9.4.2) has demonstrated that adequate processing infrastructure does not exist for such high levels of waste recovery, resulting in the need to stockpile or export recovered waste. Finally, certain individual member states object to the proposal because it might exert downward pressure on their own more ambitious recovery and recycling programs. France, Britain, and others have objected that members should not be allowed to exceed the mandated levels unless they can demonstrate that their excess waste recovery will not cause market distortions that hinder other members' efforts to comply with the minimum recycling levels.

The proposal for European Union packaging legislation also includes a proposal to require some form of take-back for consumer packaging. There is also a proposal to enact specific limits for heavy metals content of packaging similar to the U.S. toxics in packaging laws (see Section 8.3.1.2). The European proposal also includes the possibility of specific marking for packages subject to the take-back provisions.

9.2   ENGLAND

The United Kingdom is a member of the European Union, and as such will implement EU-wide environmentally conscious marketing measures. Britain has already established an authority for awarding the EU Eco-Label. Britain is also moving forward on domestic packaging waste reduction measures.

Domestically, Britain enforces truth-in-advertising legislation through its Advertising Standards Authority (ASA). While Britain has considered a domestic environmental label or seal, it appears that Britain will defer to the EU Eco-Label.

It should be noted that Ireland, also a member of the EU, has separate jurisdiction over environmentally conscious claims. Ireland has not, to date, been as active as the United Kingdom in this area.

9.2.1   *Truth in Advertising*

The truth and accuracy of environmental marketing claims in Britain are regulated by the Advertising Standards Authority (ASA). The ASA is a self-regulatory system created by British advertisers to avoid extensive government regulation. Although the ASA works closely with the British government, it is not a legislative body. A substantial majority of British advertisers, advertising agencies, and media are members of groups within the ASA and as such are bound to follow the standards set by the ASA.

Generally, the ASA acts only when a complaint is made against an advertisement. The ASA will investigate after a complaint is made and judges whether the claim contravenes the British Code of Advertising Practice (CAP). If the claim is adjudged to be a violation, the advertiser is asked to withdraw the claim, and if the advertiser refuses, all members of the ASA are asked to refuse to publish the claim.

The United Kingdom does have generally applicable truth-in-advertising laws that are mandatory and enforceable in civil and criminal courts. Legal controls, however, are much less prominent in the United Kingdom than in the United States, and advertising regulation depends primarily on the ASA system. The ASA and British truth-in-advertising laws impose similar truthfulness and substantiation standards to those imposed in the United States (see generally Section 8.1). Accordingly, environmental marketing claims for use in England should be evaluated using generally the same approach outlined in Section 8.1.

Not surprisingly, the actual experience of truth-in-advertising regulation of environmental claims in England has been similar to the U.S. experience and involves similar claims about the same products. The diaper debate, for example, has been a target of regulatory scrutiny in England. In 1992, the ASA prohibited claims that paper and cloth diapers have roughly equivalent overall environmental impact. Specifically, a diaper manufacturer made the claim: "There is little difference in overall environmental impact between paper and cloth nappies."

Although the manufacturer's claim was at least defensible based on several life cycle analysis reports, the ASA determined the claim could not be substantiated as true because of scientific disagreement over the "real" impact on the environment of the different kinds of diapers. The ASA ruled that the claim could not be substantiated because it "depended to a large extent upon subjective interpretation of the relative importance that should be attached to different criteria (e.g., energy consumption, water pollution, resource consumption)." Accordingly, the ASA ruled the manufacturer should disclose that its interpretation of the life cycle analysis was "simply one side of an ongoing argument."

The ASA's approach illustrated in the above example is remarkably similar to the approach of the U.S. Federal Trade Commission in evaluating advertising claims. It focuses on whether the claim is (1) true, (2) substantiated, and (3) misleading. The claim faltered because, although technically "true" under one, defensible interpretation of the science, the claim could arguably mislead consumers not sophisticated in the complex environmental issues implied in the claim. The remedy, too, is familiar from the United States experience, in that the ASA required qualification and disclosure ("simply one side of an ongoing argument") to prevent any deception.

9.2.2 Packaging and Recycling Initiatives

As a member of the European Union, Britain will be subject to any packaging initiatives that are implemented within the EU. In the meantime, the United Kingdom has moved domestically to initiate recycling and packaging reduction measures. The British government has established a target of achieving a recycling rate for household wastes of 25 percent by 2000. The British government has called on industry to implement a voluntary plan with a target for recovery of between 50 and 75 percent of packaging waste by the year 2000. In February 1994, the Producer Responsibility Industry Group submitted a voluntary plan to British Environment Secretary John Gummer, which pledges that industry will recover 58 percent of packaging waste by the year 2000, finance the costs of a packaging recovery through a consumer levy on the product, support incineration with energy collection at the local authority level.

9.2.3 Implementation of the EU Eco-Label in the United Kingdom

The United Kingdom has taken aggressive steps to rapidly implement the EU Eco-Labeling initiative. The United Kingdom has established an implementing body for awards of the Eco-Label, known as the United Kingdom Ecolabeling Board (UKEB). In September 1993, UKEB announced that it was prepared to accept applications for Eco-Labeling of the first product groups for which standards had been developed—washing machines and dishwashers. The UKEB expects that applications will be acted on within two months and has established an application fee of 500 pounds.

9.3 FRANCE

Although France is a member of the European Union, France has been active in developing its own domestic labeling and packaging initiatives. France has generally followed the German model, including the creation of a private waste handling system to comply with French packaging reduction and recovery laws.

9.3.1 French "NF Environnement" Emblem

France has adopted a domestic environmental emblem that is intended to work in harmony with the EU Eco-Label, known as the NF Environnement emblem. The French program is almost a mini-EU Eco-Label, with similar standards and procedures for an award of the emblem. Some view the French program as an effort by France to give itself a stronger voice in the development of the EU Eco-Label.

The French emblem may be useful in at least three situations. First, it may be available for some product groups for which no EU-wide standards have been developed, and it may, therefore, provide an option when the EU Eco-Label does not. Second, depending on how the Eco-Label and the NF Environnement emblems are received by consumers, the French emblem may generate greater demand in France even for products already bearing the EU Eco-Label. Finally, for products distributed only within France, the French emblem may prove easier or more cost effective to obtain that the EU Eco-Label, depending on how the programs develop.

The French emblem is available to products for which environmental performance criteria have been established by the French Association for Normalisation (AFNOR). Specific criteria are established by AFNOR's Committee for the Certification of Euro-Products using a life cycle analysis approach. Factors considered include: (1) the use of primary resources, (2) impact on the atmosphere, (3) noise from use of the product, (4) pollution or process waste, and (5) consumer waste from use of the product.

Application for use of the French emblem is administered by AFNOR. The application process includes submission of a written application and fee (which includes a plant inspection fee and testing fee). In addition to the initial application fee, use of the emblem requires payment of an annual contract fee plus a percentage of annual product sales. Awards are based on consideration of the written application, product testing, and facilities inspection. Criteria for paints and varnishes were established in 1992. Although criteria are scheduled to be established for a number of additional product groups, including lubricants, household cleaners, trash bags, heating appliances, cosmetics, insulation materials, and paper products, no final criteria were established in the 18 months following introduction of the emblem. The program is not available to pharmaceuticals, agricultural products, services, and automobiles.

9.3.2 French Packaging and Recycling Initiatives

France has legislatively established aggressive package recycling and recovery goals of 75 percent reduction or recovery over a 10-year period. The law is similar to Germany's strict packaging law. Under the law, packaging manufacturers and importers into France must implement programs to recover waste originating from their products. The law applies to all packaging of consumer products, essentially those that reach their disposition in households. The packaging includes any type of container or material that holds the product for purposes of transportation or presentation for sale.

The French packaging law requires anyone who produces or imports consumer products into France to provide for recovery of the packaging waste. Under the law, packaging producers may implement a take-back system or may participate in a private waste recovery organization.

A private organization has been created in France to help packaging producers comply with the packaging waste reduction law. The organization is known as Eco-Emballage and is based on the German model of the Duales System Deutschland (DSD). As of 1993, membership in Eco-Emballage required a membership fee of 50,000 francs (about $8,500) plus an additional fee of 2 cents per packaging unit placed on the market. The per-package fee is scheduled to be increased over time up to 5 cents per package. Eco-Emballage is attempting to finance local waste recovery and sorting programs in 37 test communities to facilitate waste collection and recovery by Eco-Emballage.

9.4 GERMANY

Generally, German advertising legislation has been characterized as the strictest and most specific in the world. Germany is perhaps the world leader in developing both environmental labeling programs and packaging reduction and recovery programs. Germany has had an established environmental labeling program known as the Blue Angel label since 1978. Germany's 1991 packaging ordinance, with the goal of eliminating all waste packaging, has been the subject of substantial international attention. German industry reacted to the ordinance by creating the private waste recovery organization known as the Duales System Deutschland (DSD) which oversees compliance with the packaging ordinance. The DSD has adopted the green dot as its symbol for packaging materials meeting its criteria.

9.4.1 *Blue Angel Emblem*

The Blue Angel emblem is a certification program administered by the German government. The program certifies products as environmentally beneficial based on a selected characteristic. Once the Blue Angel is awarded, the product may display the Blue Angel symbol. Currently, over 3,500 products carry the Blue Angel label.

Applying for the Blue Angel certification is somewhat cumbersome, requiring the manufacturer to request criteria for its product group. As a result, only a limited number of product categories have had standards developed. Manufacturers initiate the process of developing a prod-

uct standard by proposing consideration of their products for the Blue Angel emblem. A federal agency known as the Environmental Label Jury determines whether the product is appropriate for detailed evaluation. If the product is selected for review by the Environmental Label Jury, product standards are set by an independent organization known as the institute for product safety and labeling (RAL). As of 1991, criteria had been set for 66 product categories.

Blue Angel product criteria are set using life cycle analysis. The purpose of the product criteria is to distinguish among products in the same category, and RAL typically selects those product characteristics with the greatest environmental impact, or the greatest potential for discrimination among products in the same category, for setting the final standard. Accordingly, standards for an award of the Blue Angel emblem typically contain a limited number of performance criteria intended to identify products that have a relatively superior environmental impact to other products within the same category.

As an example, RAL has developed a standard for varnishes and glazes. To qualify for a Blue Angel emblem, a varnish or glaze must:

1. Contain no more than half the allowed concentration of certain listed hazardous substances under German law.
2. Not contain certain other chemicals including biocide agents, formaldehyde in excess of 10 mg/kilo, or lead, cadmium, and chromium.
3. Minimize volatile organic compounds.
4. Use no CFCs as propellants.
5. Meet performance standards.
6. Contain a warning that standard precautions for use of varnishes and glazes should be employed.

Certifications are valid for a three year period. There is an annual fee for use of the emblem based on a percentage of sales.

9.4.2 *German Packaging Law*

Germany's packaging law, which took effect in June 1991 and is known as the Töpfer Law, is the broadest and strictest, by far, of any packing initiatives worldwide. Essentially, it requires manufacturers and retailers of any packaging to provide that the packaging can be taken back, reused, or recycled so that the packaging will not enter the public waste stream. Manufacturers or retailers and distributors taking back packaging materials are responsible for the material recycling and disposal. The law applies both to German producers and to importers into Germany.

The packaging ordinance covers all packaging unless the packaging is specifically exempted because it constitutes a health risk or an environmental risk. The exemptions are set out in separate statutes and apply to packages that contain or are contaminated with hazardous chemicals.

The ordinance divides packaging into three categories: (1) transport packaging such as drums, pallets, and outerpacking; (2) secondary packaging (such as display boxes); and (3) sales packaging used by the consumer to transport the goods from the point of sale, including disposable cups and plates.

Transport packaging must be taken back by manufacturers and reused or recycled outside the public waste stream. This means retailers are entitled to send transport packaging back to the manufacturer if the retailers choose not to reuse or recycle the packaging. End consumers may request delivery of the product in the transport packaging, and the transport packaging then becomes subject to the requirements for sales packaging.

Secondary packaging must either (1) be removed by the retailer or (2) must be returnable by the consumer free of charge to the point of sale or a nearby collection center. Responsibility for secondary packaging falls on the retailer and not the manufacturer. The retailer is permitted to remove the secondary packaging at the point of sale, and similarly, the customer may remove secondary packaging and leave it with the retailer.

Sales packaging must be returnable by the consumer free of charge to the point of sale or at a nearby collection point. The manufacturer and retailer are jointly responsible for arranging for the reuse or the recycling of the sales packaging outside the public waste stream.

Unlike transport packaging, which must be reused or recycled, the return of secondary packaging and sales packaging is strictly voluntary by the consumer. Accordingly, a manufacturer or retailer is not responsible if a customer chooses to dispose of packaging within the public waste stream. The ordinance provides for a 50 pfennig deposit (about 35 cents) on disposable containers for beverages, detergents, and emulsion paints, as an incentive to increase the level of consumer recycling. A deposit has not been imposed on other disposable secondary and sales packaging.

The packaging ordinance has presented a number of challenges to consumers and manufacturers. At the operational level, retailers have found it difficult to prepare transport packaging for return to the manufacturer. Labor-intensive preparation may be necessary, including the separation of nails and staples from wooden packaging, separation of different materials, and similar tasks. In addition, the German recycling system has been overwhelmed by the sudden leap in supply of recyclable materials. This has led to a debate on appropriate recycling and reuse alternatives, as well as criticism of the packaging ordinance based on its bias for recycling rather than reduction. Under the ordinance, incineration is not an appropriate treatment alternative, even if it involves energy recovery. Therefore, the German system has been unable

to process all the return packaging, and accordingly a significant volume of waste is being exported or stockpiled. This has led to calls for changes in the law to encourage reduction of packaging waste rather than recycling or reuse.

Another challenge to the system is criticism by other members of the European Union that the system is a competitive barrier in violation of the free trade provisions of the community. Some products have been refused entry at German ports and border crossings because importers have been unable to demonstrate that the products comply with the ordinance. This has led to charges that the ordinance is, in effect, an illegal trade barrier. To date, however, the European Union government bodies have upheld the German packaging ordinance. As discussed in Section 9.1.2, there have also been claims that Germany has developed anticompetitive power in recycling markets.

In November 1993, the German government proposed amendments to the packaging ordinance so as to reduce exports of DSD-collected papers and plastics. The main features of the proposed amendments are (1) a reduction and postponement of the targets; (2) the introduction of separate ''return'' targets for manufacturers remaining outside of DSD; (3) recycling is required only if it is ''technically possible'' and the cost is not ''economically unreasonable'' compared with other recovery options, but this assumption does not apply to DSD (i.e., DSD must achieve its targets even if the cost is excessive); (4) the requirement that individual companies prove that they are in compliance with the law; (5) obligations to mark packaging for refillability, recyclability, recoverability, or nonrecoverability and the estimated cost of disposal; and (6) the introduction of incineration with energy recovery as a viable recycling option although limited to materials collected beyond and above the targets.

9.4.2.1 Duales System Deutschland—The green dot.

The German packaging ordinance allows manufacturers and retailers to participate in alternative, private collection systems to comply with the obligations of the packaging ordinance. To qualify, an alternative collection system must meet certain collection and sorting quotas. Utilizing this alternative, German industry has constituted the Duales System Deutschland (''DSD'') as a private collection and sorting organization to meet the obligations of the packaging law. DSD has reported that it collected 57 percent of all packaging used in Germany in 1993.

The DSD has established a system of recognizable collection facilities where consumers can drop off packaging for reuse and recycling. Manufacturers indicate their membership in DSD, and the eligibility of their packages for collection by DSD, by displaying a green dot on their packages. The green dot symbol is developing secondary meaning as an environmental marketing emblem. Some large retailers, for example, make purchasing decisions based on whether a product carries the Green Dot. There is some evidence that consumers as well are influenced by the green dot.

Packagers or their importers must apply to DSD for membership to qualify for the green dot. The application includes data on expected sales, product size, product type, and the packaging material. Producers must pay a fee to participate in the DSD. In addition, consumers pay a fee at the point of sale for a product carrying the green dot.

The green dot system applies only to sales packaging and does not handle transport or secondary packaging.

The DSD has encountered some difficulties. Foremost are financing difficulties and difficulties in processing the huge volume of waste collected by the system. The German government, which has a decided interest in the survival of the DSD, has mediated some of the financial difficulties of DSD. The DSD claims its financial difficulties are a result of difficulties in collecting fees from manufacturers for use of the green dot. The DSD claims new licensing arrangements and fee auditing procedures with sanctions and punitive fines will eliminate any financial difficulties.

The DSD's other problem is that consumers have responded with greater separation of waste than anticipated. For example, in 1993, the DSD expected to collect over 100,000 tons of plastic wrappings, but collected as much as 440,000 tons. The German National recycling capacity of 176,000 tons was overwhelmed. Under the packaging ordinance, the DSD is not permitted to incinerate or landfill the excess and must either export it or stockpile it for processing when capacity has caught up.

9.4.2.2 Plastic resin coding. German law requires many plastic containers to be labeled with the type of plastic resin used to formulate the container.

9.5 AUSTRIA

The Austrian Packaging Waste Ordinance took effect on October 1, 1993, and calls for an 80 percent reduction in packaging waste by the year 2000. The ordinance requires manufacturers and distributors that place packaging on the market to take back used packaging based on the following quotas:

Effective	Percentage Recovery
October 1, 1993	40%
July 1, 1995	50%
January 1, 1997	60%
July 1, 1998	70%
January 1, 2000	80%

Incineration with energy recovery is considered a valid means of recovery.

Alstoff Recycling Austria (ARA) has been set up by

EXHIBIT 9–1

Country	National Certification Program	EU Member	Nordic Council Member
Austria	X		
Belgium		X	
Denmark		X	X
Finland	X		X
France	X	X	
Germany	X	X	
Greece		X	
Iceland			X
Ireland		X	
Italy		X	
Japan	X		
Luxembourg		X	
Netherlands	X	X	
New Zealand	X		
Norway	X		X
Portugal		X	
Spain	X	X	
Sweden	X		X
Switzerland	X		
United Kingdom		X	
Yugoslavia	X		

packaging producers and users to assume industry's responsibility for the take-back of packaging waste. ARA collects fees from packaging producers and importers which are used to cover the costs of collecting and recycling.

9.6 JAPAN

Japan's approach to environmental regulation generally has been fragmented and localized. For example, there are as many as 28,000 locally tailored and negotiated environmental agreements between Japanese industry and local communities.

Japan has instituted a certification and emblem program administered by a nongovernmental organization under the supervision of Japan's Ministry of Environment. The emblem, known as ECOMARK, is similar to certification and emblem programs in the European Union. Performance criteria are selected for product categories based on life cycle analysis. Once criteria are developed, manufacturers can apply for an award of the emblem. The emblem is awarded for a two-year period. A fee must be paid for use of the emblem.

9.7 ADDITIONAL INTERNATIONAL CERTIFICATION PROGRAMS

A number of countries and international organizations are developing environmental certification programs (see Exhibit 9–1). Many of these programs are in the development stages.

The Nordic Council, consisting of Denmark, Finland, Iceland, Norway, and Sweden, is developing an environmental emblem to be known as the White Swan. The program is intended to be similar to the EU Eco-Label.

Other countries that are developing national certification programs include Australia, Austria, Finland, Ireland, the Netherlands, New Zealand, Norway, Portugal, Sweden, Switzerland, and Yugoslavia.

9.8 ADDITIONAL INTERNATIONAL PACKAGING PROGRAMS

The Netherlands

On June 6, 1991, the Netherlands and the Foundation on Packaging and the Environment (representing Dutch industry) signed the Dutch packaging covenant. The Covenant, which is binding only on parties to that agreement, obliges the participating companies by the year 2000 to: (1) reduce the total volume of new packaging put into circulation below the 1986 level of 2 million tons of packaging waste; (2) take back 90 percent of the waste packaging that cannot be re-used; (3) recycle 60 percent of the packaging that is not reused (giving preference to material recycling over energy recovery); and (4) stop landfilling packaging waste.

Belgium

Currently, each of the three Belgian regions—Flanders, Wallonia, and Brussels—has its own packaging waste system. The three regions are attempting to formulate uniform national legislation requiring that all product packaging waste be recycled. Belgian industry has also proposed a national DSD-like scheme (''Fost Plus'') to collect and sort packaging waste. Fost Plus would be financed through the application of a levy on individual companies proportional to the number of units of packaging placed on the market. Fost Plus has also recently announced that it would use the ''green dot'' to identify products party to the system.

Spain

In 1993, discussions began between the Spanish government and industry to formulate a voluntary scheme on the recovery and recycling of used packaging. Negotiations are underway between the Environment Ministry and industry representatives to establish targets and implement a recovery scheme.

RESOURCES

''Ecolable Criteria: UK Ecolabelling Board Newsletter,'' Numbers 3, 4, 5.

ENDS Report 216, January 1993, p. 25.

ENDS Report 206, March 1992.

''Environmental Packaging: U.S. Guide to Green Labelling, Packaging, and Recycling,'' Thompson Publising Group, p. 580.3.

International Environment Reporter (BNA), February 23, 1994, p. 195.

International Environment Reporter (BNA), July 28, 1993, p. 551.

International Environment Reporter (BNA), July 14, 1993, p. 513.

International Environment Reporter (BNA), June 2, 1993, p. 389.

International Environment Reporter (BNA), December 1, 1993, p. 894.

Nickel, Volker. ''Regulations Battle Creativity in Ads,'' *Advertising Age/Europe*, November 1979, p. 28.

Ryans, John K., Jr.; James R. Willis, Jr.; Henry Bell. ''International Advertising Regulation: A Transnational View,'' *Midwest Marketing Association 1979 Conference Proceedings*, p. 37.

Simon, Francoise L. ''Marketing Green Products in the Triad,'' *Columbia Journal of World Business*, Fall/Winter 1992, p. 268.

''Some 22 Nations Could Have 'Green Lable' Schemes by '93,'' *Toronto Star*, November 16, 1991, p. D6.

Chapter Ten

Implementing Environmentally Conscious Marketing

The previous chapters of Part II provide a background in the legal rules and business issues that define environmentally conscious marketing. This chapter outlines strategies for managing those issues. The basic strategies for managing environmentally conscious marketing challenges include the following steps:

1. Define your market opportunities.

   - Identify your product/attribute/corporate philosophy.

   - Identity your package.

   - Identify potential environmentally conscious marketing claims.

   - Identify your target areas of distribution.

   - Gather or create market research relevant to your product in your area of distribution.

   - Identify any applicable environmental marketing rules and regulations in each geographic area of distribution you have targeted.

2. Evaluate your claim options and identify challenges.

   - Evaluate whether your potential claims are allowable under the rules of your target areas.

   - Evaluate the marketing advantages of your potential claims.

   - Identify potential conflicts among applicable rules.

- Evaluate the potential backfire risk.
- Identify gaps in information necessary for implementing claims.
- Identify other considerations.

3. Reevaluate/refine your options.
   - Fill information gaps.
   - Consider modifications to your claims.
   - Consider "maximum compliance" strategy.
   - Consider using different claims in different market/regulatory regions.
   - Consider modifying your product or package.

4. Choose and implement a strategy.

5. Reevaluate/update strategy periodically.

As indicated throughout the foregoing chapters of this book, the time to begin managing environmental marketing challenges is immediately, at the very beginning of product development and design. The checklist set forth above should be applied, if possible, as a tool to help identify opportunities for new products or new markets and as a tool for designing environmental quality into the product before the first drawings or models are produced.

The following sections discuss how to implement each of the steps described in the checklist above.

10.1   IDENTIFY MARKET OPPORTUNITIES

The challenge of identifying a market demand and filling it is no less difficult in the context of environmental marketing than in any other context. The traditional ingredients of ingenuity, creativity, hard work, know-how, luck—1 percent inspiration and 99 percent perspiration—are all important. At the risk of stating the obvious, opportunities for environmental marketing are limited only by the creativity and resources of the entrepreneur. With that in mind, this discussion of "identifying your opportunities" is intended as a tool to help organize the creativity and know-how you bring to the process, not as a substitute for them.

The order in which you follow each of the steps outlined in this section is not important. Indeed, your ideas should drive the process rather than the other way around. Accordingly, the order of the steps in this section is somewhat arbitrary for the convenience of exposition and should be altered freely to suit your needs.

10.1.1   *Identify Product/Attribute/Corporate Philosophy*

In the specific context of environmental marketing, your "product" is whatever positive environmental attribute you are able to bring to the marketplace. Often, your product will be an environmentally beneficial substitute for an existing product. For example, you may be able to develop a less toxic cleaner, a package that is more easily recyclable, or a process that generates less waste.

You may develop a green product as a result of the ECM process. When you eliminate process waste or formulate your product with an environmentally preferable material, you not only reduce your production costs and compliance risks, but you also create an improved product that should give you a market edge provided you are able to communicate the benefit to your market. In such a case, the product drives the environmental marketing process, and your environmental marketing claims are shaped to fit the product.

In many cases, the process will be driven from the other end: You will identify a demand for a green attribute and create a product to meet that demand. Ideally, the process should be both supply driven and demand driven. In other words, demands in the market will lead you to alter your product, and alterations in the product will allow you to create additional demand in the market. This presents both an opportunity and a challenge. You may find that designers and operations managers can present you with a number of options for accomplishing the environmental benefit you intend to market. Such flexibility allows you to continuously refine your design and process in response to challenges or opportunities presented by legal or marketing constraints on your potential claims. Here is an opportunity. The challenge lies in integrating your team so that the trade-offs between design, operational, legal, and market needs can be balanced effectively.

For distributors and retailers, "your" product may in fact be claims you make about the products you sell even if you don't manufacture those products. Wal-Mart, for example, has been successful in promoting its use of in-store shelf labeling to call attention to green products. In many cases, it may make more sense for a local distributor or retailer, working on its own initiative or in cooperation with a manufacturer, to make environmental marketing claims. This arrangement overcomes many of the logistical problems with distribution in different areas that have different rules and regulations. Generally, when a distributor makes an environmental marketing claim, the distributor is responsible for the truth of that claim and its compliance with other environmental marketing laws and regulations.

Corporate philosophy is a much more obscure kind of "product" than a specific attribute for a specific, tangible product. Not surprisingly, a corporate philosophy tends to be that much more difficult to market. In an environment of consumer skepticism (which environmental marketing certainly is), claims of corporate commitment to the environment ring hollow absent concrete steps to address specific problems. Efforts to market a green corporate philosophy are most likely to succeed when they are linked to concrete changes in the way you do business that provide a tangible benefit you can point to.

Corporate philosophy can become a marketable product, however, when it is linked to positive changes in the way you do business. Communicating your environmental commitment should be linked to actions demonstrating the commitment, such as a move to an in-house recycling program, use of recycled paper internally and in advertisements, or similar measures. By contrast, donations to environmental groups, declarations of commitment to the environment, and investment in environmental programs unrelated to your business or the everyday lives of your customers tend not to be as effective.

10.1.2   *Identify Package*

Developments in packaging technology have created easy opportunities to reduce your production costs and the burden of your packaging on the environment. These opportunities include the reduction or abandonment of your overall packaging, use of reusable packaging, and use of recycled and recyclable materials in packaging.

Consumers have responded to marketing claims of green packaging. For example, concentrated laundry soap and refillable laundry soap bottles have made significant market gains. For companies trading in Europe, and especially Germany, reduced, reusable, or recyclable packaging is a condition of entry into the market.

Identifying your packaging is important not only because packaging provides a distinct opportunity for environmental marketing claims, but it is also necessary to distinguish between claims about your product and claims about your package under truth-in-advertising rules.

10.1.3   *Identify Potential Environmentally Conscious Marketing Claims*

A hit list of popular environmentally conscious marketing claims has already emerged. This list includes claims that a product or package is:

- Recycled
- Recyclable
- Bio/photo/degradable
- Compostable
- Phosphate free
- Reusable/refillable
- CFC free
- Reduced packaging
- Reduced waste
- Energy efficient
- Nontoxic
- Returnable
- Low-fume
- Unbleached

- Noncorrosive
- Pesticide free
- Contains less . . .

In addition to the above list, there are some common claims that are less desirable because they tend to raise compliance issues and consumer skepticism. These are general benefit claims such as "environmentally friendly," "safe," "safe for ozone," "organic," "natural," "cruelty free" and the like. As discussed in Chapter Seven, market researchers have tracked how consumers rank these common environmental marketing claims in terms of relative importance and skepticism. You can use such research to evaluate your proposed claims.

Identifying which of these common ECM claims potentially applies to your product or package is relatively simple. The universe of all possible environmentally conscious marketing claims is by no means limited to this hit list, however. Just as the terms on the hit list had relatively little importance to consumers as recently as 30 years ago or less, there are certainly other positive environmental attributes for which unidentified demand exists or can be created. While the temptation is to follow a beaten track, the greatest reward may come to the green marketer with a unique claim.

One method for identifying potential new and unique environmental marketing claims is to monitor the environmental press. The increasing focus on new environmental problems creates corresponding opportunities to solve those problems. Whenever you can contribute to the solution of a perceived problem, you have the basis for an environmental marketing claim.

There are challenges in breaking new ground in the environmentally conscious marketing context. First, you may need to educate your market to create the demand for your new benefit. If you tackle a problem that is relatively unrecognized or complicated, it may be expensive and difficult to communicate to your market exactly why your product is an important and positive environmental step, or indeed that it is even a step forward at all. Your "solution" to an unrecognized problem may be met with skepticism, and you may be perceived to be puffing or aggrandizing yourself. In addition, it may be more difficult to substantiate a new or unique environmental marketing claim, if the state of scientific research is substantially undeveloped or disputed on the issue.

You can minimize the risks of making new or unique claims. To minimize the risk of consumer skepticism or misunderstanding, you can provide background on why your claim is meaningful and commission market research to verify that it will be understood. Obviously, the extent of your research may be dictated by the resources available. At a minimum, however, a simple written survey of your actual customers may be helpful in gauging consumer reaction.

To minimize the risk of inadequate substantiation, you can (1) be conservative, as in making any claim, and (2) obtain an opinion or at least a reaction from applicable regulatory offices. Conservatism means limiting your claims to those statements you can prove and obtaining credible opinions from knowledgeable individuals to back up your claim. Once you have substantiated the claim, many regulatory authorities will provide formal or informal review of claims to determine whether they comply with applicable laws and regulations.

10.1.4   *Identify Target Areas of Distribution*

Knowing where you will distribute your product is required to determine what rules and regulations will apply to your claims. In addition, consumer preferences will vary in different geographic areas, particularly in case of distribution in foreign countries.

One caveat should be noted: The important determination is where you make your claim and not necessarily where the product itself will be distributed. If you make claims in advertising media, for example, rather than at the point of sale, environmental marketing regulations generally will apply only in those places where your claim is actually published. In addition, if your claim is made in some places where your product is not distributed, you may be less concerned because arguably there is no threat of consumer confusion where your products are not actually sold.

Identifying your areas of distribution and the areas in which you will make your claim can be as simple as making a list of each state and each foreign country in which you distribute your product. In cases in which your product is resold by others in connection with your environmental marketing claim, however, the process is much more difficult, because you may have no control over or even knowledge of where distributors finally sell your product. In such a case, you may be able to develop distribution information from your distributors. If this is impractical, it may be necessary to assume your product will be distributed globally, or at least throughout a region of the world, and to develop a compliance strategy based on worldwide compliance or compliance throughout an entire region.

10.1.5   *Gather or Create Market Research*

Typically there will be a number of options for the exact claims you can make, and for the way you can make them. At a minimum, it will be necessary to decide how much detail and how much prominence you should give your environmental marketing claims. It will also be necessary to determine whether you need balancing statements on price and performance to maintain a competitive position for a green product. To make these decisions, it is necessary to understand as much as possible about consumer expectations in your market.

The ideal method of collecting market data is to perform original market research through use of a qualified marketing consultant in your target areas. An original market study allows you to develop very specific information about the claims you care about in the markets you would like to target. In addition, such studies can be decisive in a dispute over deception or unfairness because they will show how actual consumers in your market actually understand your claim. Such studies, which typically involve surveys of 1,000 to 1,500 consumers, can be expensive.

More accessible market research exists through environmental marketing research services. Environmental marketing research is becoming a fairly active field, and a surprisingly large body of market research is published on a regular basis. Particularly when computer research services are available, these published reports can provide reasonably current and specific market research.

The important questions to consider in developing market research are:

- What environmental concerns are most important to consumers in your markets?

- How sophisticated an understanding of the environmental issue relevant to your claim do the consumers in your market have?

- Are consumers willing to pay a premium for the environmentally conscious attribute you intend to offer?

- What is the level of consumer skepticism regarding environmental claims generally, and your type of environmental claim in particular in your markets?

- What alternatives and substitutes to your product are available in your markets?

- If you intend to make a claim of recyclability, what actual recycling collection options are available in each of the markets for the materials used in your product?

- If your claim is dependent on how consumers use or dispose of your product, what is the actual consumer behavior in your markets? (For example, do consumers separate recyclable materials from their garbage?)

The more market data you can gather before evaluating your marketing options, the better. When gathering market research, it is always best if there is data on actual consumer behavior as opposed to consumer's claims about their behavior in hypothetical situations. For example, a study of how consumers actually separate their trash is more valuable than a survey cataloging consumers' responses to the question of whether they separate recyclable materials. Likewise, data on actual purchasing decisions are more reliable than consumers' responses to a survey question whether they would pay more for environmentally conscious products.

10.1.6   *Identify Applicable Rules*

Once you have completed the steps outlined in the preceding sections, you will have a list of claims you may wish to make and a list of geographic areas in which you intend to distribute your product and make your claims. The next step is to identify any applicable rules and regulations that may govern the use of those claims in each area in which you may make them. Depending on the scope of your geographic distribution and the type of claim you are considering, this exercise can be quite simple or it can be aggravatingly difficult.

As discussed previously, identifying applicable rules and regulations presents two challenges. First, regulations are developing quickly, and it is important to make sure your information is current. Second, many states, countries, and even international regions may impose their own set of regulations on environmentally conscious marketing. Accordingly, it will probably be necessary to subdivide your area of distribution into a number of regulatory/marketing regions in which different rules may apply.

Generally, the smallest geographic region that will have its own set of environmentally conscious marketing rules is a state. (It is possible that municipalities or counties could impose their own sets of local rules, but this does not seem to be a trend in the regulation of environmentally conscious marketing, except perhaps with respect to limitations on packaging materials.) Typically, the authorities that have issued environmentally conscious manufacturing regulations are state governments, national governments, and in the case of Europe, international bodies.

The best way to make sure you identify each regulatory scheme that may apply to the claim you wish to make is to simply list each state, each country, and each common market (such as the European Union) in which you intend to distribute your product and make your claims. As discussed above, if your product is distributed widely, and especially if it is distributed by individuals or companies outside your control, it may be necessary to simply assume your product is distributed throughout an entire country or region.

If you are dealing with a relatively limited area of distribution, it may be possible to simply contact the regulatory bodies for each of the states or countries for current information on the rules that may apply in each of those areas. If your area of distribution extends throughout the entire United States or internationally, it may be more efficient to get the information from an outside source. Larger law firms that practice in the area of environmentally conscious manufacturing can provide a comprehensive analysis of current regulations in each area in which you are interested. In selecting a law firm to perform this type of analysis, questions you may wish to ask as part of the selection process include the following:

1. Has the lawyer or law firm analyzed environmentally conscious marketing regulations in the past and developed a body of experience that will benefit you and avoid a potentially costly learning curve?

2. Does the attorney or law firm have an office or contacts in any foreign regions in which you may be interested?

3. Does the attorney or law firm monitor developments in the area of environmentally conscious manufacturing?

The Thompson Publishing Group offers a loose-leaf service that is updated on a regular basis. Thompson's *U.S. Guide to Green Labeling, Packaging and Recycling* provides an overview of many regulatory programs in the United States, Canada, Europe, Scandinavia, and Japan. Such loose-leaf services can be valuable in providing a current overview of many applicable regulations in the areas in which you are interested. It is best, however, to verify information in loose-leaf services by checking with the regulatory agencies responsible for the rules. In addition, the loose-leaf services are intended primarily as surveys or overviews of regulatory programs. You should not necessarily rely on them as the only source of information, particularly in determining whether your specific claim complies with those regulations that may apply to it. In addition, the services may not cover general truth-in-advertising laws or advertising self-regulation, which should also be investigated.

10.2   EVALUATE YOUR OPTIONS AND IDENTIFY CHALLENGES

Once you have gathered the information discussed in Section 10.1, you can make an initial evaluation of which claims (1) will be most effective from a marketing point of view and (2) will be most manageable from a legal and operational point of view. This initial evaluation is intended to reduce the problem to a manageable set of issues and to identify any remaining information gaps that should be filled before a final selection of your claims. In cases in which you are distributing your product in a relatively small number of marketing/regulatory regions, and in which your claims are relatively simple, you may be able to select final claims at the end of this initial evaluation. In more complicated situations, the initial evaluation will result in a short list of potential claims and a potentially longer list of additional questions.

10.2.1   *Evaluate Legality of Potential Claims*

In the best of all possible worlds, determining which claims are legal and which are not would be a simple process of comparing a list of potential claims against a list of rules in each marketing/regulatory region. If you have a limited and specific claim from the list of highly regulated claims, the process may be almost this simple. For example, if you intend only to make a claim of "recycled content" and your distribution area is the western United States, you can develop some pretty definite answers about what you may say and what you may not say. On the other hand, the process can be much more complicated. Generally, the evaluation of which claims are allowable will be more complicated if (1) you are making a new or unique claim, (2) you are making a complex claim, (3) your target distribution area includes a large number of marketing/regulatory regions, (4) you have identified a large number of potential claims, (5) you are considering a claim of "recyclability," or (6) you are evaluating claims in several different languages.

New or unique claims can be difficult to evaluate because it is always difficult to predict how regulators will apply rules in new situations. Because of the complexity of and scientific disagreements about environmental issues, there is usually room for disagreement about whether a given claim is true. When there is no track record or specific rule on a particular claim because it is new or unique, the potential for disagreement must be factored into the evaluation of whether the claim is proper under more general standards regulating the truth and fairness of claims. Accordingly, the evaluation of new or unique claims will usually result in a conclusion based on probability rather than certainty. In such a case, operational, logistical, and marketing factors will have a significant influence on the decision whether to make the claim.

Complex claims are difficult to evaluate for the same reasons as new or unique claims. In particular, complex claims may be more difficult to communicate in the rapid-fire context of marketing. Accordingly, the potential for a claim to be considered "misleading," because consumers do not fully understand it, is a greater concern in the case of complex claims. Evaluations of complex claims, therefore, will also likely lead to softer conclusions about whether regulators will consider the claims proper under the rules.

When the target distribution area includes a large number of market/regulatory regions, the evaluation is difficult simply because it involves a larger number of discrete decisions. Although this does not necessarily mean the compliance evaluation will lead to softer conclusions, the evaluation will be more time consuming and expensive than an evaluation of a more limited number of regions.

An evaluation of a large number of potential claims potentially presents both the difficulties of evaluation of a complex claim and the difficulties of evaluation of a large number of market/regulatory regions. If you are considering making a large number of claims in combination with each other, the sum of these multiple discrete claims becomes a single complex claim that may run a higher risk of consumer confusion. Depending on the nature of each discrete claim, harder conclusions may be available for the individual components of the claim, while a softer conclusion may apply to the overall claim. In addition, the evaluation of a large number of potential claims is inherently more expensive and more time consuming than the evaluation of fewer claims.

Claims of recyclability may be difficult and expensive to evaluate because evaluation may involve the issue of whether collection facilities exist for the product or package in each geographic area in which it is distributed. Many regulatory programs currently define *recyclability* in terms of whether meaningful opportunities exist for consumers to recycle the product or package with reasonable convenience. For materials that have only emerging recycling markets, an evaluation of recyclability could potentially involve an analysis of whether there is a collection facility for the material in each town or county in which consumers may dispose of the product or package. The practical difficulties of performing such an analysis may eliminate unqualified claims of recyclability from consideration. As discussed in Chapters Eight and Nine, however, this does not mean claims of recyclability must be abandoned. Instead, it may be possible to implement a take-back program or to qualify or limit the recyclability claim appropriately.

Evaluating a claim to be made in several languages is difficult both because it multiplies the number of issues for consideration and because it may also multiply the number of evaluators. Ideally, foreign language claims should be evaluated by qualified persons knowledgeable both in the foreign language and in the regulatory scheme applicable in those markets and regulatory regions where the foreign language claim will be made. That is, a German language claim should be evaluated by someone reasonably fluent in German and reasonably expert in applicable German laws. Accordingly, additional time and resources should be budgeted when foreign language claims are under consideration.

Before beginning a detailed, initial compliance evaluation for your claims under consideration, you may wish to identify whether there are some potential claims on your list that would be particularly difficult or expensive to evaluate. Before evaluating the legality of these claims, you may wish to consider whether there are logistical or marketing reasons to eliminate them from consideration. For example, if your marketing research revealed that a complex claim is not likely to be effective in your target market, you may wish to eliminate that complex claim from consideration before you spend time and money on a legal evaluation of the claim.

Potential claims should not necessarily be dismissed simply because the legal evaluation is potentially complex or expensive. Some of the greatest rewards may come from claims that are difficult to evaluate. The issue is one for business judgment of whether the potential return on the claim justifies the expense of the legal evaluation.

The goal of your initial evaluation is to classify your potential claims into three tiers for each market/regulatory region: (1) claims that are almost certainly proper under applicable rules, (2) claims that are almost certainly improper under applicable rules, and (3) claims whose status is uncertain under applicable rules. For claims that fall into the uncertain category, the initial evaluation should provide a softer evaluation of whether the claim is likely to comply, unlikely to comply, or whether it is impossible to determine.

Another likely product of the legal evaluation will be the identification of information gaps that need to be filled. For example, an evaluation of whether a claim of "recyclable content" is proper may result in a conclusion that the claim is proper so long as the percentage of post-consumer waste used in the manufacturing process is specified. Ideally, this factual information will already have been documented as part of the ECM process. Because the process is based on continuous *improvement*, however, and not perfection, it is likely that some information gaps will first come to light during the legal evaluation. As discussed below, the same is true of the marketing evaluation.

10.2.2 *Evaluate Marketing Advantages of Potential Claims*

It may be possible to prioritize certain claims you are considering because they present particularly strong marketing advantages or, by the same token, to eliminate claims that have relatively weak marketing impact. Predictions of how the market will react to your claims are at best an inexact science. Nevertheless, you should be able to develop some hierarchy of claims that are likely to perform better than others.

Important issues to consider in evaluating the marketing advantages of your claim include the following:

1. Do consumers feel the issue you are addressing is important? (For example, if you are reducing packaging wastes, do consumers in your target market feel that solid waste disposal is an important issue?)

2. Do consumers understand your claim or the issue your claim addresses?

3. Have similar claims been effective for other manufacturers or other products in your target markets?

4. For corporate philosophy marketing, is the claim directly related to your product or have you taken concrete action based on your corporate philosophy?

You should have a significant body of market research based on the steps discussed in Section 10.1. This research may consist of more general, published studies from the trade press or specialized sources. In the best case, you will have conducted specific research on your claims in your target markets.

By evaluating the market research available for each of the potential claims you have identified, you should be able to make a tentative assessment of the claims that are more likely to be effective and the claims that are least likely to be effective. This information can be combined with the legal evaluation to narrow your list of potential claims to a short list of final prospects for further evaluation and implementation.

10.2.3 *Identify Potential Rules Conflicts*

One of the most challenging parts of the legal evaluation of green marketing claims is determining whether there are true conflicts among the rules that may apply to your potential claims in each of the market/regulatory regions you have targeted. A very frustrating aspect of the patchwork of rules currently in place is that the rules of a single state or region may threaten to skew the entire legal analysis when that particular rule is dramatically out of line with the mainstream of regulation. This happens because it is typically cost prohibitive to market your product differently in different areas of distribution. Accordingly, there is a cost and logistical pressure to select a single claim that is acceptable in every market/regulatory region in which you distribute. When this happens, a roadblock in one market/regulatory region effectively eliminates your claim in all market/regulatory regions.

Proposition 65 in California is a recurring example of such a single-region roadblock. As discussed in Chapter Eight, California requires most products that contain any one of a list of hundreds of hazardous chemicals to display a health risk warning mandated by the State of California. Most companies that distribute their products throughout the United States find it impossible to isolate those products destined for California and to label them separately with the California health risk warning. As a result, some companies find themselves labeling all their products with the California health warning, no matter where the product is distributed. The same problem may occur with negative restrictions when a single jurisdiction would prohibit a claim that is otherwise allowable in all other markets/regulatory regions. The first step in dealing with this challenge is to identify any such problem regions.

The identification of conflicts may occur as a logical part of the legal evaluation. On the other hand, it may be more cost-effective to wait until you develop your short list before performing the conflict analysis in order to minimize the time and expense involved.

10.2.4 *Evaluate Backfire Risk*

As discussed in Chapter Eight, environmental manufacturing claims may present a significant risk of backfiring if they are exaggerated or unsubstantiated. For each of the claims on your short list, evaluate whether there is a significant risk of backfire:

1. Do regulators, public interest groups, and consumers in your target market consider the issue you are addressing to be an important one?

2. Is your claim true?

3. Do you have adequate written substantiation for your claim supported by individuals qualified to evaluate the claim?

4. Is the claim conservative?

5. Are there adequate limitations and disclosures accompanying the claim?

6. If similar claims have been made by others in your target markets, how have those claims been received?

7. Do you have exposure as a "polluter" on issues unrelated to your claim? What is the status of your overall environmental compliance historically and at present? Do you generate relatively large amounts of waste within the terms of any applicable permits or regulatory programs?

8. Are there competitors who are likely to challenge the claim?

10.2.5 *Identify Information Gaps*

The initial evaluation discussed in this section will probably result in identifying gaps in the information necessary to finally select and implement environmental marketing claims. There may be additional factual background necessary to complete the legal evaluation of the claims on your short list. Similarly, additional market research may be helpful in evaluating which claims on your short list are likely to be most effective. Finally, there may be substantiation that must be further developed in order to fully comply with applicable regulatory programs.

In addition to true gaps in your information, the selection of a short list may enable you to identify additional targeted research that would not be cost-effective for the entire list of potential claims that were first identified.

For example, it may not have been cost-effective to perform the market research on consumer acceptance and understanding of the 25 potential claims you first identified. However, if you narrow the list to three potential claims, it may be much less expensive and much more effective to conduct a survey of consumer preferences and understanding among those three claims. The same is true of market research and substantiation.

10.2.6 *Identify Other Challenges*

A comprehensive discussion of the challenges that may be presented by any marketing claim, whether environmental or other, is beyond the scope of this book. A general list of other challenges to all kinds of marketing claims includes the following:

1. Intellectual property laws, including the protection of copyrights and trademarks, may prevent the use of a claim or design that is already being used by someone else.

2. There may be product-specific labeling requirements:

    a. Automobiles generally must include information on fuel efficiency.

    b. Many consumer appliances require labeling with information on energy efficiency.

    c. Gasoline must disclose its octane rating.

    d. Insulating materials and windows may be required to disclose the R value.

    e. Light bulbs generally must disclose average lumens and average life.

3. Some media categories impose internal review standards. In particular, the three major broadcast networks have developed guidelines, and each network exercises prebroadcast review of all advertising. Many radio stations and publications review advertising according to their own criteria.

4. The Fair Packaging and Labeling Act requires that all consumer commodities convey information about the quantity of the contents accurately.

5. The federal Food and Drug Administration regulates the "adulteration or misbranding" of food, drugs, and medical devices.

6. The FTC and many states regulate express and implied warranties. Generally, a warranty may be created by any description of the goods or representation that reasonably could be construed as a promise or factual representation about the product.

7. The National Advertising Division (NAD) of the Council of Better Business Bureaus and its appellant branch, the National Advertising Review Board (NARB), monitor advertisements in national media for adequate substantiation.

10.3 REEVALUATE AND REFINE YOUR OPTIONS

The initial evaluation process described in Section 10.2 should result in (1) the selection of a short list of potential claims, (2) a list of information gaps for further research, and (3) a list of challenges including potential legal obstacles and risks relative to each potential claim on your short list. This section considers how to fill the information gaps, refine the claims or the environmentally conscious attributes of the product, and design a strategy for implementing your claim. Like all stages of the process, this one should involve all the members of your team, including persons knowledgeable in design, operations, marketing goals and strategies, logistics, and any other key links in the design through distribution chain.

While the discussion in this chapter is presented as a series of discrete steps, each step is interdependent on the others. Accordingly, it may be most effective to conduct this phase of the process using team meetings to ensure that changes in one aspect of the process can work with the requirements for other aspects of the process.

Brainstorming sessions or team meetings are one way to address the need for a high level of communication and integration at this point in the process. At a minimum, you should take affirmative steps toward ensuring that there is a mechanism for proposals to be communicated throughout the team and for feedback on those proposals to be recirculated promptly. Depending on the style of your organization, this communication mechanism could include relatively frequent status meetings, circulation of status memos by a team leader in charge of coordinating and communicating all developments and ideas of the team, or circulation of memos with proposals by individual team members.

10.3.1 *Fill Information Gaps*

At this point in the process, you should have a list of questions for further research as a result of the initial evaluation of options. In many cases, these issues will be more narrowly focused and more manageable than they may have been at earlier stages in the process, as a result of the selection of a limited number of potential claims and the identification of discrete issues during the legal and marketing analysis of your potential claims. Accordingly, it may be possible to frame more specific issues for further research and to focus greater resources on those issues. For example, market research studies that might be too costly as applied to a whole range of claims may be possible once the field is narrowed to two or three potential claims.

It should be obvious that the information-gathering process is rarely as simple and straightforward as identifying a limited number of discrete issues and simply finding the answers. The process is more likely to be a continuous one in which further research generates not only answers but also additional questions. Moreover, as you design your implementation strategy, it is likely that more information gaps will pop up, particularly as you refine the product or your claims. In short, the information-gathering process must provide the flexibility for continuous adaptation to your developing product and claim.

10.3.2 *Consider Claims Modifications*

Sometimes a great claim, one that you would like to keep on your short list, may not make the initial cut because of legal or marketing risks. For example, the claim may be too generalized and run the risk of being misleading or deceptive. You may not be able to fully substantiate the claim because there is disagreement in the scientific community over your issue. You may be unable to pin down the market data you need to substantiate your claim.

An alternative to discarding claims that fail to make the short list is qualification and disclosure. As discussed in Chapter Eight, a number of claims that may be considered misleading or deceptive by themselves can be rehabilitated with qualifications and disclosures.

Generally, a qualification and disclosure simply means supplying more information to make sure consumers do not misinterpret your claim. A number of examples of claims that require qualification or disclosures are contained in the Federal Trade Commission Guides for the Use of Environmental Marketing Claims. Some typical examples are:

1. Comparative claims that fail to disclose the thing being compared. For example, if a package claims it contains "X% more recycled content," it may be necessary to disclose the basis for comparison, such as "X% more than a previous package."

2. Further steps that must be taken by the consumer. For example, if a package is compostable only at commercial composting facilities, it may be necessary to disclose that the package is not suitable for home composting. Similarly, if a product must be disassembled before the parts can be recycled, it may be necessary to disclose that "the _____ part of this package is recyclable when separated prior to collection (i.e., at home) from the remaining parts."

3. Limitations on recycling opportunities. In addition to the specific statutory definitions of recycling facilities that must be available to make a recyclability claim in certain states, it is generally deceptive to represent that a product is recyclable without disclosing whether opportunities exist to recycle the product that are accessible to the consumer. It may be necessary to include a disclaimer to a recyclability claim that states, for example, "appropriate facilities may not exist in your area" or "this product can be composted where municipal solid waste composting facilities exist. There are currently [X number of] municipal solid waste composting facilities across the country."

4. Recycled content. It is best to disclose the amount of preconsumer and postconsumer waste in connection with any claim of recycled content.

The number of ways consumers potentially can misunderstand an environmental marketing claim is as endless as the possible claims that can be made. Likewise, the number of qualifications and disclosures that might be used to remedy such a risk is vast. One way to approach the process of attempting to identify claims in need of qualification and disclosure is to subject the claim to adversarial scrutiny. This simply means attempting to identify every possible interpretation a consumer might put on a claim. To be effective, the hypothetical consumer should be wildly optimistic about the potential environmental benefits of your product. At this stage of the process, do not be constrained by reasonableness; rather, attempt to identify every possible misunderstanding that could result from your claim, no matter how unreasonable. A sample examination might go like this:

Consider an engine cleaner formerly packaged in an aerosol can with a CFC propellant. The product has been repackaged in a pump spray bottle. CFCs have been eliminated, but the product still contains organic solvents that are suspected to deplete the ozone layer. Consider the claim that the product "contains no CFCs." Possible consumer interpretations include the following:

1. The product now contains no CFC propellant (a correct interpretation!).
2. No CFCs are used in the formulation of the product itself.
3. No CFCs are used in the formulation of the package.
4. The product contains no ozone-depleting chemicals of any kind.
5. The product is nontoxic.
6. The product is environmentally safe and imposes no burdens on the environment.

One reaction to these possible interpretations is to argue that most of the interpretations are substantially correct, and the remaining ones are hair-brained. While some form of legal defense based on these arguments could be crafted, the legal defenses would not head off a consumer backlash, and at the least, costly litigation would be necessary to establish the defenses.

A better alternative might be to modify the claim with appropriate qualifications and disclosures. This would not only protect against legal liability and consumer backlash, but also serve as an opportunity to further educate your market about the genuine and important benefits of your reformulated product. You might modify the claim as follows: "New package eliminates CFC propellant—use of engine cleaner creates 90 percent less ozone-depleting chemicals than previous aerosol package."

It should be apparent from the above example that adding a qualification or limitation to your claim in effect creates a different claim that must be reevaluated by a process similar to that described in Section 10.2. In the above example, it would be necessary to evaluate whether the claim as qualified is allowable under applicable laws, to evaluate the potential market benefits of the qualified claim, and to develop substantiation for the qualification (that use of the product in fact generates 90 percent less ozone-depleting chemicals than the prior package).

Another way to get problematic claims onto the short list is to reword the claims. Rewording is very similar to qualification and disclosure in that it generally involves making the claim more specific.

10.3.3 *Consider Maximum Compliance Strategy*

One strategy for cutting through the regulatory maze is simply to pick the strictest possible requirement and to comply with that requirement everywhere you do business. The idea behind this strategy is that if you comply with the strictest rule, you will necessarily be in compli-

ance with less strict rules elsewhere. Four possible advantages to such a strategy are: (1) it avoids or at least simplifies the potentially costly and long process of legal analysis, (2) it provides a margin of error in those regulatory/market regions that have less strict standards with a corresponding margin of comfort that you are in compliance, (3) you may save product costs by using a single standard, and (4) you reduce the risk of accusations that you apply a double environmental standard.

The cost and time savings of the legal analysis may be somewhat limited by the fact that it is still generally necessary to identify all applicable legal standards in order to determine which is the strictest. The principal savings, however, comes because it is necessary only to identify the standards in order to choose the strictest; it is not necessary to apply each separate standard to your proposed claims. Depending on the number of claims and number of jurisdictions you are dealing with, some savings may accrue.

The operational cost savings are potentially much more significant. These savings represent the difference in production costs between producing two or more separate packages or labels or advertisements instead of only one. To a certain extent, this opportunity is limited to distribution within same language areas of distribution, because distribution in areas of foreign language will require separate packaging or labeling or advertising regardless of the legal considerations. Accordingly, "maximum compliance" may be most effective for distribution within the United States and Canada.

There are two exceptions to the foreign language problem: packaging and recycling. With respect to packaging (at least all packaging except the outer package when it functions as labeling and trade dress in addition to its packaging function), language considerations are probably unimportant. Consumer preferences may have some influence over packaging and may sometimes provide a reason for differentiated packaging in different markets, but these preferences do not take on the mandatory character of using the local language. Accordingly, consumer preferences with respect to packaging can be weighed against operational cost savings from employing a "maximum compliance" method of packaging. Accordingly, "maximum compliance" may make sense even for international distribution, at least with respect to packaging.

The other exception to the foreign language problem is the growing implementation of take-back programs. A take-back program is essentially a method of maximum compliance with recyclability requirements because offering a take-back option will probably qualify your product as recyclable in any market/regulatory region.

Maximum compliance is not always available. Because of the patchwork of developing regulation of environmentally conscious marketing claims, different markets/regulatory regions may have fundamentally different approaches to the regulation of particular claims that do not compare or harmonize neatly through a single compliance option. This is particularly true of emblems or certifications that must be registered with local government authorities. Obviously, it is necessary to comply with any unique requirements of each authority whose emblem you desire to display.

10.3.4 *Consider Differentiation Strategy*

The differentiation strategy is the opposite alternative to maximum compliance. This strategy involves producing two or more separate packages, labels, or claims for use in different markets/regulatory regions. This strategy may be particularly well suited to situations in which differentiated packaging and labeling is already necessary because you distribute in foreign language areas. In addition, the differentiation strategy may work well if your distribution network includes local distributors that can tailor your claims to fit local regulations.

The obstacle to the differentiation strategy is probably obvious: It is expensive and difficult logistically to produce a differentiated product and to control differentiated distribution to different markets/regulatory regions. In some cases, however, there may be large market incentives for differentiated distribution because of the unusual or strict local regulations on environmentally conscious marketing.

California, for example, imposes a unique requirement of the use of warnings on most products that contain any of a number of listed chemicals. Products covered by the regulation must display language that may include a warning that the product contains a chemical known to the State of California to cause cancer or reproductive harm. These warning labels often may be required for products not traditionally considered to be a health risk, including, for example, china tableware, fruits and vegetables sprayed with herbicides and a large number of other products. Some marketers may find the use of this warning so objectionable that the cost of separately labeling products distributed in California is justified so products distributed elsewhere need not display the warning.

Another example of a strict requirement that may justify differentiated compliance is the German packaging law, which generally requires manufacturers and distributors to take back all disposable packaging and components of their products. For products difficult to ship or difficult to recycle, it may not be cost-justified to institute a worldwide take-back program. Market forces have recognized this problem, resulting in the availability of a private contracting service within Germany that can undertake the manufacturer's take-back obligation within Germany. In such a case, it may make economic sense to contract locally for compliance with the take-back requirement.

One potential disadvantage of a differentiated compliance strategy is the risk of public criticism based on a perception that it is inequitable to treat one part of the world better than another. It is risky to undertake an environmentally conscious position if you are unwilling to extend your commitment to all parts of your operations in all places in the world. Accordingly, all other things being equal, differentiated compliance strategy probably does not make sense unless there are compelling reasons based on local requirements.

One way to implement a plan of differentiated claims is to use stickers or adhesive labels with differentiated claims printed on them. The cost of printing and applying differentiated labels will usually be much less than the cost of designing, fabricating, and utilizing differentiated packaging and labeling. However, an adhesive sticker on outer packaging may impair the recyclability of the outer package. It should be obvious that stickers will not address every differentiation problem. Stickers, for example, will not solve a differentiation problem related to packaging issues or materials content.

10.3.5 *Consider Product or Package Modifications*

The process of examining potential claims may, under the principle of continuous improvement, lead you to identify opportunities to improve your product even further. This may come in the form of an opportunity, when you simply discover a way to make the product even better through the process of talking about it. On the other hand, it may come in the form of a challenge because the only way to make the claim you have selected is to modify the product to fit the claim.

It may be difficult at this point in the process to consider redesigning your product or package to the extent that redesign may involve adjustments in procurement, manufacturing, and other operations. In short, the ripple effect of redesign may be much greater at this stage in the process. Accordingly, a business judgment must be made about whether the redesign is cost-justified. To the extent your operations include continuous improvement mechanisms, this process will be that much easier.

Regulations on environmentally conscious marketing claims may provide a significant incentive to redesign. Redesign may be the only way or the best way to make your claims proper. This occurs when a specification that is a requirement for the claim has not been designed into the product. For example, if you have identified a recycled content claim as an effective one for your product, you may find that the recycled content designed into the product is insufficient to meet the threshold content requirements in one or more marketing/regulatory regions in which you intend to distribute. A simple way to solve this problem is to rewrite the material specification to include a greater percentage of postconsumer recycled material.

10.4   CHOOSE AND IMPLEMENT YOUR STRATEGY

In some cases, a clear ''winner'' will emerge from the evaluation and selection process, leaving the final selection of a marketing strategy clear. In other cases, the process will leave you with a number of roughly equal options from which to choose. In some cases, the options may seem limited and there is no choice that is clearly satisfactory.

One strategy for dealing with a situation in which your best claims remain somewhat risky is to consider whether relevant government agencies are available to review the claim and provide some comfort that it will comply with applicable requirements. Many state attorneys general or trade commissions are willing to provide review on a semiformal or informal basis. While such agency review is helpful, it does not eliminate all risks. In many cases, an agency opinion is not binding on the agency and may be reversed if the agency decides to challenge your claim. Furthermore, agency opinions will not, in most cases, provide a defense to a suit by a competitor or citizens group. Finally, agency review provides little protection from the backfire risk.

Another option for the cautious is to test-market your claim on a limited basis to get a sense of how it will perform in the market. Test-marketing simply means making the claim in a selected area or on a limited number of samples and tracking the results.

10.5   REEVALUATE AND UPDATE YOUR STRATEGY PERIODICALLY

Once the long and potentially difficult process of identifying, evaluating, and selecting an environmentally conscious marketing strategy has been completed, the temptation is to leave well enough alone. In some cases, when your claim rests on relatively well-established scientific principles and research and does not depend on market factors, it may make sense to leave well enough alone. In most cases, however, a claim that is true when it is first made can easily become misleading or false based on developments in scientific research, the regulatory environment, or the market. For this reason, it is important to continuously monitor scientific developments, regulatory developments, and market developments and to reevaluate your claim in light of those developments.

Claims that are particularly subject to reevaluation include the following:

1. Comparative claims that depend on what other alternatives are available in the market.
2. New or unique claims that may attract regulatory attention.
3. Claims that address an issue that is subject to active and ongoing scientific research.
4. Claims that depend on consumer behavior and waste management technology (such as claims of degradability, compostability, or recyclability).

The common denominator for these types of claims is that they depend on facts outside the manufacturing process that can change without your knowledge if you are not paying attention. Laws can change. Consumer behavior can change. Technology can change. Knowledge about environmental issues can change. If the truth and accuracy of your claim depends on any of these external facts, a change in the fact mandates review of your claim to ensure it is still fair and accurate.

Assume, for example, that a paper plate manufacturer makes a claim that the paper plates are recyclable. The claim is proper at the time it is made because the plates are distributed primarily in the northeastern United States, and the manufacturer has done market research documenting that its customers actually recycle the paper plates at a rate in excess of 50 percent. During the two years after the product launch, the manufacturer expands its distribution to the southeastern coast of the United States and assumes the composition of the customers in the Northeast has changed, while municipal collection of separated recyclable waste has been scaled back due to local funding deficits. The recyclability claim may well continue to be fair and accurate. The substantiation, however, which was based on the historical fact that 50 percent of the supplier's customers actually recycled the product, is no longer sufficient to demonstrate that the product meets the legal definition of recyclability. Therefore, it is necessary to conduct an updated evaluation to document continuing compliance.

Obviously, changes in the formulation and manufacturing of your product can also affect whether an environmentally conscious marketing claim continues to be fair and accurate. Any internal changes should be evaluated to determine whether they affect environmentally conscious marketing claims.

APPENDIXES

A Sample Pollution Prevention Plan

Prepared By:

XYZ CORPORATION
100 MAIN STREET
BORA BORA, MINNESOTA

CERTIFICATION
of the
POLLUTION PREVENTION PLAN
for the
XYZ Corporation
Bora Bora, Minnesota

I certify under penalty of law that I have personally examined and am familiar with the information contained in this plan, and that based on my inquiry of those individuals who prepared or are responsible for obtaining that information, I believe that the information is true, accurate and complete. I am aware that there are significant penalties for submitting false information, including the possibility of fines and imprisonment.

Facility Manager _____

Title _____ Date _____

Company Officer _____

Title _____

Date _____

CONTENTS

POLLUTION PREVENTION POLICY
FOR
XYZ CORPORATION

XYZ Corporation is committed to excellence and leadership in protecting the environment. In keeping with this policy, our objective is to work toward the elimination of hazardous wastes and emissions by modifying our products and processes. We strive to set a standard for excellence in pollution prevention. By successfully preventing pollution at its source, we can achieve cost savings, increase operational efficiencies, improve the quality of our products and services, and maintain a safe and healthy workplace for our employees.

XYZ Corporation's environmental guidelines include the following:

- Environmental protection is everyone's responsibility. XYZ is committed to being a good neighbor and to operating in complete compliance with federal, state, and local environmental laws. Meeting this commitment requires the continued efforts of all employees.

- Preventing pollution by reducing and eliminating the generation of waste and emissions at the source is a prime consideration in plant operations. XYZ is committed to identifying and implementing pollution prevention opportunities through encouragement and involvement of all employees.

- Technologies and methods that substitute nonhazardous materials and utilize other source reduction approaches will be given top priority in addressing all environmental issues.

FACILITY IDENTIFICATION

Name and Location

XYZ Corporation
100 Main Street
Bora Bora, Minnesota 55555

SIC Code

3523—Farm Machinery and Equipment

Business Activity

XYZ Corporation manufactures an automated food dispensing system for farm animals. The equipment made at this facility is a control mechanism that consists of a cabinet containing various small machined parts. Parts are machined and plated, and cabinets are formed and painted at XYZ.

Time in Business

XYZ Corporation has manufactured cabinets at this facility since 1970.

Number of Employees

Average number of employees: 234.

FACILITY DESCRIPTION

XYZ Corporation's manufacturing operations can be broken down into three areas: fabricating, painting, and plating.

The fabricating area consists of machining, press and welding operations, and also has supporting operations such as cleaning. Two types of cleaning processes are used on production parts: immersion tanks with an alkaline cleaning solution for in-process cleaning and 1,1,1, trichloroethane vapor degreasing for final cleaning. The machining operations use both straight cutting oils and water soluble coolants.

The cabinets are painted for both aesthetics and corrosion protection. The painting is done using conventional air spray application equipment in dry booths. Solvent-based compliance coatings consisting of 20 to 25 percent 1,1,1 trichloroethane make up the majority of paint used.

The plating area contains two automatic barrel noncyanide alkaline zinc dichromate plating lines and the facility wastewater pretreatment system. Carbon steel parts are plated to provide corrosion protection for the components.

MAJOR WASTE STREAMS AND EMISSION SOURCES

To establish priorities, XYZ defines a major waste stream as one that constitutes greater than 5 percent by weight of the total hazardous waste generated at the facility. Ultrafiltration waste is also included in this evaluation even though it is not a hazardous waste since XYZ is concerned that this oily waste stream could be regulated as a hazardous waste in the future. Processes that generate SARA Title III Form R reportable emissions are considered major sources of emissions. Exhibits A–1, A–2, and A–3 include block diagrams showing the raw materials used, final products produced, and hazardous wastes and emissions generated for each process.

Waste Stream	Waste Quantity (lbs)	Emissions (lbs)
Wastewater pretreatment sludge	32,000	
Cleaning wastes (trichloroethane)	17,000	50,000
Ultrafilter oil/water concentrate	61,400	
Paint wastes (trichloroethane)	15,400	20,000
Total	125,800	70,000

EXHIBIT A–1
Fabrication Process Flow Diagram

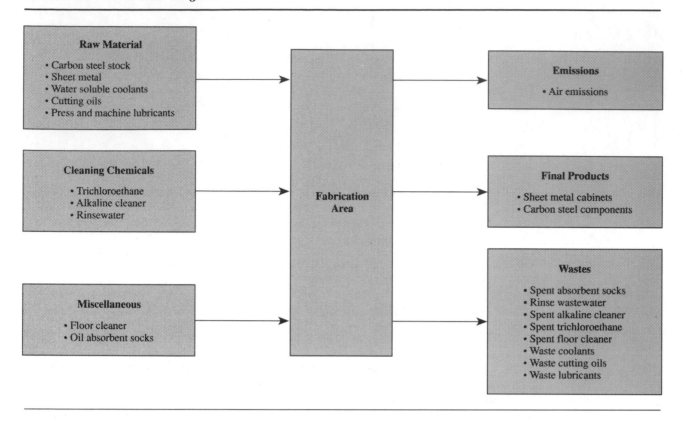

Raw Material
- Carbon steel stock
- Sheet metal
- Water soluble coolants
- Cutting oils
- Press and machine lubricants

Cleaning Chemicals
- Trichloroethane
- Alkaline cleaner
- Rinsewater

Miscellaneous
- Floor cleaner
- Oil absorbent socks

Fabrication Area

Emissions
- Air emissions

Final Products
- Sheet metal cabinets
- Carbon steel components

Wastes
- Spent absorbent socks
- Rinse wastewater
- Spent alkaline cleaner
- Spent trichloroethane
- Spent floor cleaner
- Waste coolants
- Waste cutting oils
- Waste lubricants

EXHIBIT A–2
Plating Process Flow Diagram

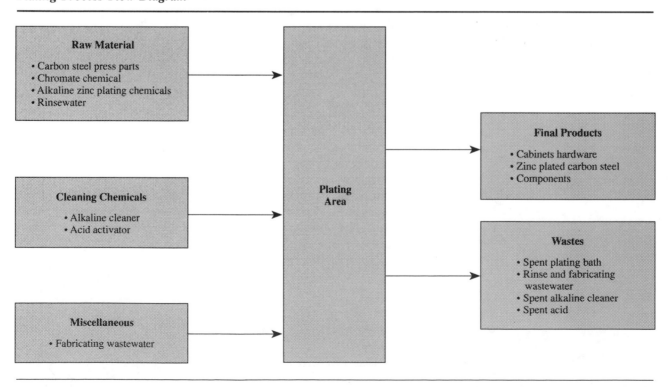

Raw Material
- Carbon steel press parts
- Chromate chemical
- Alkaline zinc plating chemicals
- Rinsewater

Cleaning Chemicals
- Alkaline cleaner
- Acid activator

Miscellaneous
- Fabricating wastewater

Plating Area

Final Products
- Cabinets hardware
- Zinc plated carbon steel
- Components

Wastes
- Spent plating bath
- Rinse and fabricating wastewater
- Spent alkaline cleaner
- Spent acid

EXHIBIT A-3
Painting Process Flow Diagram

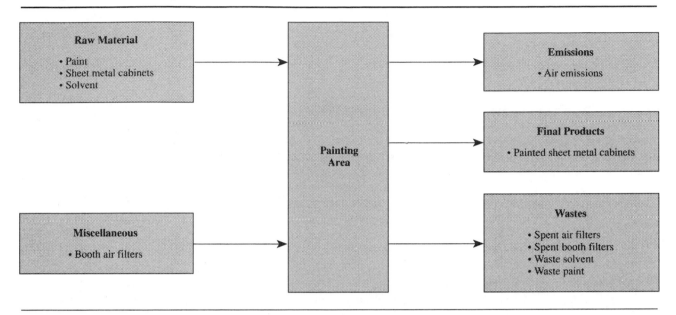

CURRENT POLLUTION PREVENTION PRACTICES

Machining

Many of the machining oils have been replaced by aqueous solutions to minimize the generation of oily wastes. All aqueous machining waste is processed through an ultrafilter to remove oil before being pumped to the wastewater pretreatment system. The wastewater from the ultrafilter is processed through the same pretreatment system used for the plating wastewater to remove any metals before final discharge.

Cleaning

A small aqueous cleaning system was purchased to remove coolants and oils before in-process inspection. This system eliminated the use of solvent cold cleaners.

The vapor degreaser is covered when it is not in use and is also equipped with a still for solvent recovery.

Painting

Painting wastes are minimized through good employee practices. These practices include keeping unused paint containers covered, minimizing the amount of solvent used for gun cleaning, and avoiding leaving excessive paint in pots, which may require disposal.

Plating

In the plating area, water is conserved by turning off rinses when the system is not in use. Flow restrictors have been installed on all rinses. Deionized water is used to make up baths to minimize sludge generation and reduce the need to dispose of baths. XYZ also converted from cyanide zinc plating to noncyanide alkaline processes. While it was hoped this would allow waste sludge to be considered nonhazardous (noncyanide zinc plating on steel is exempt from F006 classification), it is still hazardous due to the presence of leachable chrome.

The current plating wastewater treatment system consists of chrome reduction, metals precipitation, and pH adjustment to remove metals from plating and ultrafiltered fabricating wastewater.

POLLUTION PREVENTION OPTIONS

Machining

A coolant management program can be developed to review the types of coolants being used and if and how they can be recycled. Through consolidation, the number of coolants will be reduced, making recycling easier to accomplish. The processes still using straight oils can be evaluated for the feasibility of switching to water soluble machining fluids. If this cannot be done, the oils can be filtered and reused.

Several equipment options are available for the coolant recycling program. A central or satellite recycling system can be implemented. A settling tank using a drag chain can be used for chip removal or a filtration device can be used. Belts, discs, or coalescers can be used to remove contaminating tramp oils. Finally a chemical maintenance program is required for biocide addition, pH adjustment, and coolant makeup.

The key to good fluid management is to have a committed program supported by management and operations personnel. The program requires regular attention with machine and chemical maintenance. The proper fluid recycling equipment can simplify the management of such a program.

Two systems are being considered. Both options require the use of water soluble metal working fluids. Since there is no existing trenching system at XYZ's facility, satellite systems are recommended over a central system. A sump cleaner will be used to empty the machine sumps and transport the contaminated fluid to the recycling equipment. The sump cleaner will have a 1 cubic foot capacity strainer with a fabric liner to capture metal chips and fines. Overhead return pipes will return the cleaned fluid to the machines.

System One is a single-pass pasteurization unit. The used fluid is heated to 160 degrees Fahrenheit to kill most of the bacteria. The heat also improves the removal of tramp oil by coalescence. Chemical maintenance takes place in the clean fluid tank. The entire operation treats 200 gallons of fluid in eight hours.

System Two is a filtration and coalescing unit. A floating skimmer in the inlet tank removes free-floating tramp oils, and a bag filter is used to remove suspended solids. The bag filter prevents buildup of fines on the coalescer. The coalescer removes free-floating, dispersed, and loosely emulsified tramp oils. The fluid overflows from the clean tank to the dirty tank for further clarification, providing continuous processing. Chemical maintenance takes place in the clean tank.

For coolants that cannot be recycled, the plant has an unfiltration system. Improvements can be made to the system to ensure more efficient operation and reduced machine downtime. The benefit of these changes are that the improved system will reduce the amount of coolants shipped off site. The plant can more tightly control what is disposed of in the waste coolant tank. Tramp oils or other floating oil sludges and solids should not enter the system. These will cause the membranes to foul more quickly, if not damage them permanently. Oil skimmers can be used wherever possible to remove the floating oils from coolants before processing, and filtration should be used in areas like the slop room to reduce solids loading.

Cleaning

Significant waste and emission reductions can be realized in the cleaning/degreasing operations. The existing degreaser can be retrofitted and brought up to current standards or a new degreaser can be purchased. To eliminate solvent use, the degreasing process can be replaced with aqueous cleaning operations. It is important to have adequate rinsing following the aqueous cleaning to minimize spotting and poor paint adhesion.

The spent aqueous cleaner along with the spent alkaline cleaner already in use can be transferred to the ultrafiltration system for concentration before disposal. Filtration and reuse of the cleaner should be investigated to reduce the volume of spent cleaner produced.

Painting

The painting at XYZ is done using conventional air spray application equipment in dry booths where solvent-based coatings consisting of 1,1,1 trichlorethylene make up the majority of paint used. To reduce the emissions and solvent wastes, high solids and waterborne paints should be considered. Also more efficient paint application equipment can be used.

Powder coating provides a means to eliminate solvent emissions and save money by applying a 100 percent solid powder rather than a solvent-dispersed resin. Similar parts to those painted by XYZ are powder coated at other companies. For this reason, powder coating is seen as a technically feasible option. The feasibility of waterborne coatings can also be investigated. More efficient application equipment can reduce the overspray thus reducing the amount of paint used and associated paint wastes. Electrostatic spray guns and high-volume, low-pressure (HVLP) guns are both good options for this particular application.

Paint wastes can be reduced by using alternate filters in the spray booths, increasing the loading of paint on the filters before they are spent. This increases the time between filter changes and results in a reduction in the generation of wastes.

Plating

XYZ's wastewater treatment of plating rinsewaters currently produces 32,000 pounds of sludge annually. Implementation of a metals recovery process could reduce the generation of sludge by as much as 75 percent. Recovery of metals from the plating rinsewaters requires potentially expensive equipment. Options include ion exchange, reverse osmosis, electrolytic recovery, and process modifications such as dragout reduction techniques.

Another series of options may allow the waste generated to be no longer considered a hazardous waste. Low and no-chrome seal coatings may be able to replace the existing chromate coating. These coatings would reduce or eliminate chrome in the wastewater sludge, rendering it nonhazardous.

TECHNOLOGICAL AND ECONOMIC FEASIBILITY

Machining

Evaluation: Two options are outlined above. Under consideration are a pasteurization unit and a filtration unit. Both systems are easy to operate and maintain and have been widely used in industry. In removing the tramp oil and continuously recirculating the fluid, the filtration unit accomplishes the same task as the pasteurization unit. The bacterial concentration is reduced by removing its food (tramp oil), aerating the fluid thus killing the anaerobic bacteria, and adding biocide during chemical maintenance. The pasteurization unit also claims to reduce the bacterial concentration, but still leaves colonies to grow in the somewhat stagnant solution. Option One has higher capital costs and operating costs than Option Two. The payback period for Option One is 6.3 years and for Option Two is 3.0 years. An economic analysis is included in Exhibit A–4.

Selected measure: Option Two, the filtration unit was selected for the fluid management program in the machining area. It is a proven technology at reasonable cost.

Rejected measure: Option One, the pasteurization unit, was rejected for the fluid man-

EXHIBIT A-4
Machining Economic Analysis

	Option One	Option Two
Equipment cost		
Recycling equipment	$32,000	$17,000
Sump cleaner	10,000	10,000
Mechanical/piping		
Installation and materials	8,000	8,000
Electrical installation and materials	2,000	2,000
Total installed cost	$52,000	$37,000
Labor	$30,500	$30,500
Raw materials	11,000	11,000
Maintenance	6,000	6,000
Replacement parts	1,200	1,200
Utility cost	4,000	—
Total annual operating cost	$52,700	$48,700
Savings		
Material handling	$11,500	$11,500
Raw material	26,000	26,000
Waste disposal	23,500	23,500
Total annual savings	$61,000	$61,000

Economic payback

Installed equipment cost ÷ (Savings − Operating cost)

	Option One	Option Two
	$\dfrac{\$52,000}{(\$61,000-52,700)}$	$\dfrac{\$37,000}{(\$61,000-48,700)}$
	= 6.3 years	3.0 years

agement program. The costs were significantly higher than for Option Two with no significant added efficiency.

Cleaning

Evaluation: The first option is upgrading or replacing the degreaser to continue with solvent cleaning. The second option is to design and install an aqueous cleaning system.

The first option has proven its cleaning ability in the plant for over 20 years; however, the regulatory environment has changed significantly. Considering the ongoing tightening of regulations, it is not feasible to continue to operate a solvent-emitting system. The costs of solvent and solvent disposal in the future are not known, but they most certainly will rise.

The second option has been shown through testing to meet cleaning requirements. The system that was investigated is a conveyorized spray system using an alkaline cleaner. It is followed by two rinses, the second using deionized water with a rust inhibitor. Cleaner wastes will be processed through the existing ultrafiltration system to reduce oil and BOD loading in the wastewater. An economic analysis compares this system to the current system, using waste disposal costs for 1990. The anticipated payback period is 4.8 years (see Exhibit A–5).

Selected measure: The second option, aqueous cleaning, was selected for the cleaning operation. Its major advantage is that it eliminates the use of solvents. It cleans well and has a good payback period.

Rejected measure: The first option, upgrading or replacing the vapor degreaser, was rejected. Although it cleans well, there are other economically feasible options that clean equally well without the hazards of solvents.

EXHIBIT A-5
Cleaning Economic Analysis

Equipment costs
Deionizer	$ 3,000
Cleaning tank	16,000
Rinse tanks	1,000
Mechanical/piping	
Installation and materials	6,400
Electrical installation and materials	2,300
Total installed cost	$28,700

Labor	$30,500
Raw materials	5,000
Maintenance	4,000
Replacement parts	900
Utility cost	7,100
Total annual operating cost	$47,500

Savings
Material handling	$22,000
Raw material	16,000
Waste disposal	15,500
Total annual savings	$53,500

Economic payback

Installed equipment cost ÷ (Savings − Operating cost) =

$$\frac{\$28,700}{(\$53,500-47,500)}$$

$$= 4.8 \text{ years}$$

Painting

Evaluation: The options being considered are the use of water-based coatings or powder coating, along with improved application equipment. The safety hazards and emitting potential of the water-based and powder coatings are much lower than with the current solvent-based paints.

Some concerns of water-based paints are transfer efficiency and texture consistency. The problems can be reduced, but not eliminated by using better application methods. The economic analysis for a water-based coating using HVLP guns gives a payback of 4.3 years.

Transfer efficiencies for powder coating with an electrostatic spray gun are very good because the overspray can be collected and reused. The texture is also improved over the liquid spray painting applications. The payback period for the second system is 3.8 years (see Exhibit A-6).

Selected measure: Powder coating using electrostatic spray equipment appears to be the best option. It provides a good finish and minimizes waste and economic losses because it is able to be reused. Thus, this is also an economically feasible option.

Rejected measure: The use of water-based coatings was rejected due to low transfer efficiency compared to powder coating and inconsistent texture.

Plating

Evaluation: Options for waste reduction in the plating area include ion exchange, reverse osmosis, electrolytic recovery, and process modifications. Recovery technologies concentrate the metal ions, allowing for easier metals recovery through electrolytic recovery or some other process. In general, these processes have significant capital requirements relative to other chemical modification options. The option with most promise appears

EXHIBIT A–6
Painting Economic Analysis

	Powder/ Electrostatic	Waterborne/ HVLP
Equipment costs		
Paint booth	$130,000	Use Existing
Cure oven	34,000	
Part rack	1,000	
Filters	400	3,000
Installation	5,200	4,000
Electrical	4,600	1,600
Total installed cost	$175,200	$ 8,600
Operating costs		
Paint	$ 9,500	$ 43,000
Maintenance	3,000	4,500
Replacement parts	1,000	1,000
Utility	8,000	28,000
Labor	60,000	32,500
Total annual operating cost	$ 81,500	$109,000
Savings		
Material handling	$ 11,000	$ 35,500
Raw material	106,000	72,000
Waste disposal	11,000	3,000
Economic payback	$128,000	$110,500

Installed equipment cost ÷ (Savings − Operating cost) =

$$\frac{\$175,200}{(\$128,000-81,500)} \qquad \frac{\$8,600}{(\$110,500-109,000)}$$

$$= 3.8 \text{ years} \qquad\qquad = 5.7 \text{ years}$$

to be the use of no or low-chrome seals to eliminate or significantly reduce the hazardous constituent from the plating sludge, thus making it nonhazardous.

Selected measure: XYZ will evaluate, and if accepatble, implement the use of no and low-chrome seals to replace the present chromate coating. This process change has no capital cost and will elimate the hazardous constituent form this wastestream.

Rejected measures: Plating rinsewater and metal recovery systems were rejected due to cost. These options may be reevaluated in a future plan to reduce water use and discharges and to minimize generation of the nonhazardous sludge.

GOALS AND OBJECTIVES

The goal of XYZ Corporation is to reduce hazardous waste generation by 75 percent within three years and 90 percent within five years. XYZ also expects to eliminate SARA Form R reportable emissions by 1995. The baseline year is 1990. Changes in production from the baseline year will be taken into account when evaluating progress toward these goals. Hours of direct labor will be used to relate production growth to waste generation. In 1990, there were 340,000 direct manufacturing labor hours.

XYZ's objective is to accomplish its goal through material substitution, process modification, recycling and reuse, as well as by involving employees in the program. Employee involvement includes enlisting the support of existing quality circles and periodic updates on the status of environmental activities in the facility newsletter.

Sample Environmental Management Program Survey

1.0 GENERAL

1.1 Plant name: _____

1.2 Plant street address: _____

1.3 Plant mailing address: _____

1.4 City, state, and ZIP code: _____

1.5 County: _____

1.6 Facility manager or equivalent: _____

1.7 Title: _____

1.8 Environmental contact: _____

1.9 Title: _____

1.10 Phone number: _____

1.11 Fax number: _____

1.12 Number of employees: _____

1.13 Production variable (e.g., direct labor hours): _____

2.0 WASTEWATER TREATMENT

2.1 Wastewater permits

2.1.1 Does your facility have a permit to discharge wastewater?
[Y] [N] [U]

2.1.2 If yes, please include the permit number(s).
_____ _____ _____ _____

2.1.3 Are any of these permits NPDES discharge permits?
[Y] [N]   [Y] [N]   [Y] [N]   [Y] [N]

2.1.4 What are the wastewater discharge receiving streams?
[ ] Publicly owned treatment plant (POTW)
[ ] River, stream, creek, or ditch
[ ] Lake, pond, swamp, or marsh
[ ] Septic tank, leach field, or french drain
[ ] Soil or ground
[ ] Recycled or reused
[ ] Unknown

2.2 Wastewater treatment processes performed on site. (Check all that apply.)

[ ] Cyanide oxidation	[ ] Filtration
[ ] Chrome reduction	[ ] Reverse osmosis
[ ] Metal hydroxide precipitation	[ ] Ultrafiltration
[ ] Metal sulfide precipitation	[ ] Acid splitting of emulsions
[ ] Oil skimming	[ ] Lagoon or surface impoundment
[ ] Neutralization	[ ] Settling pond/sump
[ ] Ion exchange	[ ] Other: _____
[ ] Electrolytic recovery	_____
[ ] Flocculation	_____
[ ] Dissolved air flotation	_____

2.3 Is the stormwater runoff for your facility required to be permitted?

[Y] [N] [U]

If yes, did your facility operate under such a permit prior to October 1, 1992? Please list the associated permit number. _____

If no, did your facility submit a permit application by October 1, 1992? Please indicate whether submission was completed for an individual, general, or group permit, or if a notice of intent (NOI) for a general permit was filed.

[Y] [N] [U]   [ ] Individual
                   [ ] General
                   [ ] Group
                   [ ] NOI

If required, has your facility begun preparations for a stormwater pollution prevention plan?

2.4 Wastewater and stormwater permit violations

2.4.1 Was your facility found to be in violation of any conditions set forth in your wastewater or stormwater discharge permit during the past year?
[Y] [N]

2.4.2 If yes, indicate below the type of discharge involved (e.g., stormwater, discharge to POTW or surface waters), sample dates, the specific parameters in violation, and the established permit limits for each parameter. Additional information can be attached if necessary.

Discharge	Sample Date	Parameter	Discharge Level (Reported Concentration)	Permitted Limit

2.4.3 Did any of these violations result in a citation or fine being imposed?

[Y] [N]

2.4.4 If yes, please describe the current or pending action taken by the regulatory authority.

3.0 COMMUNITY RIGHT-TO-KNOW (SARA TITLE III)

3.1 Was your facility required to file either Tier I or Tier II reports with state agencies in the past year?

[Y] [N] [U]

3.2 If yes, check the appropriate box.

[ ] Tier I submitted
[ ] Tier II submitted
[ ] Neither submitted—please give a brief explanation:

3.3 If you submitted either Tier I or Tier II forms, list the substances reported.

_____	_____
_____	_____
_____	_____
_____	_____

3.4 Was your facility required to file Toxic Chemical Release Inventory Reporting Forms (Form R) in the past year?

[Y] [N] [U]

3.5 If yes, please attach copies of your completed Form Rs.

4.0 REGULATORY STATUS

4.1 RCRA identification number: EPA ID number: _____

4.2   State ID number: _____

4.3 Regulatory filing status:

[ ] Conditionally exempt small-quantity generator
      (<100 kg/month or <220 lb/month)

[ ] Small-quantity generator
      (100–1,000 kg/month or 220–2,200 lb/month)

[ ] Generator
      ($>1,000$ kg/month or $>2,200$ lb/month)

[ ] Treatment, storage, and disposal faciltiy

5.0 REGULATORY OVERSIGHT

5.1 Inspections

5.1.1 Were any inspections conducted at your facility in the past year?

5.1.2 If so, indicate on Form 2 (Exhibit B–1) the inspection date, regulatory agency conducting the inspection, the stated reason for the inspection (e.g., wastewater treatment system, air emissions, safety), and specify whether any deficiencies were noted or whether a notice of violation was issued.

5.2 Citations, deficiencies, and notices of violation

5.2.1 Were any citations, deficiencies, or notices of violations (NOV) given at your facility in the past year?

5.2.2 If so, idicate on Form 3 (Exhibit B–2) the issuance date, regulatory agency involved, and the stated reason for the citation, deficiency, or NOV. Describe the action taken to correct the deficiency. For each citation, deficiency, or NOV, state whether any fine or proposed penalty was issued or proposed to be issued. List the amount of the fine.

5.3 Penalties/fines

5.3.1 Were any penalties or fines issued at your facility in the past year?

5.3.2 If so, indicate on Form 4 (Exhibit B–3) the issuance date, regulatory agency involved, the amount of the penalty, and the stated reason for the issuance of the fine or penalty. Attach a copy of the correspondence, notice, or other documentation reflecting the amount and reason for the fine or penalty.

5.4 Consent orders

5.4.1 Were any consent orders entered between the company and a regulatory agency or other government unit in the past year?

5.4.2 If so, indicate on Form 5 (Exhibit B–4) the issuance date, regulatory agency involved, and the stated reason for the consent order.

EXHIBIT B-1
Form 2 Inspections

Inspection Date	Regulatory Agency	Reason	Result of Inspection

EXHIBIT B-2

Form 3 Citations, Deficiencies, and NOVs

Issuance Date	Regulatory Agency	Reason	Corrective Action Taken	Penalty Issued or Proposed
				[I] [P]
				[I] [P]
				[I] [P]
				[I] [P]
				[I] [P]

EXHIBIT B–3
Form 4 Penalties/Fines

Issuance Date	Regulatory Agency	Amount of Fine or Penalty	Reason

EXHIBIT B–4
Form 5 Consent Orders

Issuance Date	Regulatory Agency	Reason

6.0 RECORD-KEEPING

6.1 Does your facility have a written Hazardous Waste Management Plan?

[Y]   [N]

6.2 If so, check each completed section of your waste management plan and include the last date of revision.

Last
Review

[ ] _____ Closure plan
[ ] _____ Postclosure plan
[ ] _____ SPCC plan (spill prevention control and countermeasure plan)
[ ] _____ Contingency plan
[ ] _____ Preparedness and prevention plan
[ ] _____ Inspection manual
[ ] _____ Inspection log
[ ] _____ Training for personnel managing hazardous wastes
[ ] _____ Waste analysis plan
[ ] _____ Hazardous waste operating record

7.0 WASTE CONTRACTORS

Complete the attached Form 6 (Exhibit B–5) indicating all the off-site transporters and disposal facilities used during the past year.

8.0 WASTES DISPOSED OF OFF SITE

Complete the attached Forms 7A and 7B (Exhibit B–6) for all hazardous and nonhazardous wastes shipped off site.

9.0 SOLVENT USAGE

9.1 Hazardous air pollutants

9.1.1 Do you use any of the 189 compounds on the Initial List of Hazardous Air Pollutants, which are part of the Clean Air Act Amendments?

9.1.2 If so, identify the compound and specify how each one is used.

10.0 STORAGE TANKS

Complete a copy of Forms 9A and 9B (Exhibit B–8) for each storage tank currently on site.

11.0 WASTE/EMISSION REDUCTION UPDATE

Complete Form 10 (Exhibit B–9) for each new waste-related cost reduction project implemented or initiated in the past year.

EXHIBIT B–5
Form 6 Waste Contractors

	Company Name	Address	Phone Number	EPA ID Number / State ID Number
H1				
H2				
H3				
H4				
H5				
H6				
H7				
H8				
H9				
H10				

EXHIBIT B-6
Form 7A Wastes Shipped Off-Site

	Waste Generated	Mat'l Code	Analysis Completed **	EPA Hazard Codes	State Hazard Codes	Vendors Used			
						Hauler	Broker	Disp.
1								
2								
3								
4								
5								
6								
7								
8								
9								
10								

* Shaded areas are to be completed by database entry team.
** If yes, please attach a copy of the current analytical report.

EXHIBIT B-6
Form 7B Wastes Shipped Off-Site (Concluded)

	Total Annual Quantity **	Standard Quality	Disposal Frequency	Disposal Method	Fate Code	Transporter Cost	Disposal Cost
1							
2							
3							
4							
5							
6							
7							
8							
9							
10							

* Shaded areas are to be completed by database entry team.
** Please include units of measurement.

141

EXHIBIT B-7

Form 8 PCB-Bearing Equipment

	Inventory Number	Equipment Type	Equipment Owner	PCB Concentration	Retrofit	Year Installed	Removal Date
1					[Y] [N]		
2					[Y] [N]		
3					[Y] [N]		
4					[Y] [N]		
5					[Y] [N]		
6					[Y] [N]		
7					[Y] [N]		
8					[Y] [N]		
9					[Y] [N]		
10					[Y] [N]		
11					[Y] [N]		
12					[Y] [N]		
13					[Y] [N]		
14					[Y] [N]		
15					[Y] [N]		

EXHIBIT B-8

Form 9A Underground Storage Tanks

Inventory Number	Year Installed	Capacity (Gal.)	Leak Detection Methods	Registration	Construction Materials	In Use	Material Stored
1				[Y] [N]		[Y] [N]	
2				[Y] [N]		[Y] [N]	
3				[Y] [N]		[Y] [N]	
4				[Y] [N]		[Y] [N]	
5				[Y] [N]		[Y] [N]	
6				[Y] [N]		[Y] [N]	
7				[Y] [N]		[Y] [N]	
8				[Y] [N]		[Y] [N]	
9				[Y] [N]		[Y] [N]	
10				[Y] [N]		[Y] [N]	
11				[Y] [N]		[Y] [N]	
12				[Y] [N]		[Y] [N]	
13				[Y] [N]		[Y] [N]	
14				[Y] [N]		[Y] [N]	
15				[Y] [N]		[Y] [N]	

EXHIBIT B-8

Form 9B Aboveground Storage Tanks (Concluded)

Inventory Number	Year Installed	Capacity (Gal.)	Containment Capacity (Gal.)	Leak Detection Method	Registered	Construction Materials	In Use	Material Stored
1					[Y] [N]		[Y] [N]	
2					[Y] [N]		[Y] [N]	
3					[Y] [N]		[Y] [N]	
4					[Y] [N]		[Y] [N]	
5					[Y] [N]		[Y] [N]	
6					[Y] [N]		[Y] [N]	
7					[Y] [N]		[Y] [N]	
8					[Y] [N]		[Y] [N]	
9					[Y] [N]		[Y] [N]	
10					[Y] [N]		[Y] [N]	
11					[Y] [N]		[Y] [N]	
12					[Y] [N]		[Y] [N]	
13					[Y] [N]		[Y] [N]	
14					[Y] [N]		[Y] [N]	
15					[Y] [N]		[Y] [N]	

EXHIBIT B-9
Form 10 Cost/Waste/Emission Reductions

Project name: _____

Category:

[ ] Painting	[ ] Heat treating
[ ] Cleaning	[ ] Wastewater treatment
[ ] Metal finishing/plating	[ ] Soldering
[ ] Machining/grinding	[ ] Other

Type of change:

[ ] Raw material substitution	[ ] Treatment
[ ] Process modification	[ ] Procedural
[ ] Other	

Does this project result in a reduction in:

[ ] Costs
[ ] Waste generation
[ ] Toxic emissions
[ ] ITP emissions
[ ] Ozone depleting compounds

Implementation:

[ ] Implemented, on (date): _____
[ ] Planned, scheduled date of implementation: _____

Project description. Briefly describe the project. In this description, quantify the amount and type of reductions expected. If possible or applicable, relate these reductions to specific waste streams or emissions.

_____

_____

_____

_____

_____

_____

_____

Hazardous Waste and Materials Management Plan

IMMEDIATE RESPONSE PROCEDURES

Employees who discover an actual or potential emergency situation involving hazardous materials, other than incidental releases, must immediately report this information to the emergency coordinator or, if not immediately available, the receptionist. The receptionist is then responsible for immediately contacting the EC or a designated alternate. The emergency coordinator or alternate must then:

1. Ensure the alarm system has been sounded.
2. Order evacuation of involved zones—this order must be repeated three times.

   Evacuation order: _____

   _____

   _____

3. Report to the site and begin record-keeping.
4. Evaluate the extent of the emergency, the nature, source, amount, and type of material released, and any actual and/or potential hazards.
5. Implement the emergency plan.
6. Inform guards or other designated individuals to contact outside assistance, if needed.
7. Determine if other immediate contacts are necessary: National Response Center (NRC), state and local authorities or agencies, corporate office.
8. Determine if any portions of plant must be shut down.
9. Arrange for any cleanup contractors needed.
10. Supervise any decontamination required.
11. Submit a written report summarizing the incident and actions taken AS SOON AS POSSIBLE to state and local authorities, as required.
12. Submit the final written report within 15 days to regional EPA office, as required.
13. Revise the emergency plan, if necessary; submit the revised plan to proper authorities.

FACILITY AND LOCAL EMERGENCY PHONE NUMBERS

Emergency coordinator:
    Name: Office: _____

    Address: Home: _____

Alternates:
    Name: Office: _____

    Address: Home: _____

Name: _____   Office: _____
                                              _____
Address: _____   Home: _____
                                              _____

Outside emergency response team:
    Contact names: _____ _____
                     _____ _____
                     _____ _____

Plant first aid:
    Contact names: _____ _____
                     _____ _____
                     _____ _____

    Fire department   _____
    Police department   _____
    State patrol   _____
    Hospital   _____
    Ambulance service   _____
    Poison control center   _____
    Publicly owned treatment works (POTW)   _____

Corporate contacts: _____
    Contact names: _____ _____
                     _____ _____
                     _____ _____

Other: _____ _____
       _____ _____

FEDERAL AND STATE CONTACTS

                                         Phone Numbers

Local Emergency Planning Committee

(LEPC) contact: _____   _____

State Emergency Response Commission

(SERC) contact: _____   _____

National Response Center (NRC)   1-800-424-8802

State Department of Natural Resources
    (DNR or Pollution Control Agency)   _____

Regional Environmental Protection
    Agency (EPA)   _____

EPA RCRA/Superfund Hotline   1-800-424-9346

Other: _____   _____

EMERGENCY RESPONSE INITIAL NOTIFICATION FORM

Agency name and contact: _____

Agency phone number: _____

Facility address:

_____

_____

_____

Time/date: _____

Location of incident:

_____

_____

Type of incident (i.e., fire, explosion, spill): _____

_____

Extent: _____

_____

_____

_____

Cause of incident: _____

_____

Type and quantity of material(s) involved: _____

_____

Extent of injuries, if any: _____

_____

_____

Possible hazards to human health or the environment outside of the facility: _____

_____

_____

_____

Agency response/action/comments: _____

_____

_____

_____

EMERGENCY RESPONSE PROCEDURES FOR THE EMERGENCY COORDINATOR

The emergency coordinator (EC) is responsible for directing all emergency response activities implemented during any actual or potential emergency situation. Either the EC or a designated alternate must be on site or on call at all times. The EC must be thoroughly familiar with all aspects of the contingency plan, plant operations and activities, the location and characteristics of hazardous substances and wastes, the location of records, and the plant layout. The EC has the authority to commit the resources needed to carry out the emergency plan or other appropriate emergency response measures.

Whenever there is an emergency situation, the emergency coordinator or alternate will decide whether to implement the emergency plan. If the plan is implemented, the EC will:

1. Activate internal facility alarms or communication systems to notify facility personnel.

2. Order evacuation of the plant.

3. Notify appropriate state or local agencies if their help is needed. Such help could include police, fire, rescue squads, emergency response contractors, and/or the National Response Center.

4. Identify the characteristics, source, amount of material involved, and the extent of the problem. The EC will do this by observation or review of facility records, material safety data sheets, manifests, and, if necessary, by chemical analysis.

5. Assess possible hazards to human health and the environment that may result from a release caused by fire, spill, or explosion. This assessment will consider both direct and indirect effects of the release. Direct effects could include release of toxic gases; indirect effects could include runoff from water or chemical agents used to control the fire or explosion.

6. Determine whether evacuation of the surrounding area is required and advise the local authorities of this decision. Additionally, the EC will be available to assist agency officials with the evacuation.

7. Use all reasonable measures necessary to ensure that fires, explosions, and releases do not occur, recur, or spread to other hazardous materials at the facility. If the facility must stop operations, the EC will continue to monitor for leaks, pressure buildup, gas generation, or ruptures in valves and pipes.

8. Ensure all appropriate company and corporate personnel and regulatory authorities have been contacted.

9. Note in the operating log the time, date, and details of any incident that requires implementing the contingency plan.

10. Immediately following an emergency, provide for treatment, storage, or disposal of recovered waste, contaminated soil or surface water, and decontamination of equipment or any other materials that result from the emergency.

11. After the emergency, notify the Regional Administrator and appropriate state and local authorities that the facility is in compliance before operations are resumed. The EC must also certify that all emergency equipment listed in the Emergency Plan is cleaned and fit for its intended use before operations are resumed. Ensure all necessary follow-up reports are completed and filed with federal, state, and local agencies.

The EC and designated alternates must be familiar with emergency reporting requirements as specified by federal, state, and local regulations. Chemical spills that result in a release to the environment in excess of established Reporting Quantities (RQ) must immediately notify the National Response Center (NRC), the State Emergency Planning Commissions (SERC), and the Local Emergency Planning Committee (LEPC). When notifying these agencies, the EC must be prepared to provide the following information:

- Name of person making phone report and phone number.
- Name and address of facility.
- Time and type of incident.

- Name and quantity of material involved.
- Extent of injuries, if any.
- Possible hazards to human health and the environment inside and outside the facility.

The SERC and LEPC require written reports of follow-up and updated information AS SOON AS POSSIBLE and as often as necessary; for releases reported to the NRC, a written follow-up report must be filed within 15 days of the incident. Written follow-up reports must include the following information:

- Date, time, and type of incident (e.g., fire, explosion).
- Extent of injuries, if any.
- Assessment of actual or potential hazards to human health or the environment, if applicable.
- Actions taken during the event and any ongoing remediation activities.
- Estimated quantity and related disposition of any recovered materials.
- Revisions to the Emergency Plan or Emergency Response procedures as a result of this incident.

EMERGENCY RESPONSE PROCEDURES FOR PLANT PERSONNEL

1. Move into a safe area.
2. Immediately contact the Emergency Coordinator. If the EC is unavailable, contact the receptionist or, during off-hours, a security guard who can contact the EC or designated alternate.
3. Activate the alarm system or otherwise notify other personnel in the area of the emergency situation.
4. Await further instruction by the EC regarding evacuation or other emergency response actions.

EMERGENCY RESPONSE PROCEDURES FOR GUARDS

1. Notify the Emergency Coordinator or the designated alternate.
2. Sound alarm and announce evacuation order over the intercom. Repeat evacuation order three times.
3. Notify plant medical team, if required.
4. Call for outside fire, police, and rescue if required.
5. Make sure gates are open for arriving fire and rescue crews, direct traffic.
6. After danger has passed, sound the all clear signal.

MEDICAL TEAM MEMBERS AND RESPONSIBILITIES

1. Medical Team Members:
   - List names of the medical team member.
   - Training (registered nurse, Red Cross CPR, etc.).
   - Specific capabilities or duties during an emergency.
2. Responsibilities:
   - Upon sounding of the first alarm, the medical team shall report to the scene of the emergency. Identify the equipment that each member is responsible for bringing.
   - Medical team shall administer the necessary first aid.
   - When trained rescue personnel arrive at the scene, the medical team shall immediately evacuate the plant site.

VERIFICATION OF LOCAL ARRANGEMENTS

Date: _____

Police Department
Fire Department
Hospital
Emergency Response Team
Emergency Response Contractors

Dear (   ):

The corporate policies of Company Y require facilities that generate and/or store hazardous materials and wastes to develop a plan to prepare for and respond to emergency situations, such as a fire or chemical release. As part of this plan, each facility is required to inform and pre-arrange emergency response procedures with those local authorities (police, fire, etc.) that may be required to respond to such an emergency.

Enclosed is a copy of the Management Plan for our facility that outlines the preparations and procedures established to respond to a potentially hazardous situation. Please review this information in relation to the services that you may be called upon to provide.

Please sign this letter and return it to our facility to acknowledge that you have received and reviewed our Management Plan. If you would like to tour this facility or discuss any information in the Management Plan in greater detail, please do not hesitate to contact me at (phone number).

Sincerely,

(Name)
(Title)

I acknowledge receiving the emergency procedures and plan from
_____(Company Name)_____   on   _____(Date)_____ .

Signed: _____

Title: _____

LIST OF INCOMPATIBLE MATERIALS

In the lists below, the mixing of a Group A material with a Group B material may have the potential consequence as noted.

Group 1–A	*Group 1–B*
Acetylene sludge	Acid sludge
Alkaline caustic liquids	Acid and water
Alkaline cleaner	Battery acid
Alkaline corrosive liquids	Chemical cleaners
Alkaline corrosive battery	Electrolyte, acid

Caustic wastewater lime sludge and other corrosive alkalies	Etching acid liquid or solvent
Lime wastewater	Pickling liquor and other corrosive acids
Lime and water	Spent acid
Spent caustic	Spent mixed acid, spent sulfuric acid

Potential consequences: Heat generation; violent reaction.

Group 2–A	*Group 2–B*
Aluminum	Any material in Groups 1–A or 1–B
Beryllium	
Calcium	
Lithium	
Magnesium	
Potassium	
Sodium	
Zinc powder	
Other reactive metals and metal hydrides	

Potential consequences: Fire or explosion; generation of flammable hydrogen gas.

Group 3–A	*Group 3–B*
Alcohols and water	Any concentrated material listed in Groups 1–A or 1–B
	Calcium
	Lithium
	Metal hydrides
	Potassium
	Sulfur dioxide, chlorine, thionyl chloride, phosphorus trichloride, acetylene, silicon tetrachloride
	Other water-reactive materials

Potential consequences: Fire, explosion, or heat generation; generation of flammable or toxic gases.

Group 4–A	*Group 4–B*
Alcohols	Concentrated Group 1–A or 1–B materials
Aldehydes	Concentrated Group 1–A or 1–B materials
Halogenated hydrocarbons	Concentrated Group 1–A or 1–B materials
Nitrated hydrocarbons	Group 2–A materials

Unsaturated hydrocarbons	Group 2–A materials
Other reactive organic compounds and solvents	Group 2–A materials

Potential consequences: Fire, explosion, or violent reaction.

Group 5–A	*Group 5–B*
Spent cyanide and sulfide solutions	Group 1–B wastes

Potential consequences: Generation of toxic hydrogen cyanide or hydrogen sulfide gas.

Group 6–A	*Group 6–B*
Chlorates	Acetic and other organic acids
Chlorine	Acetic and other organic acids
Chlorites	Concentrated mineral acids
Chromic acid	Concentrated mineral acids
Hypochlorites	Group 2–A materials
Nitrates	Group 4–A materials
Nitric acid, fuming	Other flammable and combustible materials
Permanganates	Flammable and combustible materials
Perchlorates	Flammable and combustible materials
Peroxides	Flammable and combustible materials
Other strong oxidizers	Flammable and combustible materials

Potential consequences: Fire, explosion, or violent reaction.

ENDNOTES

Title 40, Code of Federal Regulations, Part 265, Appendix V.

Guides for the Use of Environmental Marketing Claims

The Application of Section 5 of the Federal Trade Commission Act to Environmental Advertising and Marketing Practices

TABLE OF CONTENTS

A. STATEMENT OF PURPOSE:

These guides represent administrative interpretations of laws administered by the Federal Trade Commission for the guidance of the public in conducting its affairs in conformity with legal requirements. These guides specifically address the application of Section 5 of the FTC Act to environmental advertising and marketing practices. They provide the basis for voluntary compliance with such laws by members of industry. Conduct inconsistent with the positions articulated in these guides may result in corrective action by the Commission under Section 5, if, after investigation, the Commission has reason to believe that the behavior falls within the scope of conduct declared unlawful by the statute.

SOURCE: Federal Trade Commission, July 1992.

B. SCOPE OF GUIDES:

These guides apply to environmental claims included in labeling, advertising, promotional materials and all other forms of marketing, whether asserted directly or by implication, through words, symbols, emblems, logos, depictions, product brand names, or through any other means. The guides apply to any claim about the environmental attributes of a product or package in connection with the sale, offering for sale, or marketing of such product or package for personal, family or household use, or for commercial, institutional or industrial use.

Because the guides are not legislative rules under Section 18 of the FTC Act, they are not themselves enforceable regulations, nor do they have the force and effect of law. The guides themselves do not preempt regulation of other federal agencies or of state and local bodies governing the use of environmental marketing claims. Compliance with federal, state or local law and regulations concerning such claims, however, will not necessarily preclude Commission law enforcement action under Section 5.

C. STRUCTURE OF THE GUIDES:

The guides are composed of general principles and specific guidance on the use of environmental claims. These general principles and specific guidance are followed by examples that generally address a single deception concern. A given claim may raise issues that are addressed under more than one example and in more than one section of the guides.

In many of the examples, one or more options are presented for qualifying a claim. These options are intended to provide a "safe harbor" for marketers who want certainty about how to make environmental claims. They do not represent the only permissible approaches to qualifying a claim. The examples do not illustrate all possible acceptable claims or disclosures that would be permissible under Section 5. In addition, some of the illustrative disclosures may be appropriate for use on labels but not in print or broadcast advertisements and vice versa. In some instances, the guides indicate within the example in what context or contexts a particular type of disclosure should be considered.

D. REVIEW PROCEDURE:

Three years after the date of adoption of these guides, the Commission will seek public comment on whether and how the guides need to be modified in light of ensuing developments.

Parties may petition the Commission to alter or amend these guides in light of substantial new evidence regarding consumer interpretation of a claim or regarding substantiation of a claim. Following review of such a petition, the Commission will take such action as it deems appropriate.

E. INTERPRETATION AND SUBSTANTIATION OF ENVIRONMENTAL MARKETING CLAIMS:

Section 5 of the FTC Act makes unlawful deceptive acts and practices in or affecting commerce. The Commission's criteria for determining whether an express or implied claim has been made are enunciated in the Commission's Policy Statement on Deception.[1] In addition, any party making an express or implied claim that presents an objective assertion about the environmental attribute of a product or package must, at the time the claim is made, possess and rely upon a reasonable basis substantiating the claim. A reasonable basis consists of competent and reliable evidence. In the context of environmen-

---

1. *Cliffdale Associates, Inc.,* 103 F.T.C. 110, at 176, 176 n. 7, n. 8, Appendix, *reprinting* letter dated Oct. 14, 1983, from the Commission to The Honorable John D. Dingell, Chairman, Committee on Energy and Commerce, U.S. House of Representatives (1984) ("Deception Statement").

tal marketing claims, such substantiation will often require competent and reliable scientific evidence. For any test, analysis, research, study or other evidence to be "competent and reliable" for purposes of these guides, it must be conducted and evaluated in an objective manner by persons qualified to do so, using procedures generally accepted in the profession to yield accurate and reliable results. Further guidance on the reasonable basis standard is set forth in the Commission's 1983 Policy Statement on the Advertising Substantiation Doctrine. 49 Fed. Reg. 30,999 (1984); *appended to Thompson Medical Co.*, 104 F.T.C. 648 (1984). These guides, therefore, attempt to preview Commission policy in a relatively new context—that of environmental claims.

F. GENERAL PRINCIPLES:

The following general principles apply to all environmental marketing claims, including, but not limited to, those described in Part G below. In addition, Part G contains specific guidance applicable to certain environmental marketing claims. Claims should comport with all relevant provisions of these guides, not simply the provision that seems most directly applicable.

1. *Qualifications and Disclosures:* The Commission traditionally has held that in order to be effective, any qualifications or disclosures such as those described in these guides should be sufficiently clear and prominent to prevent deception. Clarity of language, relative type size and proximity to the claim being qualified, and an absence of contrary claims that should undercut effectiveness, will maximize the likelihood that the qualifications and disclosures are appropriately clear and prominent.

2. *Distinction Between Benefits of Product and Package:* An environmental marketing claim should be presented in a way that makes clear whether the environmental attribute or benefit being asserted refers to the product, the product's packaging or to a portion or component of the product or packaging. In general, if the environmental attribute or benefit applies to all but minor, incidental components of a product or package, the claim need not be qualified to identify that fact. There may be exceptions to this general principle. For example, if an unqualified "recyclable" claim is made and the presence of the incidental component significantly limits the ability to recycle the product, then the claim would be deceptive.

Example 1: A box of aluminum foil is labeled with the claim "recyclable," without further elaboration. Unless the type of product, surrounding language, or other context of the phrase establishes whether the claim refers to the foil or the box, the claim is deceptive if any part of either the box or the foil, other than minor, incidental components, cannot be recycled.

Example 2: A soft drink bottle is labeled "recycled." The bottle is made entirely from recycled materials, but the bottle cap is not. Because reasonable consumers are likely to consider the bottle cap to be a minor, incidental component of the package, the claim is not deceptive. Similarly, it would not be deceptive to label a shopping bag "recycled" where the bag is made entirely of recycled material but the easily detachable handle, an incidental component, is not.

3. *Overstatement of Environmental Attribute:* An environmental marketing claim should not be presented in a manner that overstates the environmental attribute or benefit, expressly or by implication. Marketers should avoid implications of significant environmental benefits if the benefit is in fact negligible.

Example 1: A package is labeled, "50% more recycled content than before." The manufacturer increased the recycled content of its package from 2 percent recycled material to 3 percent recycled material. Although the claim is technically true, it is likely to convey the false impression that the advertiser has increased significantly the use of recycled material.

Example 2: A trash bag is labeled "recyclable" without qualification. Because trash bags will ordinarily not be separated out from other trash at the landfill or incinerator for recycling, they are highly unlikely to be used again for any purpose. Even if the bag is technically capable of being recycled, the claim is deceptive since it asserts an environmental benefit where no significant or meaningful benefit exists.

Example 3: A paper grocery sack is labeled "reusable." The sack can be brought back to the store and reused for carrying groceries but will fall apart after two or three reuses, on average. Because reasonable consumers are unlikely to assume that a paper grocery sack is durable, the unqualified claim does not overstate the environmental benefit conveyed to consumers. The claim is not deceptive and does not need to be qualified to indicate the limited reuse of the sack.

4. *Comparative Claims:* Environmental marketing claims that include a comparative statement should be presented in a manner that makes the basis for the comparison sufficiently clear to avoid consumer deception. In addition, the advertiser should be able to substantiate the comparison.

Example 1: An advertiser notes that its shampoo bottle contains "20% more recycled content." The claim in its content is ambiguous. Depending on contextual factors, it could be a comparison either to the advertiser's immediately preceding product or to a competitor's product. The advertiser should clarify the claim to make the basis for comparison clear, for example, by saying "20% more recycled content than our previous package." Otherwise, the advertiser should be prepared to substantiate whatever comparison is conveyed to reasonable consumers.

Example 2: An advertiser claims that "our plastic diaper liner has the most recycled content." The advertised diaper does have more recycled content, calculated as a percentage of weight, than any other on the market, although it is still well under 100% recycled. Provided the recycled content and the comparative difference between the product and those of competitors are significant and provided the specific comparison can be substantiated, the claim is not deceptive.

Example 3: An ad claims that the advertiser's packaging creates "less waste than the leading national brand." The advertiser's source reduction was implemented sometime ago and is supported by a calculation comparing the relative solid waste contributions of the two packages. The advertiser should be able to substantiate that the comparison remains accurate.

G. ENVIRONMENTAL MARKETING CLAIMS:

Guidance about the use of environmental marketing claims is set forth below. Each guide is followed by several examples that illustrate, but do not provide an exhaustive list of, claims that do and do not comport with the guides. In each case, the general principles set forth in Part F above should also be followed.[2]

1. *General Environmental Benefit Claims:* It is deceptive to misrepresent, directly or by implication, that a product or package offers a general environmental benefit. Unqualified general claims of environmental benefit are difficult to interpret, and depending on their context, may convey a wide range of meanings to consumers. In many cases, such claims may convey that the product or package has specific and far-reaching environmental benefits. As explained in the Commission's Ad Substantiation Statement, every express and material, implied claim that the general assertion conveys to reasonable consumers about an objective quality, feature or attribute of a product must be substantiated. Unless this substantiation duty can be met, broad environmental claims should either be avoided or qualified, as necessary, to prevent deception about the specific nature of the environmental benefit being asserted.

---

2. These guides do not address claims based on a "lifecycle" theory of environmental benefit. Such analyses are still in their infancy and thus the Commission lacks sufficient information on which to base guidance at this time.

Example 1:   A brand name like "Eco-Safe" would be deceptive if, in the context of the product so named, it leads consumers to believe that the product has environmental benefits which cannot be substantiated by the manufacturer. The claim would not be deceptive if "Eco-Safe" were followed by clear and prominent qualifying language limiting the safety representation to a particular product attribute for which it could be substantiated, and provided that no other deceptive implications were created by the context.

Example 2:   A product wrapper is printed with the claim "Environmentally Friendly." Textual comments on the wrapper explain that the wrapper is "Environmentally Friendly because it was not chlorine bleached, a process that has been shown to create harmful substances." The wrapper was, in fact, not bleached with chlorine. However, the production of the wrapper now creates and releases to the environment significant quantities of other harmful substances. Since consumers are likely to interpret the "Environmentally Friendly" claim, in combination with the textual explanation, to mean that no significant harmful substances are currently released to the environment, the "Environmentally Friendly" claim would be deceptive.

Example 3:   A pump spray product is labeled "environmentally safe." Most of the product's active ingredients consist of volatile organic compounds (VOCs) that may cause smog by contributing to ground-level ozone formation. The claim is deceptive because, absent further qualification, it is likely to convey to consumers that use of the product will not result in air pollution or other harm to the environment.

2. *Degradable/Biodegradable/Photodegradable*: It is deceptive to misrepresent, directly or by implication, that a product or package is degradable, biodegradable or photodegradable. An unqualified claim that a product or package is degradable, biodegradable or photodegradable should be substantiated by competent and reliable scientific evidence that the entire product or package will completely break down and return to nature, *i.e.*, decompose into elements found in nature within a reasonably short period of time after customary disposal.

Claims of degradability, biodegradability or photodegradability should be qualified to the extent necessary to avoid consumer deception about: (a) the product or package's ability to degrade in the environment where it is customarily disposed; and (b) the rate and extent of degradation.

Example 1:   A trash bag is marketed as "degradable," with no qualification or other disclosure. The marketer relies on soil burial tests to show that the product will decompose in the presence of water and oxygen. The trash bags are customarily disposed of in incineration facilities or at sanitary landfills that are managed in a way that inhibits degradation by minimizing moisture and oxygen. Degradation will be irrelevant for those trash bags that are incinerated and, for those disposed of in landfills, the marketer does not possess adequate substantiation that the bags will degrade in a reasonably short period of time in a landfill. The claim is therefore deceptive.

Example 2:   A commercial agricultural plastic mulch film is advertised as "Photodegradable" and qualified with the phrase, "Will break down into small pieces if left uncovered in sunlight." The claim is supported by competent and reliable scientific evidence that the product will break down in a reasonably short period of time after being exposed to sunlight and into sufficiently small pieces to become part of the soil. The qualified claim is not deceptive. Because the claim is qualified to indicate the limited extent of breakdown, the advertiser need not meet the elements for an unqualified photodegradable claim, *i.e.*, that the product will not only break down, but also will decompose into elements found in nature.

Example 3:   A soap or shampoo product is advertised as "biodegradable," with no qualification or other disclosure. The manufacturer has competent and reliable scientific evidence demonstrating that the product, which is customarily disposed of in sewage systems, will break down and decompose into elements found in nature in a short period of time. The claim is not deceptive.

3. *Compostable:* It is deceptive to misrepresent, directly or by implication, that a product or package is compostable. An unqualified claim that a product or package is compostable

should be substantiated by competent and reliable scientific evidence that all the materials in the product or package will break down into, or otherwise become part of, usable compost (*e.g.*, soil-conditioning material, mulch) in a safe and timely manner in an appropriate composting program or facility, or in a home compost pile or device.

Claims of compostability should be qualified to the extent necessary to avoid consumer deception. An unqualified claim may be deceptive: (1) if municipal composting facilities are not available to a substantial majority of consumers or communities where the package is sold; (2) if the claim misleads consumers about the environmental benefit provided when the product is disposed of in a landfill; or (3) if consumers misunderstand the claim to mean that the package can be safely composted in their home compost pile or device, when in fact it cannot.

Example 1:   A manufacturer indicates that its unbleached coffee filter is compostable. The unqualified claim is not deceptive provided the manufacturer can substantiate that the filter can be converted safely to usable compost in a timely manner in a home compost pile or device, as well as in an appropriate composting program or facility.

Example 2:   A lawn and leaf bag is labeled as "Compostable in California Municipal Yard Waste Composting Facilities." The bag contains toxic ingredients that are released into the compost material as the bag breaks down. The claim is deceptive if the presence of these toxic ingredients prevents the compost from being usable.

Example 3:   A manufacturer indicates that its paper plate is suitable for home composing. If the manufacturer possesses substantiation for claiming that the paper plate can be converted safely to usable compost in a home compost pile or device, this claim is not deceptive even if no municipal composting facilities exist.

Example 4:   A manufacturer makes an unqualified claim that its package is compostable. Although municipal composting facilities exist where the product is sold, the package will not break down into usable compost in a home compost pile or device. To avoid deception, the manufacturer should disclose that the package is not suitable for home composting.

Example 5:   A nationally marketed lawn and leaf bag is labeled "compostable." Also printed on the bag is a disclosure that the bag is not designed for use in home compost piles. The bags are in fact composted in municipal yard waste composting programs in many communities around the country, but such programs are not available to a substantial majority of consumers where the bag is sold. The claim is deceptive since reasonable consumers living in areas not served by municipal yard waste programs may understand the reference to mean that composting facilities accepting the bags are available in their area. To avoid deception, the claim should be qualified to indicate the limited availability of such programs, for example, by stating, "Appropriate facilities may not exist in your area." Other examples of adequate qualification of the claim include providing the approximate percentage of communities or the population for which such programs are available.

Example 6:   A manufacturer sells a disposable diaper that bears the legend, "This diaper can be composted where municipal solid waste composting facilities exist. There are currently [X number of] municipal solid waste composting facilities across the country." The claim is not deceptive, assuming that composting facilities are available as claimed and the manufacturer can substantiate that the diaper can be converted safely to usable compost in municipal solid waste composting facilities.

Example 7:   A manufacturer markets yard waste bags only to consumers residing in particular geographic areas served by county yard waste composting programs. The bags meet specifications for these programs and are labeled, "Compostable Yard Waste Bag for County Composting Programs." The claim is not deceptive. Because the bags are compostable where they are sold, no qualification is required to indicate the limited availability of composting facilities.

4. *Recyclable:* It is deceptive to misrepresent, directly or by implication, that a product or package is recyclable. A product or package should not be marketed as recyclable

unless it can be collected, separated or otherwise recovered from the solid waste stream for use in the form of raw materials in the manufacture or assembly of a new package or product. Unqualified claims of recyclability for a product or package may be made if the entire product or package, excluding minor incidental components, is recyclable. For products or packages that are made by both recyclable and non-recyclable components, the recyclable claim should be adequately qualified to avoid consumer deception about which portions or components of the product or package are recyclable.

Claims of recyclability should be qualified to the extent necessary to avoid consumer deception about any limited availability of recycling programs and collection sites. If an incidental component significantly limits the ability to recycle the product, the claim would be deceptive. A product or package that is made from recyclable material, but, because of its shape, size or some other attribute, should not be marketed as recyclable.

Example 1: A packaged product is labeled with an unqualified claim, "recyclable." It is unclear from the type of product and other context whether the claim refers to the product or its package. The unqualified claim is likely to convey to reasonable consumers that all of both the product and its packaging that remain after normal use of the product, except for minor, incidental components, can be recycled. Unless each such message can be substantiated, the claim should be qualified to indicate what portions are recyclable.

Example 2: A plastic package is labeled on the bottom with the Society of the Plastics Industry (SPI) code, consisting of a design of arrows in a triangular shape containing a number and abbreviation identifying the component plastic resin. Without more, the use of the SPI symbol (or similar industry codes) on the bottom of the package, or in a similarly inconspicuous location, does not constitute a claim of recyclability.

Example 3: A container can be burned in incinerator facilities to produce heat and power. It cannot, however, be recycled into new products or packaging. Any claim that the container is recyclable would be deceptive.

Example 4: A nationally marketed bottle bears the unqualified statement that it is "recyclable." Collection sites for recycling the material in question are not available to a substantial majority of consumers or communities, although collection sites are established in a significant percentage of communities or available to a significant percentage of the population. The unqualified claim is deceptive since, unless evidence shows otherwise, reasonable consumers living in communities not served by programs may conclude that recycling programs for the material are available in their area. To avoid deception, the claim should be qualified to indicate the limited availability of programs, for example, by stating, "Check to see if recycling facilities exist in your area." Other examples of adequate qualifications of the claim include providing the approximate percentage of communities or the population to whom programs are available.

Example 5: A soda bottle is marketed nationally and labeled, "Recyclable where facilities exist." Recycling programs for material of this type and size are available in a significant percentage of communities or to a significant percentage of the population, but are not available to a substantial majority of consumers. The claim is deceptive since, unless evidence shows otherwise, reasonable consumers living in communities not served by programs may understand this phrase to mean that programs are available in their area. To avoid deception, the claim should be further qualified to indicate the limited availability of programs, for example, by using any of the approaches set forth in Example 4 above.

Example 6: A plastic detergent bottle is marketed as follows: "Recyclable in the few communities with facilities for colored HDPE bottles." Collection sites for recycling the container have been established in a half-dozen major metropolitan areas. This disclosure illustrates one approach to qualifying a claim adequately to prevent deception about the limited availability of recycling programs where collection facilities are not established in a significant percentage of the population. Other examples of adequate qualification of the claim include providing the number of communities with programs, or the percentage of communities or the population to which programs are available.

Example 7:   A label claims that the package "includes some recyclable material." The package is composed of four layers of different materials, bonded together. One of the layers is made from the recyclable material, but the others are not. While programs for recycling this type of material are available to a substantial majority of consumers, only a few of those programs have the capability to separate out the recyclable layer. Even though it is technologically possible to separate the layers, the claim is not adequately qualified to avoid consumer deception. An appropriately qualified claim would be, "includes material recyclable in the few communities that collect multi-layer products." Other examples of adequate qualification of the claim include providing the number of communities with programs, or the percentage of communities or the population to which programs are available.

Example 8:   A product is marketed as having a "recyclable" container. The product is distributed and advertised only in Missouri. Collection sites for recycling the container are available to a substantial majority of Missouri residents, but are not yet available nationally. Because programs are generally available where the product is marketed, the unqualified claim does not deceive consumers about the limited availability of recycling programs.

5. *Recycled Content:* A recycled content claim may be made only for materials that have been recovered or otherwise diverted from the solid waste stream, either during the manufacturing process (pre-consumer), or after consumer use (post-consumer). To the extent the source of recycled content includes pre-consumer material, the manufacturer or advertiser must have substantiation for concluding that the pre-consumer material would otherwise have entered the solid waste stream. In asserting a recycled content claim, distinctions may be made between pre-consumer and post-consumer materials. Where such distinctions are asserted, any express or implied claim about the specific pre-consumer or post-consumer content of a product or package must be substantiated.

It is deceptive to misrepresent, directly or by implication, that a product or package is made of recycled material. Unqualified claims of recycled content may be made only if the entire product or package, excluding minor, incidental components, is made from recycled material. For products or packages that are only partially made of recycled material, a recycled claim should be adequately qualified to avoid consumer deception about the amount, by weight, of recycled content in the finished product or package.

Example 1:   A manufacturer routinely collects spilled raw material and scraps from trimming finished products. After a minimal amount of reprocessing, the manufacturer combines the spills and scraps with virgin material for use in further production of the same product. A claim that the product contains recycled material is deceptive since the spills and scraps to which the claim refers are normally reused by industry within the original manufacturing process, and would not normally have entered the waste stream.

Example 2:   A manufacturer purchases material from a firm that collects discarded material from other manufacturers and resells it. All of the material was diverted from the solid waste stream and is not normally reused by industry within the original manufacturing process. The manufacturer includes the weight of this material in its calculations of the recycled content of its products. A claim of recycled content based on this calculation is not deceptive because, absent the purchase and reuse of this material, it would have entered the waste stream.

Example 3:   A greeting card is composed 30% by weight of paper collected from consumers after use of a paper product, and 20% by weight of paper that was generated after completion of the paper-making process, diverted from the solid waste stream, and otherwise would not normally have been reused in the original manufacturing process. The marketer of the card may claim either that the product "contains 50% recycled material," or may identify the specific pre-consumer and/or post-consumer content by stating, for example, that the product "contains 50% total recycled material, 30% of which is post-consumer material."

Example 4:   A package with 20% recycled content by weight is labeled as containing "20% recycled paper." Some of the recycled content was composed of material collected from consumers after use of the original product. The rest was composed of overrun newspaper stock never sold to customers. The claim is not deceptive.

Example 5: A product in a multi-component package, such as a paperboard box in a shrink-wrapped plastic cover, indicates that it has recycled packaging. The paperboard box is made entirely of recycled material, but the plastic cover is not. The claim is deceptive since, without qualification, it suggests that both components are recycled. A claim limited to the paperboard box would not be deceptive.

Example 6: A package is made from layers of foil, plastic, and paper laminated together, although the layers are indistinguishable to consumers. The label claims that "one of the three layers of this package is made of recycled plastic." The plastic layer is made entirely of recycled plastic. The claim is not deceptive provided the recycled plastic layer constitutes a significant component of the entire package.

Example 7: A paper product is labeled as containing "100% recycled fiber." The claim is not deceptive if the advertiser can substantiate the conclusion that 100% by weight of the fiber in the finished product is recycled.

Example 8: A frozen dinner is marketed in a package composed of a cardboard box over a plastic tray. The package bears the legend, "package made from 30% recycled material." Each packaging component amounts to one-half the weight of the total package. The box is 20% recycled content by weight, while the plastic tray is 40% recycled content by weight. The claim is not deceptive, since the average amount of recycled material is 30%.

Example 9: A paper greeting card is labeled as containing 50% by weight recycled content. The seller purchases paper stock from several sources and the amount of recycled material in the stock provided by each source varies. Because the 50% figure is based on the annual weighted average of recycled material purchased from the sources after accounting for fiber loss during the production process, the claim is permissible.

6. *Source Reduction:* It is deceptive to misrepresent, directly or by implication, that a product or package has been reduced or is lower in weight, volume or toxicity. Source reduction claims should be qualified to the extent necessary to avoid consumer deception about the amount of the source reduction and about the basis for any comparison asserted.

Example 1: An ad claims that solid waste created by disposal of the advertiser's packaging is "now 10% less than our previous package." The claim is not deceptive if the advertiser has substantiation that shows that disposal of the current package contributes 10% less waste by weight or volume to the solid waste stream when compared with the immediately preceding version of the package.

Example 2: An advertiser notes that disposal of its product generates "10% less waste." The claim is ambiguous. Depending on contextual factors, it could be a comparison either to the immediately preceding product or to a competitor's product. The "10% less waste" reference is deceptive unless the seller clarifies which comparison is intended and substantiates that comparison, or substantiates both possible interpretations of the claim.

7. *Refillable:* It is deceptive to misrepresent, directly or by implication, that a package is refillable. An unqualified refillable claim should not be asserted unless a system is provided for: (1) the collection and return of the package for refill; or (2) the later refill of the package by consumers with product subsequently sold in another package. A package should not be marketed with an unqualified refillable claim, if it is up to the consumer to find new ways to refill the package.

Example 1: A container is labeled "refillable x times." The manufacturer has the capability to refill returned containers and can show that the container will withstand being refilled at least x times. The manufacturer, however, has established no collection program. The unqualified claim is deceptive because there is no means for collection and return of the container to the manufacturer for refill.

Example 2:   A bottle of fabric softener states that it is in a "handy refillable container." The manufacturer also sells a large-sized container that indicates that the consumer is expected to use it to refill the smaller container. The manufacturer sells the large-sized container in the same market areas where it sells the small container. The claim is not deceptive because there is a means for consumers to refill the smaller container from larger containers of the same product.

8. *Ozone Safe and Ozone Friendly:* It is deceptive to misrepresent, directly or by implication, that a product is safe for or "friendly" to the ozone layer. A claim that a product does not harm the ozone layer is deceptive if the product contains an ozone-depleting substance.

Example 1:   A product is labeled "ozone friendly." The claim is deceptive if the product contains any ozone-depleting substance, including those substances listed as Class I or Class II chemicals in Title VI of the Clean Air Act Amendments of 1990, Pub. L. No. 101–549, or others subsequently designated by EPA as ozone-depleting substances. Class I chemicals currently listed in Title VI are chlorofluorocarbons (CFCs), halons, carbon tetrachloride and 1.1.1.-trichloroethane. Class II chemicals currently listed in Title VI are hydrochlorofluorocarbons (HCFCs).

Example 2:   The seller of an aerosol product makes an unqualified claim that its product "Contains no CFCs." Although the product does not contain CFCs, it does contain HCFC-22, another ozone depleting ingredient. Because the claim "Contains no CFCs" may imply to reasonable consumers that the product does not harm the ozone layer, the claim is deceptive.

Example 3:   A product is labeled "This product is 95% less damaging to the ozone layer than past formulations that contained CFCs." The manufacturer has substituted HCFCs for CFC-12, and can substantiate that this substitution will result in 95% less ozone depletion. The qualified comparative claim is not likely to be deceptive.

SUMMARY OF FTC LAW ENFORCEMENT ACTIONS IN THE ENVIRONMENTAL MARKETING AREA

The Federal Trade Commission has brought charges against nine companies in the last few years in connection with allegedly misleading or deceptive claims about the environmental attributes or benefits of their products. In all instances, the companies have agreed to settle the charges under consent agreements designed to prevent similar misleading claims in the future.[1] In addition to the cases listed below, the Commission has several other investigations underway.

Cases Involving Degradability Claims:

1. RMED International, Inc., the maker of "TenderCare" disposable diapers, agreed to settle FTC charges that the firm made unsubstantiated claims the diapers are biodegradable and offer a significant environmental benefit compared to other disposable diapers when disposed of in a landfill. The Commission issued the consent order in final form on May 14, 1992.

2. First Brands Corporation settled FTC charges that it failed to have an adequate basis for claims that its "Glad" trash bags, when disposed of as trash, will completely break down, decompose and return to nature in a reasonably short period of time, and that they offer a significant environmental benefit compared to ordinary plastic bags. The Commission issued the consent order in final form on Jan. 3, 1992.

3. American Enviro products, Inc., settled FTC charges that it made unsubstantiated claims that its "Bunnies" disposable diapers, when disposed of in a landfill, will decompose and return to nature within 3–5 years or "before your child grows up," and that they offer a significant environmental benefit compared to other diapers. The Commission issued the consent order in final form on March 26, 1992.

---

1. A consent agreement is for settlement purposes only and does not constitute an admission of a law violation. When the Commission issues a consent order on a final basis, it carries the force of law with respect to future actions. Each violation of such an order may result in a civil penalty of $10,000.

DISSENTING STATEMENT OF COMMISSIONER MARY L. AZCUENAGA
CONCERNING ISSUANCE OF COMMISSION GUIDES
ON
ENVIRONMENTAL MARKETING CLAIMS

Today the Commission issues guides on environmental marketing claims. The guides should prove useful to the business and law enforcement communities and to consumers, that is, to all those who make, analyze or rely on environmental claims in the advertising and marketing of goods and services. In an area that seems always to prove more difficult than initial impressions suggest, the Commission should be commended for producing a clear, careful and balanced document.

It has been my pleasure to work with my colleagues and Commission staff in this important and difficult endeavor and with the government agencies and other concerned groups and individuals who have participated so generously and constructively in this process. With regret, I nevertheless find I must dissent.

Basic to the exercise of the responsibility of my office is the obligation to act within the authority conferred on that office and, as I understand that obligation, it is not satisfied by forecasting that a challenge is unlikely or by deferring to the courts to decide on review whether the exercise lies within the bounds of the authority, but rather is my obligation to decide in the first instance and without regard to the prevailing political climate in which that decision will be received. As I read the law, the Commission has no authority to issue these guides, as written, without first employing the rulemaking procedures of Section 18(b)(1) of the FTC Act, which it has not done.

Section 18(a)(1) of the FTC Act, 15 U.S.C. § 57a(a)(1), provides that the Commission may prescribe:

(A) interpretive rules and general statements of policy with respect to unfair or deceptive acts or practices . . ., and

(B) rules which define with specificity acts or practices which are unfair or deceptive acts or practices. . . .

Section 18(b)(1) directs that ''[w]hen prescribing a rule under subsection (a)(1)(B),'' the Commission is to proceed in accordance with the notice and comment requirements of section 553 of the Administrative Procedure Act and shall also follow the more extensive procedures set forth in Section 18 that often are referred to as ''Magnuson-Moss rulemaking.''

As the guides expressly state, the majority of the Commission does not view its guides as having the force and effect of law but as explanations of existing statutory terms and obligations. Under the Administrative Procedure Act, 5 U.S.C. § 553, and under Section 18 of the FTC Act, therefore, the Commission apparently would categorize its guides as ''interpretive'' (or ''interpretative'') rules or policy statements rather than ''legislative'' rules or ''rules which define with specificity . . . deceptive acts or practices.'' I cannot agree.

By stating definitively, for example, that a particular act ''is deceptive'' or that particular conduct ''would be deceptive,'' or that under specified circumstances, firms, ''must'' or ''should'' act in a particular way, language that appears throughout the document,[1] I believe that the document has ''defined with specificity'' a deceptive act or practice as set forth in Section 18(a)(1)(B). Since the enactment of the Magnuson-Moss Act in 1975, the Commission has been empowered to take such an action only if it first adheres to Magnuson-Moss rulemaking procedures.

If the Commission in issuing its guides were relying on a body of past precedent, I might be persuaded that my colleagues were correct in their assessment, and that the decisive ''guidance'' in the document simply explicates existing Commission case law and policy. In issuing its Deception Statement in 1983, for example, the Commission reviewed decided cases to synthesize principles, but that is not the case here. The Commission's case law on environmental claims consists almost entirely of consent agreements and orders issued without adjudicative records or admissions of liability. These agreements and orders may convey to the public some sense of what the Commission is likely to do in other similar situations, but they are not binding precedent.

Were I entirely alone in my concern over the need to distinguish between interpretive and legislative rules in issuing some form of guidance on environmental claims, I might be inclined to accede to the position of the majority. Again, this is not the case. Although the courts, particularly in the District of Columbia Circuit, have not instructed agencies unambiguously on how they should distinguish interpretive and legislative rules, recent decisions suggest that my concern is not without validity. At the least, they reflect judicial concern that agencies attend to this question with care in reaching their regulatory decisions and judicial unwillingness

---

1. Guides and trade practice rules issued before the enactment of § 18 and before the judicial decisions discussed below contain similarly didactic language.

blindly to acquiesce in agencies' characterizations of their actions. In short, saying that these are guides and not rules does not make it so.

Even in the presence of express language disavowing agency intent to bind either itself or the public, courts in this circuit have considered whether allegedly interpretive rules are sufficiently mandatory and definitive to render them legislative in nature. *See Community Nutrition Institute* v. *Young*, 818 F.2d 943, 946 (D.C. Cir. 1987) (noting that it is appropriate to ''give some, 'albeit not overwhelming' deference to an agency's characterization of its statement'' and refusing to sustain FDA rules because the agency failed to follow the appropriate rulemaking process); *Arrow Air, Inc.* v. *Dole*, 784 F.2d 1118, 1122 (D.C. Cir. 1986) (listing agency intent as only one among other factors differentiating interpretive and legislative rules); *General Motors Corp.* v. *Ruckelshaus*, 742 F.2d 1561, 1565 (D.C. Cir. 1984) (*en banc*), *cert. denied*, 471 U.S. 1074 (1985) (upholding agency's interpretation but finding agency's own label relevant but not dispositive).[2]

The likelihood, in whatever degree, that what the Commission calls guides are in fact rules under Section 18(a)(1)(B) could easily have been avoided without diminishing the basic guidance the Commission seeks to offer. The Horizontal Merger Guidelines recently issued by the Commission and the Department of Justice, for example, refrain from definitive conclusions about what does or does not violate the law in various ways, one of which is by using the qualifier ''likely.'' For example, in discussing the significance of post-merger market concentration measured by the Herfindahl-Hirschman Index (''HHI''), the Merger Guidelines say, ''Where the post-merger HHI exceeds 1800, it will be presumed that mergers producing an increase in the HHI of more than 100 points *are likely* to create or enhance market power or facilitate its exercise.'' 1992 CCH Trade Cas. ¶ 13, 104 at 20,573-6 (emphasis added).[3]

A similar approach could be used here. Instead of saying that a particular claim ''is'' or ''is not'' deceptive, the environmental guides could have said that a particular claim ''is likely'' or ''is unlikely'' to be deceptive. Although adding the qualifiers ''likely'' or ''unlikely'' sounds more tentative, if that language were used throughout the document, the basic message of the guides, which is to indicate the Commission's likely response in various hypothetical situations, would remain. If the Commission prefers the more definitive language because indeed it wants to be definitive about what is or is not deceptive, then it seems to me that the Commission runs squarely into the problem that it is in fact issuing rules rather than guides. I confess some puzzlement about whether the Commission intends to be definitive (and issue rules) or to indicate what it is likely to do (and issue guides), but, even more than that, I regret that the Commission has not seen fit to make this single change, which would have enabled me to join in making this a unanimous document.

Second, I differ from the Commission in its decision not to place the guides on the public record for a short period of time to enable the public to comment on them. Although we have sought to obtain accurate information and to consider the issues thoroughly, it is conceivable, nevertheless, that someone outside the agency might offer useful observations and suggestions for improvement. The Commission has obtained comment on the merits of issuing guidance and on the issues that such guides should address, but it has not provided to those affected by the guides an opportunity to assess the economic benefits and costs of the actual provisions or to call to our attention provisions that may cause unintended effects. A short, appropriately focused comment period on the guides could have coincided with the public comment period on the Environmental Assessment that is required under the National Environmental Policy Act of 1969, 42 U.S.C. § 4321, as amended.

---

2. Although, as already noted, the law of the circuit is not settled, there is a serious possibility, and in my opinion likelihood, that the Court of Appeals for the District of Columbia Circuit, at least, would find that portions, if not all, of the guides just issued are legislative rules rather than interpretive rules or policy statements. *Compare The Fertilizer Institute* v. *EPA*, 935 F.2d 1303, 1307–08 (D.C. Cir. 1991), *quoting General Motors Corp.* v. *Ruckelshaus, supra, and Citizens To Save Spencer County* v. *EPA*, 600 F.2d 844, 876 and n. 153 (D.C. Cir. 1979) (distinction between interpretive and legislative rules depends on whether document ''simply states what the administrative agency thinks the statute means, and only 'reminds' affected parties of existing duties'' or demonstrates that ''the agency intends to create new law, rights or duties''), *with Alaska* v. *DOT*, 868 F.2d 441, 446–47 (D.C. Cir. 1989), *and Community Nutrition Institute* v. *Young, supra* at 947–49, (distinction depends on several factors including use of mandatory language, inclusion of exception process, practical application and limitations placed on agency discretion).

3. Magnuson-Moss rulemaking procedures do not apply to antitrust rules, but the notice and comment rulemaking requirements in the Administrative Procedure Act (''APA'') apply and presumably would have precluded the Commission and the Department from issuing the merger guidelines had they purported to bind the government or the public by requiring or prescribing particular conduct without first providing for public notice and comment. When it recently issued revisions to the so-called Fred Meyer Guides (Guides for Advertising Allowances and Other Merchandising Payments and Services, 55 Fed. Reg. 33,651 (Aug. 17, 1990)), under the antitrust laws, the Commission employed the appropriate APA rulemaking procedures.

The Green Report II: Recommendations for Responsible Environmental Advertising

1.9 *Source Reduction Claims.* [New]

Source reduction claims should be specific, and where possible include percentages. Comparisons should be clear and complete.

2. *Claims Should Reflect Current Solid Waste Management Options.* [Changed]

Environmental claims relating to the disposability or potential for recovery of a particular product (e.g., "compostable" or "recyclable") should be made in a manner that clearly discloses the general availability of the advertised option where the product is sold.

2.1 *Use of Terms "Degradable," "Biodegradable," and "Photodegradable."* [Slightly modified]

Products that are currently disposed of primarily in landfills or through incineration—whether paper or plastic—should not be promoted as "degradable," "biodegradable," or "photodegradable."

2.2 *Use of Term "Compostable."* [Changed]

Unqualified compostability claims should not be made for products sold nationally unless a significant amount of the product is currently being composted everywhere the product is sold. In all other cases, compostability claims should be accompanied by a clear disclosure about the limited availability of this disposal option. If a claim of degradability is made in the context of a product's compostability a disclosure should be made that the product is not designed to degrade in a landfill.

2.3 *Use of Term "Recyclable."* [Changed]

Unqualified recyclability claims should not be made for products sold nationally unless a significant amount of the product is being recycled everywhere the product is sold. Where a product is being recycled in many areas of the country, a qualified recyclability claim can be made. If consumers have little or no opportunity to recycle a product, recyclability claims should not be made.

2.4 *Safe for Disposal.* [Changed—Replaces "Safe for Incineration"]

Vague safety claims concerning disposability should be avoided. Instead, products should specifically disclose those environmentally dangerous materials or additives that have been eliminated.

3. *Claims Should Be Substantive* [Unchanged]

Environmental claims should be substantive.

3.1 *Trivial and Irrelevant Claims.* [Unchanged]

Trivial and irrelevant claims should be avoided.

3.2 *Single Use Products.* [Unchanged]

Single use disposable products promoted on the basis of environmental attributes should be promoted carefully to avoid the implication that they do not impose a burden on the environment.

4. *Claims Should Be Supported.* [Unchanged]

Environmental claims should be supported by competent and reliable scientific evidence.

EXECUTIVE SUMMARY

There is no more pressing issue facing our society than the need to change the way we live, work and do business in order to protect our natural environment. This is a monumental task and will take the combined efforts of government, business and an actively engaged public. Our free market economic system is a powerful tool that can be used towards achieving that end. As consumers demand products that are safer for or have a less harmful impact on the environment, industry will strive to meet that demand. But consumer clout can become a major motivating force for improving environmental policies only if the public receives accurate, specific and complete information about the environmental effects of the goods and services they buy. Unfortunately, attempts to take advantage of consumers' increasing interest in the environment have led some companies to make environmental advertising claims that are trivial, confusing and misleading.[1]

In November, 1989, the Attorneys General of California, Massachusetts, Minnesota, Missouri, New York, Texas, Utah, Washington and Wisconsin (who have since been joined by the Attorneys General of Florida and Tennessee) formed an *ad hoc* task force to study this advertising trend. The Task Force held a public forum on environmental marketing in March, 1990, and in November, 1990, issued "The Green Report," which was a comprehensive overview of the problem.

At its Spring, 1990, meeting the National Association of Attorneys General adopted a resolutions that called on the federal government to work with the states to develop uniform national standards for environmental advertising. In "The Green Report," the Task Force reiterated the need for such action and called specifically for federal definitions of environmental marketing claims, federal testing protocols for terms that have a technical basis such as "degradable," and strong federal involvement in the process of developing methods for conducting lifecycle assessments for product evaluation. Of course, by recommending a national regulatory scheme for environmental claims, the Task Force was not recommending that states be preempted from regulating in this area. In addition to proposing federal action, the Report made a series of preliminary recommendations for responsible environmental advertising to guide industry until national standards are put into place.

In December, 1990, after issuing "The Green Report," the Task Force held follow-up hearings offering industry, environmental groups, and consumers an opportunity to respond (either orally or in writing) to the Task Force's preliminary findings and recommendations. The overall response to "The Green Report" was very positive. It was recognized as an excellent first step toward the goal of developing a uniform national program to ensure truthful, nondeceptive environmental advertising. However, some of the Task Force's specific recommendations were criticized as untenable, unfair, or ill-advised.

The Task Force has carefully reviewed all of the testimony and written submissions concerning "The Green Report," rethought some of its initial recommendations, and where appropriate has revised some of its preliminary recommendations. These final recommendations of the Task Force are intended to serve as interim guidance to the business community and to provide a framework upon which more concrete definitions and standards can be developed. In sum, the Task Force recommends that:

- Environmental claims should be as specific as possible, not general, vague, incomplete or overly broad.

- Environmental claims relating to the disposability or potential for recovery of a particular product (e.g., "compostable" or "recyclable") should be made in a manner that clearly discloses the general availability of the advertised option where the product is sold.

- Environmental claims should be substantive.

- Environmental claims should, of course, be supported by competent and reliable scientific evidence.

The most significant changes from the preliminary Task Force recommendations can be found in Section B 2, the section dealing with the advertising of solid waste management options. In the preliminary report the Task Force recommended that claims, such as recyclability and compostability, be location specific. On rethinking this issue, the Task Force has determined that, if a product can be recycled or composted in many, but not all, communities where the product is sold, a properly qualified claim can be made for that product nationally. (See Sections B 2.2 and 2.3.) With respect to claims regarding recycled content, the Task Force now recommends that only post-consumer material be referred to as "recycled" and that recaptured pre-consumer or industrial materials that have been diverted from the waste stream be referred to by some other term such as

---

1. The term "advertising" as used throughout, includes advertisements in any media—radio, television or print, or any promotional materials such as coupons or flyers distributed to the public to induce the purchase of a product, and, of course, the product label or packaging itself. "Environmental claims" means "environmental advertising claims."

"recovered [or reprocessed] industrial material." The Task Force continues to recommend that separate percentages be disclosed for each and that factory waste that is routinely fed back into the manufacturing process not be included at all. (See Section B 1.5.) Because of the increasing interest in environmental certification and seal of approval programs, the Task Force has added two new recommendations regarding the use of environmental certifications in advertising and the use of lifecycle assessment claims. (See Sections B 1.7 and 1.8.) Finally, the Task Force has clarified its recommendation concerning pre-existing, but previously unadvertised, environmental attributes (See Sections B 1.7 and 1.8) and added a recommendation on source reduction claims. (See Section B.1.9.)

The Task Force firmly believes that there is a need for the development of a national regulatory scheme for environmental advertising. (See Section A.) Members of the Task Force are actively pursuing every opportunity to work with the Federal Trade Commission, the Environmental Protection Agency, the U.S. Office of Consumer Affairs, and Congress to realize that goal. But until concrete and specific federal definitions and standards are in place, the Task Force offers its own recommendations to the business community as immediate guidance on how environmental claims can be made in a manner that is most likely to be consistent with state laws.

It is important to note that these recommendations are not laws, do not have the force and effect of law, and are not an attempt at rulemaking. They, in fact, do nothing to change pre-existing state laws prohibiting false, misleading, or deceptive advertising. These recommendations are intended simply to provide some guidance to industry so that companies can avoid making environmental advertising claims that violate the current deceptive advertising laws of various states. The Task Force believes that, if companies conform their advertisements to these recommendations, consumer confusion over environmental advertising claims will be substantially reduced. We want to emphasize that the recommendations made by the Task Force are not intended to be mandatory or prescriptive. Any company that believes it can make a claim that meets the goals of these recommendations should not refrain from doing so simply because that specific claim has not been addressed.

TASK FORCE RECOMMENDATIONS FOR
RESPONSIBLE ENVIRONMENTAL ADVERTISING

A. RECOMMENDATIONS FOR FEDERAL ACTION

In "The Green Report," the Task Force called upon the federal government to adopt national standards for environmental marketing claims used in the labeling, packaging and promotion of consumer products. The Task Force now reaffirms its initial recommendation. Indeed, as more and more manufacturers turn to environmental claims to market their products, the need for federal standards to control and regulate these claims is more important than ever.

Specifically, uniform definitions are needed for terms such as "degradable," "compostable," "recycled" and "recyclable" to ensure that marketers know what properties their products must possess before they can make these claims and to ensure that consumers get the accurate information they need to make informed purchasing decisions based on environmental considerations. The federal regulatory program also must include testing protocols and standards for industry to follow in determining whether a product satisfies a particular environmental definition. This regulatory scheme should be developed with input from, and after consultation with, the states, industry, environmental and consumer organizations and all other interested members of the public.

Finally, the "product life" or "lifecycle" issue must be addressed by the federal government. There is a general consensus that evaluating a product's total environmental impact—from the harvesting of raw materials to final disposal of waste—will prove valuable in conserving resources and reducing the volume of waste generated. At this time, however, no reliable and meaningful methodology exists for conducting complex product life assessments. The federal government should therefore work on developing acceptable methodologies for conducting product life assessments, and should restrict product life claims until federal standards are in place.

It should be noted that industry groups have also come forward recently and called for the federal government to develop uniform environmental marketing standards. The National Food Processors Association led a coalition of 11 industry trade associations in filing a petition with the FTC requesting that the Commission adopt ''Industry Guides'' governing the use of environmental claims in labeling, advertising and other marketing materials. In addition, separate petitions calling for environmental guidelines have been filed with the FTC by several individual companies and trade associations.

It is essential that any federal programs developed to govern environmental marketing claims be enforceable not only by federal regulators, but also by the states. Enforcement of consumer protection and false advertising laws is an essential function of the states' general police powers. The federal regulatory scheme therefore should supplement, rather than supplant, existing state law governing false advertising and deceptive practices. The states must continue to share with the federal government the authority and responsibility for taking action against companies that violate standards developed to govern environmental marketing claims. The states would, accordingly, vigorously oppose any statute or regulation that proposes preemption of states' rights in this area.

Since ''The Green Report'' was issued in November, 1990, there have been several significant developments on the federal level. The Federal Trade Commission, the Environmental Protection Agency, and the U.S. Office of Consumer Affairs have informed a joint agency working group to study environmental marketing claims. In addition, the FTC has announced plans to hold public hearings on environmental marketing and is expected to publish notice of these hearings in the Federal Register. And, with respect to the lifecycle issue, the EPA is studying a methodology for developing this type of complex product assessment.

In addition to these agency efforts, Congress is also moving forward in the area of environmental marketing claims. Senator Frank Lautenberg of New Jersey and Representative Gerry Sikorski of Minnesota have introduced the Environmental Marketing Claims Act of 1991. This legislation proposes that the EPA establish regulations to define environmental marketing terms. Significantly, this bill, as proposed, provides for the states as well as EPA to enforce the marketing standards developed under this legislation.

The Task Force commends the efforts undertaken to date by the FTC, EPA and OCA on this issue, as well as the work of Senator Lautenberg and Representative Sikorski, in moving forward with federal legislation. Although the Task Force recognizes that there may be some differences of opinion on certain issues among the federal agencies working on environmental marketing issues, it is vital that these agencies continue to work together toward development of national environmental marketing standards. At the same time, the Task Force will continue to support federal legislation that mandates development of national environmental marketing standards, as long as such legislation preserves the right of the states to enforce these standards and to take action under their own consumer protection laws against companies which engage in false, deceptive and misleading environmental marketing campaigns.

B. RECOMMENDATIONS TO INDUSTRY[2]

1. Claims Should Be Specific. [Unchanged]

Environmental claims should be as specific as possible, and not general, vague, incomplete or overly broad.

Commentary: Not only are terms such as ''environmentally friendly'' or ''safe for the environment'' too vague to be meaningful but, because of the inherent complexity of environmental issues, simplified statements have a tendency to be inaccurate. Moreover, vague and incomplete claims do not permit consumers to make meaningful comparisons between products. Providing more specific information allows consumers to evaluate environmental attributes for themselves

---

2. This report contains the full text of all of the final Task Force recommendations to industry for responsible environmental advertising. Some recommendations are unchanged from our initial Green Report, some have been notified, and some are new. In brackets, next to each recommendation is a notation that indicates whether the recommendation is ''Unchanged,'' ''Slightly Modified,'' ''Changed,'' or ''New'' in comparison to the preliminary recommendation.

and makes an important contribution to consumer education. Specific claims also prevent the misunderstanding that is probable when a more generalized term or phrase is used because such term or phrase may be subject to more than one reasonable interpretation. Finally, specific information minimizes the risk that consumers will attach a broader significance to the product's actual environmental attributes than is warranted.

1.1 Use of Terms "Environmentally Friendly" and "Safe for the Environment." [Changed]
Generalized environmental claims which imply that a product has no negative or adverse impact on the environment should be avoided. Instead, claims should be specific and state the precise environmental benefit that the product provides.

Commentary: In the absence of standards for comparing the environmental impacts of products throughout a product's lifecycle, it is very difficult, if not impossible to substantiate claims of generalized environmental benefit. Moreover, generalized environmental benefit claims may create an unwarranted impression that a product is good for the environment in all respects. As stated in our preliminary report, the production and use of products necessarily have adverse environmental consequences. For these reasons, such claims should be avoided altogether. Instead, companies should make truthful, narrowly-drawn claims that specify the precise environmental attribute of a product. Such claims are much more useful to consumers and avoid the potential for deception.

1.2 Preexisting Environmental Attributes. [Changed]
The promotion of a previously-preexisting but previously-unadvertised positive environmental attribute should not create, either explicitly or implicitly, the perception that the product has been recently modified or improved.

Commentary: There was objection to the preliminary recommendation of the Task Force that when promoting a previously-existing but previously unadvertised positive environmental attribute an advertiser make clear that the product had not been modified or improved. Companies commented that an affirmative disclosure that the product had not been modified would place a company that had long made a product with a positive environmental attribute at a competitive disadvantage to a company that only now reformulates its product to achieve the same attribute.

This was not the intended result of the Task Force. Therefore, the Task Force has modified the recommendation to make our intention clear. The final recommendation is that promotion of a previously-existing but previously-unadvertised positive environmental attribute of a product should not create, either explicitly or implicitly, the impression that the product has been recently modified or improved. Companies should not misinterpret this modification as encouragement to promote previously-unadvertised attributes in an irresponsible or deceptive manner. Companies must ensure that such claims do not mislead, even implicitly; of course, claims must also be literally true.

For example, some companies have used recycled paper in their packaging materials for years. Consumers are now sensitive to the environmental benefits of recycling and base their purchasing decisions, in part, on whether product packaging is made from recycled materials. The company that has been doing the responsible thing for years may promote that fact, provided that such promotion does not mislead. Clearly, it would be deceptive to promote a product that has been packaged in recycled paper for 10 years by saying "Now! Recycled package!" It would not be deceptive, on the other hand, to say, "We have used 100% recycled paper for years," as long as that claim is true and does not otherwise deceive.

1.3 Removal of Harmful Ingredient. [Slightly Modified]
In promoting the removal of a single harmful ingredient or a few harmful ingredients from a product or package, care should be taken to avoid the impression that the product is good for the environment in all respects.

Commentary: This is a problem that came to light when the Task Force began examining claims being made for aerosol spray products. Some aerosol spray products made without CFCs are advertised as "safe for the environment" or "ozone friendly," but they may contain other ingredients that contribute to destruction of the stratospheric ozone layer, such as 1, 1, 1, -trichloroethane. Certain of these products identify the chemical additives on the label; others do not. Promoting a product which contains ozone depleting ingredients as "ozone friendly" is clearly misleading. The Task Force is also concerned that stating that such a product "contains no CFCs" may also mislead because the phrase "no CFCs" may mean "safe for the ozone" to many consumers.

Labeling an aerosol spray product that does not contain any ozone depleting chemicals as "safe for the environment" may also be misleading because many of these products contain volatile organic compounds that are linked to the creation of ground level ozone, a component of smog. A more appropriate, less confusing claim for such a product would be one which states "contains no ozone-depleting ingredients" or "does not contribute to ozone depletion." In addition, since concerned consumers are becoming more sophisticated in their knowledge about chemicals and their impact on the environment, companies should list the ingredients used in their products so that consumers can avoid other potentially harmful propellants or ingredients if they choose.

1.4 Benefits of Products versus Packaging. [Unchanged]
A clear distinction should be made between the environmental attributes of a *product* and the environmental attributes of its *packaging*.

Commentary: The testimony of a county recycling official at the Public Forum in Minnesota illustrates the problems that can arise when such distinction is not clearly made. A manufacturer of disposable diapers placed a sticker on the plastic wrapper containing its diapers which states "RECYCLABLE" in large capital letters. Below the word "recyclable" in smaller print, were the words, "This softpac is recyclable where plastic bag recycling facilities exist." The county recycling official reported that shortly after the sticker appeared on the wrapper, a consumer dropped off a pile of the plastic wrappers *and* a garbage bag full of dirty diapers. Evidently, the consumer thought that the diapers, as well as the plastic wrapper, were recyclable. In fact, the recycling facility accepts only milk and soft drink bottles—not plastic bags and certainly not used disposable diapers.

1.5 Use of Term "Recycled." [Changed]
Recycled content claims should be specific and separate percentages should be disclosed for post-consumer and pre-consumer materials. To avoid the potential for deception, the Task Force recommends that only post-consumer materials be referred to as "recycled" material. Recaptured factory material should be referred to by some other term, such as "reprocessed [or recovered] industrial material."

Commentary: There is clearly a need to set national standards and definitions for the term "recycled" because the term is being used to refer to several different types of material. Consumers and solid waste managers should be able to discern, from a label, the sources for the recycled content of a product. Until national definitions are in place, the Task Force believes that full disclosure of both the source and the percentage of recovered materials is critical to avoid misleading consumers.

Realistically, when consumers think about recycling, they are thinking only about post-consumer waste—the trash they leave at the curb. The Task Force is of the opinion that consumers commonly believe that products labeled "recycled" contain material that consumers have recycled, i.e., household waste, that has been separated out by the consumer for separate collection by a recycler and reused in creating new products. Because solid waste managers are often unable to locate markets for materials that consumers discard, state policymakers have sought to stimulate these markets by requiring that specific amounts of post-consumer material be incorporated into products before they can be labelled as ''recycled.''[3] Consumers can only support markets for such recyclable post-consumer material, thus improving the chances that more of their waste will actually be incorporated into new products, if they can determine which products are made from post-consumer materials.

This is not to say that other forms of internal industrial recycling of industry-generated waste are unimportant for the environment. They are important. National figures indicate that the amount of industrial waste far surpasses the amount of waste generated by households. However, industry does not need to rely on advertising to stimulate the routine recycling of factory scraps back into the manufacturing process. Industry already has a strong financial incentive to recycle this type of material. When industry recycles its own by-products, it makes more internally-efficient use of raw materials and, presumably, becomes more cost efficient because it has conserved both natural materials and reduced its own internal disposal costs. Industry may, however, need an incentive to recycle factory wastes that are generally landfilled or incinerated. The Task Force believes that a distinction must be made between factory waste that is routinely fed back into the industrial process and factory waste that is routinely discarded. Only those industrial by-products that are actually diverted from the waste stream should be promoted as ''recovered content'' when used to make new products or packaging.

Because both consumers and policymakers have an interest in differentiating between the source of materials that can be included in recycled content, a distinction should be made between pre- and post-consumer materials in advertising, and a separate percentage should be listed for each. Further, because consumers generally understand ''recycled content'' to mean only post-consumer materials, to avoid deception a different word or phrase, such as ''reprocessed [or recovered] industrial material,'' is recommended to describe factory waste that has been diverted from the waste stream. For example, if a company elects to advertise only the post-consumer material content of a product, it could advertise ''made from 50% recycled fibers.'' If a company wants to advertise both pre- and post-consumer content it might say ''Our package is made from 50% recycled paper and 50% recovered industrial material.''

The Task Force recognizes that a number of companies are seeking to have new technologies or technological modifications, such as the use of sawdust in making paper, designated as ''recycling'' technologies because of their desire to advertise their product as made from recycled materials. While innovative and environmentally-sound technologies such as these should be acknowledged, the use of the term ''recycled'' to describe the resulting products would only add confusion where confusion already abounds. Advertising such a product as ''made from sawdust'' or ''made from reprocessed industrial material'' more accurately describes the technological process involved.

1.6 Comparative Claims. [Unchanged]
Only complete and full comparisons should be made; the basis for the comparison should be stated.

Commentary: Any specific claim that includes a comparative statement such as ''better for the environment'' should only be used if a complete and full comparison is made and the basis for the comparison is stated. Such a comparison might be: ''This product is better than [our former product] [our competitor's product] because. . . .''

1.7 Product Life Assessments. [New]
The results of product life assessments should not be used to advertise or promote specific products until uniform methods for conducting such assessments are developed and a general consensus is reached among government, business, environmental and consumer groups on how this type of environmental comparison can be advertised nondeceptively.

---

3. California, New York, and New Hampshire all require that a product include some percentage of post-consumer waste before it can be labeled ''recycled.'' Rhode Island requires separate disclosure of pre-consumer and post-consumer waste content for materials labelled ''recycled.''

Commentary: Although product life assessments or cradle-to-grave product analyses are expected to be extremely useful for evaluating the overall environmental effects of various manufacturing processes and products, the methodology for this type of assessment has not yet been fully developed. Experts in many fields are now working together to develop a consensus on how to conduct these complex and costly comparisons.[4]

Promotional materials that refer to product life assessments demonstrate the problems with using information from such assessments at this time. Problems include comparisons of information that technically cannot be compared; references to only the positive environmental aspects of one product and only the negative aspects of the competing product; and misuse of such assessments by third parties who do not know how to interpret the results. Moreover, the few product life assessments that have been conducted by the business community have come out in favor of the manufacturer who paid for the assessment and against that manufacturer's "target" competitor. For these reasons, the Task Force believes that, at the current time, it is misleading to use the results of product life assessments in advertisements.

However, nothing in this section should be read to discourage industry from using product life assessments to determine what products to manufacture or how to modify already existing products so as to lessen their adverse impact on the environment. In light of our growing environmental crisis, every possible tool should be employed by industry to protect our natural environment.

1.8 Third-Party Certifications and Seals of Approval. [New]
Environmental certifications and seals of approval must be designed and promoted with great care, to avoid misleading the public.

The use of certifications, seals of approval and other third-party evaluation has become a more pressing issue than it was when the Task Force began its inquiry. The last year has seen the emergence of a variety of programs in which products are evaluated by an independent third party, and the evaluation is then used in advertising and labeling.

Most conspicuous among these arrangements are several private seal programs that will, for a fee, certify one or more environmental attributes of qualifying products, and then allow the manufacturer to display the program's seal on certified products.[5] Other third-party certification programs are also proliferating. These include the governmentally sponsored environmental seals already in place in Canada, Germany, Japan, and the Scandinavian countries; a new seal being developed by the European Economic Community; and governmental seals for recycled products and "organic" foods being developed in several states. In addition, several major retailers have developed in-house programs to highlight ostensibly "environmentally superior" products—whether by a seal, shelf-labeling, in-store displays, or distinctive private label packaging.[6] In theory, there is no reason that third-party assessments cannot play an important role in the environmental advertising area. Consumers would undoubtedly benefit from detailed environmental product comparisons if those comparisons were based on testing conducted with the appropriate

---

4. Experts in many fields have been working cooperatively with the support of the EPA to reach a consensus on the methods to be used to conduct the most accurate and useful product life assessments. Most parties agree that using such assessments now to make comparisons between products is inappropriate because a great deal of technical work remains to be done before agreements on methods for conducting such assessments can be reached.

5. One member of this Task Force, Attorney General Hubert H. Humphrey III of Minnesota serves as an uncompensated member of the Board of Directors of one such organization, the nonprofit corporation, Green Seal.

6. Many companies are also pursuing less direct methods of legitimizing their products by associating them with respected third-party environmental organizations—for example, by establishing tie-in arrangements in which the company makes a financial contribution to an environmental organization for each product sold. While not certification programs in the strict sense, these arrangements are similar, in that their purpose is to establish the environmental credentials of a product by associating it with a respected third party—often by featuring the third party's name or logo in the product's packaging or advertising. These tie-in arrangements raise many of the same problems discussed in this section with regard to certification programs. Thus, those using such tie-in arrangements should take into account the discussion regarding certification programs when developing or participating in these promotions.

safeguards. However, the use of these comparisons, when communicated through advertising and on product packaging, can present significant problems.

For example, the Task Force is concerned about the criteria grantors of "seals" will use to select product categories and to determine whether a product qualifies for a "seal of approval." The criteria used are critical in determining whether the "seal" is meaningful or, on the other hand, potentially confusing and deceptive.

Another concern is the danger that financial considerations may lead programs to choose product categories and evaluation criteria that are actually at odds with environmental goals because those manufacturers most willing to pay for a "seal" may have products that are environmentally suspect or environmentally inferior to alternatives not included in the seal program. For example, the ongoing public debate about the relative merits of paper and plastic bags might tempt a program to award seals in a "grocery bag" category. However, experts agree that, whenever possible, shoppers should avoid disposable bags and carry their own reusable bags. This option would be difficult to include in establishing the product category criteria because it is unlikely to generate certification fees for the seal program.

The Task Force is equally concerned that certification programs may award seals on the basis of a single criterion that may be arbitrary, trivial or even intrinsically deceptive. For example, a program might award a seal to an environmentally harmful product simply on the basis of the recycled content of the outer package. Or a seal could be awarded to single-use paper towels, made of virgin wood products by a process using chlorine bleach, wrapped in non-recycled and non-recyclable plastic, solely on the basis of the partial recycled content in the towel's inner cardboard core. As these examples demonstrate, when a product is awarded a certificate or seal on the basis of a single criterion or a few criteria, it is critical that those criteria be carefully chosen to reflect the product's environmental impact.

In some cases—for example, where there are no clearly dominant environmental benefits—meaningful product evaluation may require an in-depth lifecycle assessment, and for this reason, such a product should not at this time receive any seal of approval or certification that is used for advertising purposes. As discussed in Section 1.7 above, there simply is not enough information available today to draw reliable conclusions about the cradle-to-grave environmental impact of most products.

Even if a program uses appropriate criteria in granting its certification, problems may still arise when the certification logo is used as a sales tool. No matter how laudable a seal program's purpose may be, if the manufacturers who pay for the use of the seal advertise it in a confusing and deceptive manner, its implementation may present more problems for consumers than solutions.

The use of seals on packages and in advertising must therefore be done with the utmost of care. In particular, advertisers must not overstate the meaning or importance of the logo to their product, either explicitly or by the configuration of their labels and advertisements. For example, seals based on only one attribute of a product must be very clear about that fact. Seals based on a constellation of "key" factors should be clearly explained, to avoid the easily-fostered misperception that a prominent seal constitutes an absolute, cradle-to-grave endorsement of the product. Nor should any environmental claim be made in proximity to a seal on a package, unless that claim is in fact certified as true by the seal grantor.

There is also a danger that a manufacturer will use a seal to imply that its product is superior to products that lack the seal, when in fact the other products may have no seal simply because their manufacturers chose not to pay to participate in the certification program, or could not afford to do so. This may be so even though their products may be superior for the environment. One safeguard for this problem would be disclosure, on products and elsewhere, that fees are paid to use the seal.

The Task Force believes that the seal grantors have an independent duty to effectively monitor the use of their seals in order to prevent deception, and may themselves be subject to legal action if they permit their seals to be used deceptively. This duty extends not only to the manner in which logos are displayed on packages, but to all advertising by licensees. It also includes an affirmative duty to communicate to the public the true significance of each seal, whether through in-store information, separate advertising by the grantor, or information on the logo.

For all of these reasons, the Task Force sees a serious potential for deception unless certification programs are designed, promoted, and monitored very carefully. If properly implemented, certifications may offer real benefits, but opportunities for missteps abound. Manufacturers and seal grantors alike should therefore proceed with great caution.

1.9 Source Reduction Claims. [New]
Source reduction claims should be specific, and where possible include percentages. Comparisons should be clear and complete.

Commentary: The Task Force recognizes that source reduction can provide a significant environmental benefit. Some companies took the Task Force's silence on source reduction claims to mean that such claims are unacceptable. This is not correct.

Companies that have made strides towards reducing packaging (for example, by using fewer layers of packaging or smaller containers/boxes), companies that have designed products that encourage consumers to reuse their original containers (for example, by selling concentrated refills in small paper containers that consumers can reconstitute in their original plastic container), and companies that have significantly reduced the actual size of their products certainly can advertise their efforts at source reduction so long as the claims made are truthful and accurate.

However, to avoid the possibility of deception, such claims should be specific and, where possible, include exact percentages for the reduction in weight or volume (e.g., "Now 10% less packaging than before"). Source reduction claims should only be made for a relative short period of time—six months to one year—immediately following the implementation of the change in size. Comparisons should always be complete (e.g., "10% less volume than our previous package"). Size reduction comparisons should be made only to the previous version of the manufacturer's product on the market unless there is a clear disclosure that a comparison is being made to a different product(s) (e.g., "10% less packaging than the leading brand").

2. Claims Should Reflect Current Solid Waste Management Options.
   [Changed]

Environmental claims relating to the disposability or potential for recovery of a particular product (e.g., "compostable" or "recyclable") should be made in a manner that clearly discloses the general availability of the advertised option where the product is sold.

In the preliminary report, the Task Force recommended that all solid waste management claims be location specific. While environmental and consumer groups heartily endorsed this approach, the opposition from industry was nearly unanimous. Representatives from the business and advertising communities argued that it is impossible to make location-specific disposability or recovery claims for products sold nationally. Moreover, they argued, if industry stops making such claims, consumers and the environment will suffer because: (1) consumers will not be informed as to which products are potentially recyclable or compostable and therefore will not be motivated to push for recycling or composting facilities that will accommodate these specific types of products in their communities; and (2) where composting or recycling facilities are available, consumers will not dispose of products properly if they are not clearly labeled as compostable or recyclable.

The Task Force is not persuaded by the first argument. It is true that there is a need to promote the creation of recycling and composting facilities, but there are many ways to achieve this goal without misleading consumers. For example, straightforward, informational advertisements that tell the full story are likely to accomplish more than ambiguous labeling buzzwords. Even on labels, companies can easily give a clear, non-deceptive message. For example, one company currently puts the following message on its plastic laundry product containers: "We are now using technology that can include recycled plastics in our bottles at levels of 25–35%. But to do so consistently, we need more recycled plastic. So please encourage recycling in your community.'' Such a statement, by indicating that plastic can be recycled and used by this manufacturer, encourages consumers to recycle, or to advocate for recycling facilities where none exist, without being misleading.

The controversy arises not because industry cannot write non-misleading labels, but rather because non-misleading labels are often less-effective sales tools. The Task Force believes, that if a disposability or recovery claim cannot be made without misleading consumers in a number of communities, then it should not be made at all. Such claims must be clarified to ensure that the public is well-informed rather than deceived. As discussed in the sections that follow, the Task Force believes there is a middle ground that will achieve that goal and foster the emergence of alternative solid waste management facilities. The worst possible solution, the Task Force is convinced, would be to continue the use of the unqualified terms ''recyclable'' and ''compostable'' on products that are not widely recyclable or compostable until consumers become so disillusioned, annoyed and frustrated that they lose interest in recycling and composting generally.

The Task Force still firmly believes that degradability claims should not be made for products that are likely to be disposed of in landfills or incinerators. During the December, 1990, hearings in San Diego, representatives from the plastic and paper industries generally conceded that degradable products provide virtually no environmental benefit when disposed of in landfills. During the past year, the Task Force has witnessed substantial movement on the part of manufacturers toward removing degradability claims from products destined for landfills or incinerators because of the confusion such claims created for consumers. Because of industry's willingness to discontinue such claims it appears that, for the most part, this particular controversy regarding the use of the word ''degradable'' has been resolved.

However, both plastic and paper manufacturers indicated that they were interested in promoting degradable products as ''compostable'' or ''degradable if deposited in a composting facility.'' This also presents potential deception problems. Although several communities now compost yard trimmings, few compost other types of municipal solid waste. Several questions are now being raised about the environmental soundness of composting inorganic and organic materials together because of the danger of contaminating the resulting compost. These problems may stall efforts to develop general municipal solid waste composting facilities.

At the current time, composting is not an available option for the vast majority of consumers in the United States. The advertising today of an environmental attribute that cannot be realized until some uncertain time in the future is confusing and misleads the consumer. Consumers are purchasing products and packaging that must be disposed of in short order. To avoid potential deception, companies that elect to make claims such as ''compostable'' or ''degradable if deposited in a composting facility'' must also clearly disclose the current limited availability of this disposal option and the fact that the product is not designed to degrade in a landfill. (See Section B 2.2. for examples.)

Recyclability claims present similar problems. Only those nationally sold products that are generally recyclable everywhere should carry an unqualified ''recyclable'' claim. Other products that are recyclable in some communities, but not in others, should only make qualified recyclability claims that inform the public that the product is potentially recyclable without misleading consumers to believe that the product is recyclable everywhere it is sold. (See Section B 2.3 for examples.)

2.1 Use of Terms ''Degradable,'' ''Biodegradable,'' and ''Photodegradable.'' [Slightly modified]

Products that are currently disposed of primarily in landfills or through incineration—whether paper or plastic—should not be promoted as ''degradable,'' ''biodegradable,'' or ''photodegradable.''

Commentary: While there was some debate over the environmental benefits of disposing of degradable products in landfills at the Public Forum in March, 1990, by December,

1990, at the hearings in San Diego, virtually all the companies that testified conceded (or had accepted) that degradability claims should not be made for products likely to be disposed of in landfills.[7]

The Task Force notes, however, that it may be appropriate to make claims about the "biodegradability" of a product when that product is disposed of in a waste management facility that is designed to take advantage of biodegradability (such as a municipal solid waste composting facility) *and* the product at issue will *safely* break down at a sufficiently rapid rate and with enough completeness when disposed of in that system to meet the standards set by any existent state or federal regulations. (Section B 2.2.)

2.2 Use of Term "Compostable." [Changed]
Unqualified compostability claims should not be made for products sold nationally unless a significant amount of the product is currently being composted everywhere the product is sold. In all other cases, compostability claims should be accompanied by a clear disclosure about the limited availability of this disposal option. If a claim of degradability is made in the context of a product's compostability a disclosure should be made that the product is not designed to degrade in a landfill.

Commentary: A product sold nationally should not be promoted as "compostable" or "compostable where composting facilities exist" until a significant amount of the product is composted nationally. There are currently very few locations in the United States where anything other than yard trimmings is composted. Thus, at the present time, promoting most nationally sold products as "compostable" is meaningless at best and potentially deceptive. Because of the extremely limited availability of municipal solid waste composting facilities, the Task Force believes that compostability claims for most products sold nationally are premature.

Further, there is a great deal of controversy over the compostability of the materials (i.e., plastic bags, paper bags, boxes, etc.) in which items for composting are delivered to composting facilities. Before making compostability claims about any such materials, companies should have competent and reliable scientific evidence to show that the product will safely breakdown at a sufficiently rapid rate and with enough completeness to render the resulting compost useful and saleable for its intended purpose without requiring an additional step in the composting procedure to remove the residue of that material. For all of these reasons, companies should proceed with extreme caution in promoting this product attribute.

To minimize the risk of deception, a compostability claim should be both qualified and as specific as possible. For example, "This product is potentially compostable, however, less than 1% of the U.S. population has access to composting facilities. To find out if there is a composting facility near you, call (800) xxx-xxxx" would be an informative, non-deceptive claim. In addition, if only a portion of the product is compostable that fact and the percentage that is compostable should be clearly disclosed. Other material facts regarding the dangers of composting certain products should also be made explicit. One such fact, for example, is that disposable diapers should not be composted in backyard

---

7. As detailed in the Preliminary Report, there is good reason why such claims should not be made. Even products specifically modified to degrade more rapidly than conventional products do not degrade at any appreciable rate in landfills. Clearly they are not breaking down fast enough to extend the useful life of our landfills. Moreover, in a report submitted to Congress in February 1990 entitled "Methods to Manage and Control Plastic Wastes," the EPA stated that degradable plastics will not help solve the landfill capacity problems facing many communities in the United States. Indeed, to the extent degradable plastics do break down over some extended period of time, their degradability may actually be harmful, because breakdown can create toxic leachates and dangerous methane gas—precisely the result that modern landfills are designed to inhibit. Further, degradability claims on plastic products may send the message that it is all right to litter such products. The Task Force believes that consumers are confused about the environmental effects of degradability for products disposed of in landfills or incinerators, and that the widespread use of deceptive degradability claims in advertising and packaging have both taken advantage of and exacerbated consumer confusion about the environmental benefits of degradability.

compost bins designed for food scraps and/or yard trimmings or in municipal lawn and leaf composting operations because of health and sanitation problems. Another is that many solid waste managers do *not* recommend that paper or plastic packaging materials be composted at home or in municipal yard trimming composting facilities.

Finally, because of the widespread misunderstanding among consumers about the benefits of "degradability" for products disposed of in land fills, the label of any product promoted as "degradable if disposed of in a composting facility" should clearly and prominently disclose that the product is not designed to degrade quickly in landfills.

2.3 Use of Term "Recyclable." [Changed]
Unqualified recyclability claims should not be made for products sold nationally unless a significant amount of the product is being recycled everywhere the product is sold. Where a product is being recycled in many areas of the country, a qualified recyclability claim can be made. If consumers have little or no opportunity to recycle a product, recyclability claims should not be made.

Products sold nationally should not be promoted with the unqualified claim "recyclable" unless the product is currently being recycled in a significant amount *everywhere* the product is sold. Thus, for example, aluminum cans, which are recyclable virtually everywhere in this country, could carry the unqualified claim "Recyclable," or the phrase "Please recycle."[8]

Where a nationally sold product can be recycled in many communities, but not everywhere it is sold, recyclability claims should be qualified to avoid the potential for deception. The Task Force strongly recommends that companies desiring to promote their products' recyclability set up 800 numbers so that consumers can find out if recycling facilities exist near them. Such a company could advertise saying, "Recyclable in many communities. Call us at (800) xxx-xxxx to find out if there is a recycling facility that accepts this product near you. Support recycling." Such a claim informs consumers that the product can be recycled in some communities and tells them how to find out if they can recycle it in their community, but avoids giving the impression that the product is being recycled everywhere.[9] In addition, if only a portion of the product (for example, the paper wrapper) is recyclable, that fact should be clearly disclosed.

However, if a product is only technically capable of being recycled and is, in fact, only being recycled at a few test sites, then no recyclability claim should be made because there is no real opportunity to recycle. If a company marketing such a product wants to promote the fact that the product has the potential to be recycled, or that is technologically possible to recycle it, then that company should clearly disclose all of the materials facts, including at least: (1) the fact that the technology is in the early stages, or that there are only "pilot" recycling programs, if that is the case; (2) the number of locations where the product is being recycled; (3) the types of collection sites if there is no curbside pick-up available (e.g. "at school cafeterias"); and (4) the number of states in which the collection and recycling facilities are located. Finally, a product should not be promoted as "recyclable" if it contains additives or other materials that make the product problematic or unsuitable for recycling.[10]

---

8. Both New York and California, by statute, restrict the use of the terms "recycled" and "recyclable" based on how widespread the recycling opportunities are in each state, and companies wishing to use these terms in those states must, of course, comply with the requirements of those states' laws.

9. The availability of recycling depends on a combination of technical and economic feasibility, market demand, and the existence of collection and separation facilities. If the manufacturer is not recycling the packaging it sells itself, the economics of recycling tend to vary by geographic location and the physical composition of materials. For example, the economics of aluminum recycling are settled and can be quantified by the value assigned to scrap aluminum. The economics of polystyrene recycling, on the other hand, are not yet known. While demonstration and pilot projects exist, it is by no means clear that recycling polystyrene is economically feasible.

10. Nothing in this section should be read to apply to the Society of Plastic Industry (SPI) code, which consists of a design of arrows in a triangular shape with a number or letter code that identifies its component plastic resin. This symbol alone, usually placed inconspicuously on the bottom of the container, is not promotional and appears for the sole purpose of permitting plastic recyclers to easily sort plastics for recycling.

2.4 Safe for Disposal. [Changed—Replaces "Safe for Incineration"]
Vague safety claims concerning disposability should be avoided. Instead, products should specifically disclose those environmentally dangerous materials or additives that have been eliminated.

Commentary: Federal and state solid waste managers need to control the components of the waste stream which enter their incinerators and landfills in order to reduce specific adverse environmental impacts associated with their waste disposal method. Thus, some waste districts may prohibit the incineration of a specific product, packaging material or the like, even though, in the manufacturer's opinion, it can be "safely" incinerated. Simply because a product meets a federal safety standard does not mean that its disposal in a landfill or incinerator is risk free or has no adverse impact on the environment. To prevent confusion, manufacturers should not promote products as "safe for incineration" or "landfill safe." If a product does not contain materials or additives that are known to be problematic for environmentally benign disposal, the manufacturer should simply state that the product does not contain (e.g., "Our packaging material contains no cadmium").

3. Claims Should Be Substantive. [Unchanged]

Environmental claims should be substantive.

Commentary: Nonsubstantive claims are widespread in "green marketing" today. These trivial and irrelevant environmental claims create a false impression of a product's overall environmental soundness. They also contribute to consumer confusion. Although this may, in the short run, aid the sale of a given product, it reflects an irresponsible attitude toward the environment and may be misleading.

3.1 Trivial and Irrelevant Claims. [Unchanged]
Trivial and irrelevant claims should be avoided.

Commentary: The broad variety of trivial and irrelevant environmental claims being made today almost defies description. Examples include products promoted as "degradable" that will be disposed of in landfills or incinerators, and trash bags, which are highly unlikely to be used again for any purpose, advertised as "recyclable." An example of a technically accurate but irrelevant claim is a polystyrene foam cup that claims to "preserve our trees and forest." It is simply irrelevant, and perhaps deceptive, to suggest that a product made of petroleum products, a scarce nonrenewable natural resource, provides an environmental benefit because it does not use trees, the natural renewable resource that would have been used if the cup had been made of paper instead of polystyrene.

3.2 Single Use Products. [Unchanged]
Single use disposable products promoted on the basis of environmental attributes should be promoted carefully to avoid the implication that they do not impose a burden on the environment.

Commentary: Many products that are designed to be thrown away after a single use, such as disposable diapers, paper plates, or shopping bags, sport claims that imply environmental soundness. Such claims convey an implicit message that disposal of a single use item—perhaps the most environmentally distressing aspect of the product—does not contribute to the overall solid waste disposal problem. These claims therefore run the risk of leading consumers to ignore or reject more durable alternatives to single use products, such as close diapers, reusable plates or reusable shopping bags. Advertisements for single use products should not convey the message that they impose no burden on the environment.

4. Claims Should Be Supported. [Unchanged]

Environmental claims should be supported by competent and reliable scientific evidence.

Commentary: Of course, this recommendation does not set forth a new legal concept. Instead it restates what has always been required under state and federal law—that advertising claims must be supported by tests, analysis, research or studies conducted or evaluated in an objective manner by persons qualified to do so using procedures generally accepted by others in the profession to yield accurate and reliable results.

The Task Force notes that in the course of its investigation of environmental claims, some companies have attempted to minimize their responsibility for claims that appear on their products by pointing to information provided by suppliers of the constituent materials. Companies that fail to independently confirm substantiation provided to them by suppliers do so at their peril when they incorporate such claims into their advertising and packaging labels. Aside from potential legal liability, a company that does not independently confirm the accuracy and completeness of claims made by suppliers abdicates its duties to its consumers and the environment.

In addition to ensuring that environmental claims are adequately supported, business can make a significant contribution toward the public debate about environmental problems by making the substantiation for environmental claims available to the public, regulators and experts. Although the Task Force recognizes that companies will often feel it necessary to keep certain information confidential, the Task Force urges companies to make information about the composition and the environmental effects of their products, and the substantiation for their environmental claims, available to the public to the greatest extent possible.

Principles and Guidelines for Environmental Labelling and Advertising

CONTENTS

PREAMBLE

Canadian consumers are concerned about the environmental effects of materials that are being used in the products and services they purchase. This concern has resulted in new selection criteria being applied: consideration for the environmental effect of materials as well as for the environmental behaviour of companies that produce or use these materials. Faced with this trend, industry has responded with what is known as "Green Marketing" or environmental labelling and advertising.

Green marketing empowers consumers to translate their growing environmental awareness into practical environmental action. For industry, the purchasing power of well-informed consumers provides the incentive to develop materials and processes that place less burden on the environment. However, this market dynamic is predicated on the availability of objective, credible and truthful information which can be readily acquired and understood. With the emergence of a broad range of descriptors, logos, vignettes

---

SOURCE: This document has been developed in partnership and consultation with associations representing public interest groups, manufacturers, distributors, retailers and advertisers. A list of participants is included in Exhibit A. March 1994.

and other representations used to describe or imply environmental features of consumer products and services, action is required to ensure responsible labelling and advertising in this regard.

Industry Canada, within its mandate to promote the fair and efficient operation of the Canadian marketplace, has undertaken to examine this issue in consultation with interest groups. Recognizing that Canadian industry associations were working independently to bring some order to green marketing, Industry Canada assembled a working group to develop this document: guiding principles to be used as a common reference by industry. The working group, as listed in Exhibit A (page 196), has worked closely and cooperatively to formulate these principles, to develop the advice and guidance based on the principles, and to adopt and promote them within their industry sectors.

In developing these principles, recognition has been given to the important role played by all participants in the marketplace, whether they are producers, distributors, advertisers or consumers. It is also recognized that while detailed technical information is not always readily available, there are a number of sources for credible information that will assist consumers to become informed on the various waste reduction and recycling programs available within their communities.

SCOPE

This document does not attempt to establish environmental standards or definitions, nor does it provide definitive solutions to complex scientific and technological issues relating to the environment.

The principles and guidelines outlined herein were developed within the context of the *Consumer Packaging and Labelling Act* (CPLA) and the *Competition Act* (CA), both of which contain broad prohibitions against false and misleading representations. Adherence to the advice contained in this document will enhance the provision of meaningful information to consumers, and will assist industry in its obligation to ensure compliance with the applicable legislation. In keeping with the intent and spirit of the CPLA, all environmental claims should be in English and French.

This document was specifically developed to ensure flexibility, recognizing that much of the knowledge in this area is based on an evolving science and technology. Background information and guidance on the use of claims that are ambiguous or profess general benefit to the environment are addressed, as are the more specific, commonly-used representations.

All parties accept that it is necessary to address consumer information issues relating to environmental labelling and advertising. There will be periodic reviews and updating of the guidance contained herein, based on changes in technology, marketplace needs and enhanced awareness of environmental concerns. The reviews will, of necessity, draw on the broadest possible range of interests. Originally issued in May 1991, this revised version attempts to address issues that have arisen since then and also to elaborate on past issues.

The guidance presented here applies equally to the labelling and advertising of products as well as services. For the sake of simplicity the text will refer to products with the understanding that services are also included.

APPLICATION

The members of the working group have brought together the most current information available to ensure a realistic and workable approach to the use of environmental claims, and will implement the principles and guidelines within their respective areas of influence. Accordingly, industry associations will recommend and encourage their members to follow the recommendations outlined to ensure that environmental claims are truthful and accurate and, where applicable, are premised on appropriate and reliable scientific and technological information that can be verified.

The advertising industry is a major user of environmental claims in the promotion of consumer products and services. The important role and responsibility of advertisers in this regard is reflected in the principles and guidelines contained in this document, which will become a supplement to the Canadian Code of Advertising Standards administered by the Standards Division of the Canadian Advertising Foundation through the Advertising Standards Council and le Conseil des normes de la publicité.

Environment Canada is responsible for initiatives and programs that have implications for environmental labelling and advertising. Two current programs are the Environmental Choice Program that enables consumers to identify products and services that reduce environmental burden (Exhibit B, page 197), and the National Packaging Protocol that targets a 50% reduction in packaging waste by the year 2000 (Exhibit C, page 198). Close liaison with Environment Canada is critical to ensure that those issues requiring a common approach are appropriately addressed and coordinated. In this regard, the glossary of terms developed by the National Task Force on Packaging, chaired by Environment Canada, will be adopted for the purposes of this document. A copy of these terms and their definitions is attached as Exhibit D (page 199).

Ultimately, Industry Canada will administer and enforce the relevant statutes controlling misrepresentation and misleading labelling and advertising. In all cases, a thorough investigation of the product and its specific claims and representations will form the basis of enforcement and/or compliance action under the *Consumer Packaging and Labelling Act* and/or the *Competition Act*. Relevant excerpts from these Acts are provided in Exhibit E (pages 200–202).

GUIDING PRINCIPLES

It is recognized that all participants in the marketplace have an important role to play in reducing the environmental impacts resulting from the production, use, and disposal of products. Environmental claims and/or representations should provide accurate and relevant information to allow meaningful comparisons to be made, thereby enabling consumer purchasing power to influence the marketplace through informed product selection.

1. *Those making environmental claims are responsible for ensuring that any claims and/or representations are accurate, and in compliance with the relevant legislation.*

2. *Consumers are responsible, to the extent possible, for appropriately using the information made available to them in labelling and advertising, thereby enhancing their role in the marketplace.*

3. *Environmental claims and/or representations that are ambiguous, vague, incomplete, misleading, or irrelevant, and that cannot be substantiated through credible information and/or test methods should not be used.*

4. *Claims and/or representations should indicate whether they are related to the product or the packaging materials.*

The advice and guidance which follow are an indication of the interpretation that will be used by officers of Industry Canada when examining claims under the *Consumer Packaging and Labelling Act* or the *Competition Act*. Environmental claims which follow this guidance are not likely to raise questions under the legislation. Positions relating to acceptability for any claim will only be taken following a complete review of the relevant facts, but final determination of acceptability will rest with the appropriate court of law.

GUIDELINES FOR ENVIRONMENTAL CLAIMS

General Claims/Representations

Recognized standards.   Claims should be based on **recognized standards or prevailing scientific principles**. In the absence of adequate evidence demonstrating that the subject of the claim or representation reduces or minimizes environmental impact, unqualified terms or representations that imply environmental benefit should not be used.

Vague statements such as "environmentally friendly," "ozone friendly," or "green" are meaningless statements that imply that the identified products or services provide some environmental benefit. In reality, these products do not benefit the environment. The message being conveyed speaks of reduced environmental impact as a result of specific product characteristics. Vague statements or catch phrases cannot clearly indicate specific benefits and should therefore not be used.

Known substances.   Some substances are currently known to have significant adverse environmental impacts. The more common substances that have been identified in consumer products are CFC's, VOC's, heavy metals and excessive levels of phosphates. In an attempt to address consumer concerns, companies remove or reduce the quantity of these substances. However, the replacement substances may not be benign, and may also be damaging in some way, or the reduction may not provide a significant benefit. Any claims relating to the removal of substances that are known to be harmful should therefore take into account the net effect of the change that has been made.

Example 1:   If an ozone-depleting blowing agent for expanded foam is replaced with an inert substance, an unqualified claim of reduced environmental impact may be acceptable; however, for the sake of clarity, the claim should be qualified.

Example 2:   If one ozone-depleting propellant in an aerosol is replaced with another, the benefit of removing one substance may be offset by the replacement substance and any claim of reduced environmental impact is not appropriate.

If current scientific opinion indicates that the replacement substance is reasonably benign or significantly less harmful, then any claim of reduced environmental impact should indicate that the offending substance has been replaced by a less harmful substance. The replacement substance should be named unless doing so would cause additional confusion. If the replacement substance is known to contribute to environmental degradation, even though the impact may be in a different area than the substance it is replacing, then claims relating to the revised product should **not** make reference to broad environmental benefit or reduced impact.

Brand name or trademark.   On occasion, a company may use a brand name or trademark that appears to relate to the environment. In these cases, where it is determined that the name or mark seems to designate environmentally sensitive products, or gives the impression that the product is environmentally sensitive, the name or mark will be considered an environmental claim. As such, the use of the name or mark will be subject to the appropriate legislation and is expected to be in accordance with the principles contained in this Guide, and, in particular, is to be accompanied by appropriate clarification or substantiation.

Life cycle analysis.   When making reference to reduced environmental impacts, it would be useful to be able to make comparisons among materials or product classes based on life cycle analysis, or LCA. Some common examples could compare paper and plastic carryout bags or disposable and reusable products. LCAs are generally based on the quantity and nature of resources used and substances released. While these quantities can be determined in most cases, for comparisons among different materials and/or processes, there may be no correlation between resources used, substances released into the environment and the relative environmental impacts of one product over another.

Many organizations, including national, international and private standards writing bodies are working to develop standards and guidelines for LCAs, which may become appropriate tools for evaluating the relative environmental impacts among materials. It must be noted, however, that a consensus on how to conduct these assessments has yet to be developed. Until such time, it is inappropriate to make public claims regarding the comparative overall environmental performance of various products and materials based on LCAs.

Where claims on a specific environmental component of a product or a material's performance is contemplated, it is important that the evaluation be based on widely accepted LCA methods. The claim should also follow logically and clearly from the analysis conclusions. In addition, any information on which the LCA claim is based should be readily available to the public.

Safe.   The word "SAFE" has broad implications that may inadvertently be perceived as relating to personal safety or security and should therefore not be as used part of an environmental claim.

Absolute Claims/Representations

Absolute claims.   An absolute claim or representation leaves no room for doubt. When used as part of a claim, some symbols and terms imply conditions that may not exist. In the absence of qualifying information, this type of claim will be interpreted to mean "100%."

Symbols.   The use of any symbol or statement relating to recyclability, without a qualifying statement, will be interpreted to mean completely recyclable (100%), or universally recyclable.

Example 1:   The recycled content symbol, either alone or with a vague statement such as "Printed on recycled paper," "Made from recycled paper" or "Made with recycled paper" creates the impression that the end product is made entirely from recycled paper, and will be interpreted as such.

Example 2:   An unqualified claim of recyclability may be justified in a market area for materials which have established deposit and return systems in place, such as beverage containers, and the returned containers are recycled. Such a system has the potential to collect all identified containers, effectively providing universal access to all purchasers in that market area.

Example 3:   Descriptors such as " . . . Free" or "Contains No . . ." are absolute in that they claim the total absence of the stated substance.

Möbius Loop.   Because the two Möbius Loop recycling symbols are still not widely understood, it is recommended that the symbols always be used with explanatory text.

Slogans.   If short slogans, catch phrases or symbols are used to indicate reduced environmental burden, and these do not clearly convey the intended message, then there should be further qualification in the claim to ensure that the message is clear and understandable.

Source Reduction of Materials

Source reduction is an important means of waste management, and one that is increasingly being used and talked about. Waste management principles in Canada identify source reduction as having the highest priority for managing waste and substances with high environmental impacts. The benefit of source reduction is derived from reduced levels of materials being used, and ultimately sent for disposal. For this reason, claims in this area must relate to the **reduction of total materials,** or **reduced levels of toxicity**.

Comparisons.   There are many ways to reduce the amount of materials used, whether it is through eliminating the package, "lightweighting," using alternative materials, or reformulating the product. Whichever method is used, however, claims of source reduction should be comparative claims. This means that any reference to reduction must be made relative to that product or package at a relevant predetermined reference point. Extreme caution should be used when making comparisons with competing products because any subsequent similar reduction in the competitor's product could make the comparison invalid resulting in a misleading claim.

Example: If a manufacturer changed the packaging of a product from a combination of inner wrap, outer box and cellophane (total weight 10 g) to a single box (weight 6 g), the claim could be "40% reduced from our previous package."

When making comparisons, it is important to ensure that the comparison is made for equivalent amounts of product.

Example: If comparing the relative benefit of packaging materials between two different products such as liquid and powdered laundry detergent, the comparison should be based on a common measure such as the number of uses, e.g., this product requires X% less packaging on a per use basis than product Y.

Distribution reduction.   Source reduction may also relate to the entire production and/or distribution system for the product.

Example: If a product or its container has not changed, but changes have been made in the distribution system to eliminate a significant amount of waste, such as using reusable pallets instead of single use pallets, then it may be appropriate to identify that initiative. It should, however, be very clear that the identified reduction has taken place in the distribution of the product and not the product or its container ("This product is now shipped from the factory on reusable pallets, resulting in an estimated waste reduction of XX tonnes per year").

Quantify.   In all cases, claims of source reduction should include a quantified indication of the reduction.

Reusable/Refillable Containers or Products

Reuse opportunities.   Many products and packages are reusable or refillable by consumers, and they are generally aware of that fact. It is important, therefore to identify new reuse opportunities that may not be readily apparent.

Example: Glass, plastic and corrugated containers have generally been reusable by the consumer after the product has been consumed. To identify these containers as reusable for that reason alone may not be appropriate.

Infrastructure.   There are many cases where a claim of reusability may be appropriate. The following examples are illustrative only and not meant to exclude other

EXHIBIT F-1

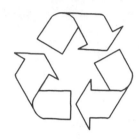

legitimate claims. In all cases, however, a reuse "infrastructure" must exist or facilities or products must exist that allow the end user to directly reuse or refill the item, and where it is not obvious, the claim must explain how the container or product is to be reused or refilled, e.g., with the purchase of a refill pouch, by returning for cleaning or restoration.

Example 1:   The container can be reused/refilled for its original purpose without reprocessing beyond normal cleaning operations.

Example 2:   A product is supplied in a durable reusable container as well as a lightweight refill.

Example 3:   A product is reformulated to a concentrated version so that it can be packed in a smaller container, but would be reconstituted in a larger reusable container for normal use.

Example 4:   A manufacturer replaces the product packaging with a durable storage container such as might normally be found for sale at retail.

Example 5:   The design of a product is modified to facilitate restoring it to an "as new" condition for reuse, e.g., printer ribbons and toner cartridges, and the facilities exist for such restoration to occur.

Recyclable Materials

Benefits of recycling.   The use of the Möbius Loop Recycling symbol (Exhibit F-1) to designate recyclable products deals to a large extent with the potential benefit that can be realized through diversion from waste and reduced resource utilization. This potential cannot be realized unless there are systems and facilities in place to recycle products and packaging. The benefit comes from reduced dependency on resources by separating the product from the waste stream and recovery of materials by directing them to a recycling facility, either through curbside collection programs or drop off sites.

Recyclable claims.   Recycling of materials is a rapidly growing industry. It is recognized that claims of recyclability may assist consumers in making choices which will ensure the availability of materials for recycling facilities. The use of the symbol may also provide an impetus to increase the development of appropriate infrastructures for recycling. However, the use of recyclable claims for materials which are not commonly recycled may have a negative impact on the viability of these infrastructures because of increased costs associated with sorting and rejection of materials that are contaminated.

Infrastructure.   The National Task Force on Packaging has defined the term "recyclable" for use in promoting materials. The definition is structured around a minimum threshold value for availability, to the Canadian population, of recycling or collection

EXHIBIT F–2
Where Facilities Accept "X"
(Là Où les Installations Acceptent le "X")

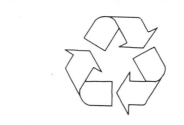

facilities. A claim of recyclability is reasonably justified if at least one-third (1/3) of the population in the area where the product is distributed has convenient access to collection or drop-off facilities for recycling. Calculation of the threshold figure will depend on the expected product distribution (e.g.: approximately 9 million for a nationally distributed product, or one-third of the local population for a regional product). There is no clear definition for "convenient access," but it is recognized that convenience is very subjective and will vary considerably depending on the demographic makeup of communities. Where no curb-side collection exists, convenience of drop off depots should be determined according to local conditions.

Claims of recyclability are inappropriate when used with products or materials for which a recycling infrastructure does not currently exist. Claims of recyclability by industry (retailers, manufacturers, and distributors) should not be made simply because the material is technically recyclable or anticipated to be recyclable in the future. **Where a material or a product cannot be recycled, or is not reasonably expected to be recycled, no claim of recyclability may be made.**

Example 1:   Plastic garbage bags are made of polyethylene, which is technically recyclable, but when used as intended will invariably end up in disposal facilities such as landfill and not be recycled. Therefore claims that plastic garbage bags are recyclable should not be made.

Example 2:   Some materials, such as old corrugated containers (occ) are increasingly collected and recycled from commercial establishments and households. Therefore it may be appropriate to label these materials as recyclable when used for commercial applications, but until these materials also reach the one-third accessibility threshold for households, they cannot claim to be recyclable when used in retail consumer goods. For example, a corrugated shipping container used and disposed of by commercial or industrial enterprises may be recyclable; but a corrugated container used as a package for retail consumer goods, and sold to households for which insufficient recycling infrastructure exists may not claim to be recyclable.

Qualifying claims.   Where a material or product is being recycled, and one-third of the population has access to collection facilities for that material or product, a claim of recyclability may be made if the claim is qualified by an explanatory statement.

Example 1:   "Recyclable in programs that accept (material name)";

Example 2:   "Where facilities exist"

There should also be a clear indication that the claim relates to the product, the packaging materials or both. Exhibit F–2 shows the use of the qualified recyclable symbol. The same message may be conveyed by the use of the statement without the symbol: "Recyclable where facilities accept (material name)."

EXHIBIT F-3

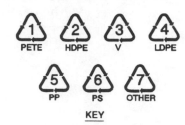

KEY

1– PETE (polyethylene terephthalate) (PET)
2– HDPE (high density polyethylene)
3– V(vinyl/polyvinyl chloride) (PVC)
4– LDPE (low density polyethlene)
5– PP (polypropylene)
6– PS (polystyrene)
7– Other

Recognition of infrastructure development initiatives.   It is recognized that recycling infrastructures for all materials are not fully developed. In establishing these infrastructures, public awareness can play a significant role in the success or failure of a program.

In some cases, a material may be collected from a limited number of locations, or the infrastructure has not been sufficiently developed to provide access to recycling as described above. Clear, accurate and meaningful information relating to these programs should not be discouraged. However, in providing information to consumers that recycling has begun for some materials, there should not be any confusion relating to the extent that these materials are collected.

In these exceptional cases, where the one-third threshold may not have been reached, but it is desirable to advise consumers of the new initiative, it may be appropriate to make a very specific and restrictive claim.

Example 1:   ''Recyclable in (city XX)''

Example 2:   ''Please return to (store name) for recycling''

Example 3:   ''For information on recycling (material name), please call 800-XXX-XXXX''

Example 4:   ''Recyclable in limited communities, call 800-XXX-XXXX for details''

Specific instructions.   If a product or package is identified as recyclable, and it is composed of several materials which cannot be recycled together, there should be specific instructions on the proper preparation of the material for deposit in the recycling infrastructure.

Example: ''Remove metal staples before recycling.''

Möbius Loop.   Because the two Möbius Loop recycling symbols are still not widely understood, it is recommended that the symbols always be used with explanatory text.

Plastic sorting codes.   The Society of the Plastic Industry (SPI) has developed a set of symbols that identifies various plastic resins for the purpose of sorting materials

EXHIBIT F–4

(Exhibit F–3). The use of these symbols is not intended to be a claim of recyclability and has enhanced the industry's attempts to address the technical problems associated with the collection, sorting, and recycling of plastics. In some instances, municipal collection programs will advise consumers to separate their plastics by SPI code numbers to ensure minimal contamination when only one or two types of plastic are collected.

Use of sorting codes.   In most cases these codes are applied on the bottoms or sides of containers as recommended by SPI. However, when the symbols appear conspicuously printed next to environmental claims, the association with these claims and the similarity to the Möbius Loop recycling symbol could give the general impression that the container is recyclable.

Example:

SPI code next to "NO CFC" symbol on expanded polystyrene foam products.

In these cases, the SPI code will be considered to be a claim of recyclability.

Recycled Content in Materials

Source of recycled content.   The recycled component of a product can originate from many sources: industry, commercial and institutional establishments, and households. Generally, any diversion of materials from the waste stream provides positive solid waste reduction benefits. However, the recycling process is related specifically to the reprocessing of materials which have already served their intended purpose. For this reason, any reference to recycled content **excludes** the primary (virgin) portion of "in-house" materials that are routinely reprocessed on the premises as part of the manufacturing process and would never have been sent for disposal. For example, in the production of plastic goods, waste materials and trimmings are normally reground and reincorporated in the production line, whether for the same product, or another product manufactured by that company. Similar operations occur in the glass, paper, steel and aluminum industries.

Claims of recycled content may be made using the Möbius Loop symbol shown in Exhibit F–4 with a recycled content disclosure as a percentage, by weight, of the product or total material. Where a claim of recycled content is made, it will be assumed that the recycled components are "post-use" materials as described in the NAPP definitions (Exhibit D). That is, they are materials from articles that have served their intended purpose and which have been diverted from the waste stream for reprocessing into new articles. This may include waste materials diverted from household, institutional, commercial or industrial sources.

EXHIBIT F-5

The **calculation of recycled content** should be based on the relative proportions of materials that make up the final product. That is, based on the total weight of raw materials required for the production, the calculations are based on the percentage of post-use materials that are being recycled.

Example: If a production run requires 1,000 kg of raw materials, and it is composed of 700 kg virgin materials and 300 kg post-use materials, then the correct claim is 30% recycled material. Calculating the recycled content in this manner provides confidence that the declared content is an accurate reflection of the materials used.

Exhibit F-5 shows the recycled content symbol used to claim 30% recycled content. The same information may be conveyed in words (30% recycled content) adjacent to the symbol. In the absence of recycled content disclosure, the claim will be interpreted to mean 100% recycled materials.

Möbius Loop.   Because the two Möbius Loop recycling symbols are still not widely understood, it is recommended that the symbols always be used with explanatory text.

Average recycled content.   Almost all industries experience variability of supply, and cannot be sure that the recycled content is consistent between production runs. Therefore, it is permissible to indicate the average recycled content based on an average over a period of 3 to 12 months depending on the production profile of the industry involved. The exact time frame will be negotiated with each affected industry and its appropriateness will be reviewed at each revision of this document. At the present time, under the Environmental Choice Program, fine paper products are operating under a three month weighted average from a single mill location.

Quantifying claims.   Because consumers are concerned with the specific benefits attributable directly to end user return of materials for recycling, industries may want to indicate the quantity of recycled materials originating from specific sources such as curbside collection systems, drop-off depots, fine paper from commercial sources, etc. This can be achieved by being specific in the claim.

Example 1:   This product contains 50% recycled material, including 10% from household collection programs

Example 2:	Made with 65% recycled post-consumer materials
Example 3:	100% recycled content—80% commercial, 20% household
Example 4:	100% recycled content—60% post-consumer, 40% pre-consumer

Degradable Materials

Definition.   A degradable material is one which breaks down in such a way that the resulting materials can be easily assimilated into the environment without having any significant negative impact on the environment. Degradability claims on package labels usually refer to biodegradability and/or photodegradability. While many materials are ultimately degradable, the conditions under which these materials are disposed of, usually through landfill, may hinder the degradation process. Degradation can take place in air, on land and in the water. Generally, in order to biodegrade, there must be biological action under specific conditions. In order to photodegrade, light must be present.

Supportable claims.   An unqualified claim that a product or package is degradable, biodegradable or photodegradable should be supportable by competent and reliable scientific evidence that the entire product or package will completely break down (i.e. decompose) after customary disposal and create no significant toxic residues. That is, biodegradable materials should be capable of undergoing decomposition into carbon dioxide, methane, water, inorganic compounds or biomass in which the predominant mechanism is the enzymatic action of micro-organisms. Photodegradable materials should be capable of undergoing a significant loss of properties after exposure to representative amounts of sunlight. The capability of these materials to degrade should be measurable by standardized tests, in a specified period of time, reflecting available disposal conditions.[1] Any evidence provided to support such claims should be recognized by the scientific community at large, and not just by a small group or individual company.

Limitation to claims.   Materials exposed to the open air may degrade through the action of light and/or biological processes. There are, however, limitations on the types of products that could legitimately claim to be degradable through atmospheric exposure. The product would normally need to be used outdoors, with the period of degradation sufficiently long that the useful life of the product would not be affected.

Example: A soil cover that acts as an artificial mulch for gardens and flower beds that will survive a growing season, but will degrade sufficiently so that it may be tilled into the soil for the next season.

Waste in a sanitary landfill site is deprived of air, moisture, and light, retarding significantly any degradation of the product. In the context of consumer items destined for landfill, current scientific opinion indicates that degradability claims as an environmental benefit may not be supportable. Products or packaging materials that are not diverted from the waste stream will invariably end up in disposal facilities such as landfill so any claim of degradability would not be appropriate.

Conditions found in liquid disposal facilities such as municipal sewage systems or septic systems will usually provide adequate conditions for biodegradation to occur. Claims of degradability may be appropriate for products normally disposed of through the wastewater system providing the by-products of degradation and/or the products in question create no significant toxic residues, and will not harm the sewage collection and treatment facility.

---

1. Derived from May 14, 1993, Main Committee Ballot of American Society for Testing and Materials (ASTM) committee on Terminology for Environmental Labelling of Packaging Materials and Packages. (D10.46)

Qualifying claims.   If a claim of degradability is used, it should be accompanied by a statement indicating the conditions under which degradation will occur, or a recognized test method that was used to determine the degradability.

Example: "Degradability determined in accordance with OECD test No. 301B"

There are a number of establishments that provide test methods and facilities for verifying claims of degradability.

Nonexaggeration of benefits.   In addition to the technical aspects of degradability, claims should not exaggerate the perceived benefit. If a product is only partly degradable, those ingredients that are identified as degradable should comprise a significant part of the product. Making degradable claims for minority ingredients could create the erroneous impression that the degradability of these ingredients is significant in relation to the waste generated by the total product and, thus, would not be appropriate.

Compostable Materials

Definition.   Similar to biodegradation, composting is the biological breakdown of organic materials that produces a soil-like product that is used to enhance, amend, or replace soil. While biodegradation may take place as a result of the disposal of a material, composting may require active participation by the end user who must separate organic from inorganic waste for municipal or large scale composting programs or, alternately, compost the material in a household composter after separation.

Supportable claims.   A claim that a product or package is compostable should be supportable by competent and reliable scientific evidence that the entire product or package will completely break down (i.e. decompose) after customary disposal. That is, it will undergo physical, chemical, thermal or biological decomposition in a compost site such that the material and its byproducts in the finished compost are not visually distinguishable and ultimately biodegrade into carbon dioxide, water, inorganic compounds, or biomass, at a rate consistent with known compostable materials and generate no significant toxic by-products.[2] This evidence should be one that is recognized by the scientific community at large, and not just a small group or individual company.

Kinds of composting.   There are significant differences between the types of materials that may be composted commercially and those suitable for composting at home. Commercial composting operations, such as those operated by municipal or large scale facilities, function under controlled conditions to better handle a larger variety of materials, that can be easily composted in the home.

Qualified claims.   When a claim of compostability is made for a material, and the material cannot be composted in a household composter:

- The claim should indicate that the material must be separated for municipal or large scale composting, and if there are special procedures to be followed for proper separation of the materials,
- There should be reasonable access to composting facilities. "Reasonable" will be determined according to the same general principles as for recycling; that is, one-third of the

---

2. Derived from May 14, 1993, Main Committee Ballot of ASTM committee on Terminology for Environmental Labelling of Packaging Materials and Packages (D10.46)

population in the area where the product is distributed has convenient access to collection or drop-off facilities for composting, and

- The claim should be qualified.

Example 1:	"Separate from waste for composting, where facilities accept X"
Example 2:	"Separate plastic materials before depositing in curbside composting program"
Example 3:	"This package is suitable for home or commercial composting—cut into 1 cm strips for use in home composters"

Unqualified claims.   If there is no indication that the material is intended for municipal or large scale composting, then the claim will be taken to mean that the material is suitable for composting in a household composter, and will break down with no significant toxic residues or other harmful products, under the normal conditions of household composting. In all cases, if the entire product or package is not compostable, the claim should identify the compostable components and/or provide instructions for proper separation of materials.

Recognition of infrastructure development initiatives.   It is recognized that composting infrastructures are not fully developed. In establishing these infrastructures, public awareness can play a significant role in the success or failure of a program.

In some cases, a material may be collected from a limited number of locations, or the infrastructure has not been sufficiently developed to provide general access to composting in a given market area. Clear, accurate and meaningful information relating to these programs should not be discouraged. However, in providing information to consumers that composting has begun for some materials, there should not be any confusion relating to the extent that these materials are collected.

In these exceptional cases, where the one-third threshold may not have been reached, but it is desirable to advise consumers of the new initiative, it may be appropriate to make a very specific and restrictive claim.

Example 1:	"Compostable in (City XX)"
Example 2:	"Please return to (facility/establishment X) for composting"
Example 3:	"For information on composting (material name), please call 800-XXX-XXXX"
Example 4:	"Compostable in limited communities, call 800-XXX-XXXX for details."

EXHIBIT A
Members of the Working Group

Association of Canadian Advertisers

Canadian Advertising Foundation

Canadian Cosmetic, Toiletry and Fragrance Association

Canadian Council of Grocery Distributors

Canadian Council of Ministers of the Environment National Task Force on Packaging

Canadian Manufacturers of Chemical Specialties Association

Canadian Pulp and Paper Association

Composting Council of Canada

Industry Canada
   Consumer Products Branch
   Marketing Practices Branch

Consumers' Association of Canada

Environment Canada
   Environmental Choice Program

Grocery Products Manufacturers of Canada

Packaging Association of Canada

Paper & Paperboard Packaging Environmental Council

Retail Council of Canada

Note: In addition to the above list, consultation was undertaken with a broad range of environmental groups, legal firms, industries, foreign governments and all levels of government in Canada.

EXHIBIT B
Environmental Choice Program

EXHIBIT F-6

The EcoLogo is an Official Mark of
Environment Canada used here with
the permission of Environmental
Choice.

Environmental Choice was created by Environment Canada to help consumers identify products and services that ease the burden on the environment. The EcoLogo is the symbol of certification that appears on products and services that meet the Environmental Choice product/ service-specific criteria. The symbol and the words "Eco-Logo" and "Environmental Choice" are Official Marks of Environment Canada and cannot be used without a license issued, or special permission granted, by Environmental Choice.

The product/service-specific criteria are set by the Environmental Choice Board, a body appointed by the federal Minister of the Environment, operating at arm's length from Government. For each set of criteria developed, a life-cycle study of the product or service is conducted. The purpose of the study is to identify which aspects of the life-cycle (manufacture, transport, use or disposal) offer opportunities to reduce significantly the product's or service's negative impacts on the environment.

Each set of product/service-specific environmental criteria is contained in a draft guideline that is made available for a public review period of 60 days. During this period, manufacturers, environmentalists, consumers and the general public may submit suggestions for the improvement of the criteria. The guideline is then finalized.

Once a guideline is final, manufacturers of products and purveyors of services may submit their products or services for testing against the criteria. An independent technical agency is under contract with Environmental Choice to test products and services against the criteria and to undertake periodic inspections to ensure they continue to measure up. Manufacturers and purveyors of services are required to pay for the initial test, as well as an annual fee to use the EcoLogo that is based on gross annual sales of the licensed product or service.

As of May 1993, over 800 products and services from some 140 companies bear the EcoLogo. The EcoLogo can be issued under the following final guidelines: re-refined motor oil, insulation from recycled wood-based cellulose fibre, products made from recycled plastic, zinc-air batteries, non-rechargeable batteries, reduced-pollution water-based paints, fine paper from recycled paper, miscellaneous products from recycled paper, newsprint from recycled paper, heat recovery ventilators, reusable cloth diapers, reduced-pollution solvent-based paints, ethanol-blended gasoline, residential composters, reusable utility bags, diaper services, water-conserving products, energy-efficient lights, and major household appliances. Other new guidelines are under development. All draft guidelines must receive Board as well as Ministerial approval before they are issued in final form.

Environmental Choice
107 Sparks Street, 2nd Floor
Ottawa, Ontario
K1A 0H3

Telephone: (613) 952-1120
FAX: (613) 952-9465

EXHIBIT C
National Packaging Protocol (NAPP)

In April 1989, the Canadian Council of Ministers of the Environment (CCME) recognized the magnitude of the waste management problem in Canada and set a goal of 50 percent reduction in waste generation by the year 2000. In order to address this problem, CCME commissioned a National Task Force on Packaging to develop a national policy for the management of packaging. After preparing an extensive technical database on packaging and conducting Canada-wide consultations, the Task Force produced the National Packaging Protocol, which recommends six packaging policies for Canada. Canadian Environmental Ministers endorsed the protocol in March 1990.

The six packaging policies constitute a plan of action, with specific waste reduction targets and schedules, that will reduce the burden of packaging waste through three achievable targets: 20 percent in 1992, 35 percent in 1996, and 50 percent by the year 2000.

To meet the milestone targets, the National Packaging Protocol recommends six policies for Canada:

Policy 1: All packaging shall have minimum effects on the environment.

Policy 2: Priority will be given to the management of packaging through source reduction, reuse and recycling.

Policy 3: A continuing campaign of information and education will be undertaken to make all Canadians aware of the function and environmental impacts of packaging.

Policy 4: These policies will apply to all packaging used in Canada, including imports.

Policy 5: Regulations will be implemented as necessary to achieve compliance with these policies.

Policy 6: All government policies and practices affecting packaging will be consistent with these national policies.

Canada's National Packaging Protocol is being implemented by a broad range of actions that will address the many facets of packaging issues. Implementation actions are closely linked to the specific statements that make up the Protocol.

For more information, contact:

Environment Canada Office of Waste Management
Conservation and Protection
Ottawa, Ontario
K1A 0H3

Telephone: (819) 997-3060

EXHIBIT D
Glossary of Terms (NAPP)

Compostable: A material that is capable of biodegradation in a composting system.

Degradable: Used to denote the ability of a material to break down significantly in a land or water eco-system.
    Bio—the prefix denoting ability to break down biologically by means of micro-organisms.
    Photo—the prefix denoting ability to break down through the action of light.

Package/packaging: A material or item that is used to protect, contain, or transport a commodity or product. A package can also be a material or item that is physically attached to a product or its container for the purpose of marketing the product or communicating information about the product.

Post-use material: Material which can no longer be used for its intended purpose and is separate from the waste stream for recycling. Post-use material **includes**:
    Post-consumer material—Material generated by commercial and institutional facilities, or households, which can no longer be used for its intended purpose; and
    Industrial scrap—Material left over from an industrial process that is not capable of being reused or reprocessed within the same plant or process. (Example: boxboard trim)
    Post-use material excludes the in-plant reutilization of materials, such as re-work, re-grind, re-pulp, or scrap materials generated within the plant and capable of being reused within the process that generated it.

Recovery: The process of reutilizing material or energy from solid waste.

Recyclable: Packages made from materials which after use can be diverted from the waste stream and recycled into a new product or package.
    For the purposes of promoting a material as recyclable, a material may be deemed recyclable where at least one-third (1/3) of the population across Canada has convenient access to collection or drop-off facilities for recycling the material, or where the material is produced for a regional market, that one-third (1/3) of the population in that market has convenient access to collection or drop-off facilities for recycling.

Recycled content: The portion of a package's weight that is composed of reprocessed post-use materials.

Reycling: A process through which post-use materials are separated from the waste stream, collected and processed for transformation into new products.

Reuse: The direct reapplication of a package, for the same or different purpose in its original form.

Source reduction: The elimination of packaging or reduction of the weight, volume or toxicity of packaging.

Waste: Any material, product or by-product which is discarded for final disposal.

Waste minimization: The reduction of waste through source reduction, reuse and recycling activities.

Note: This glossary has been developed in the context of reducing packaging waste. When used for labelling and advertising claims, any reference to "packaging" in this glossary may be used to apply to materials or products as well.

EXHIBIT E

Excerpts from the **Consumer Packaging and Labelling Act** *and the* **Competition Act**

A. *Consumer Packaging and Labelling Act*—An Act respecting the packaging, labelling, sale, importation and advertising of prepackaged and certain other products; Sections 2, 7 and 9.

   2. In this Act,

   "**Advertise**" means make any representation to the public by any means whatever, other than a label, for the purpose of promoting directly or indirectly the sale of a product;

   "**Dealer**" means a person who is a retailer, manufacturer, processor, or producer of a product, or a person who is engaged in the business of importing, packaging or selling any product;

   "**Label**" means any label, mark, sign, device, imprint, stamp, brand, ticket or tag;

   "**Prepackaged product**" means any product that is packaged in a container in such a manner that it is ordinarily sold to or used or purchased by a consumer without being repackaged.

   "**Product**" means any article that is or may be the subject of trade or commerce but does not include land or any interest therein; and

   "**Sell**" includes

   (a) Offer for sale, expose for sale and have in possession for sale, and

   (b) Display in such manner as to lead to a reasonable belief that the substance or product so displayed is intended for sale.

Representations relating to prepackaged products

   7. (1) No dealer shall apply to any prepackaged product or sell, import into Canada or advertise any prepackaged product that has applied to it a label containing any false or misleading representation that relates to or may reasonably be regarded as relating to that product.

Definition of "false or misleading representation"

   (2) For the purposes of this section, "false or misleading representation" includes

   (a) Any representation in which expressions, words, figures, depictions or symbols are used, arranged or shown in a manner that may reasonably be regarded as qualifying the declared net quantity of a prepackaged product or as likely to deceive a consumer with respect to the net quantity of a prepackaged product;

   (b) Any expression, word, figure, depiction or symbol that implies or may reasonably be regarded as implying that a prepackaged product contains any matter not contained in it or does not contain any matter in fact contained in it; and

   (c) Any description or illustration of the type, quality, performance, function, origin, or method of manufacture or production of a prepackaged product that may reasonbly be regarded as likely to deceive a consumer with respect to the matter so described or illustrated.

Containers of prepackaged products

   9. (1) No dealer shall sell, import into Canada or advertise any prepackaged product that is packaged in a container that has been manufactured, constructed or filled or is displayed in such a manner that a consumer might reasonably be misled with respect to the quality or quantity of the product.

B. *Competition Act*—An Act to provide for the general regulation of trade and commerce in respect of conspiracies, trade practices and mergers affecting competition; Sections 52 and 53

Misleading advertising

   52. (1) No person shall, for the purpose of promoting, directly or indirectly, the supply or use of a product or for the purpose of promoting, directly or indirectly, any business interest, by any means whatever,

   (a) Make a representation to the public that is false or misleading in a material respect;

EXHIBIT E *(continued)*

(b) Make a representation to the public in the form of a statement, warranty or guarantee of the performance, efficacy or length of life of a product that is not based on an adequate and proper test thereof, the proof of which lies on the person making the representation;

(c) Make a representation to the public in a form that purports to be

    (i) A warranty or guarantee of a product, or

    (ii) A promise to replace, maintain or repair an article or any part thereof or to repeat or continue a service until it has achieved a specified result

if the form of purported warranty or guarantee or promise is materially misleading or if there is a no reasonable prosect that it will be carried out; or

(d) Make a materially misleading representation to the public concerning the price at which a product or like products have been, are or will be ordinarily sold, and for the purposes of this paragraph a representation as to price is deemed to refer to the price at which the product has been sold by sellers generally in the relevant market unless it is clearly specified to be the price at which the product has been sold by the person by whom or on whose behalf the representation is made.

Deemed representation to public

(2) For the purposes of this section and section 53, a representation that is

(a) Expressed on an article offered or displayed for sale, its wrapper or container,

(b) Expressed on anything attached to, inserted in or accompanying an article offered or displayed for sale, its wrapper or container, or anything on which the article is mounted for display or sale,

(c) Expressed on an in-store or other point-of-purchase display,

(d) Made in the course of in-store, door-to-door or telephone selling to a person as ultimate user, or

(e) Contained in or on anything that is sold, sent, delivered, transmitted or in any other manner whatever made available to a member of the public,

shall be deemed to be made to the public by and only by the person who caused the representation to be so expressed, made or contained and, where that person is outside Canada, by

(f) The person who imported the article into Canada, in a case described in paragraph (a), (b), or (e), and

(g) The person who imported the display into Canada, in a case described in paragraph (c).

Idem

(3) Subject to subsection (2), every one who, for the purpose of promoting, directly or indirectly, the supply or use of a product or any business interest, supplies to a wholesaler, retailer or other distributor of a product any material or thing that contains a representation of a nature referred to in subsection (1) shall be deemed to have made that representation to the public.

General impression to be considered

(4) In any prosecution for a contravention of this section, the general impression conveyed by a representation as well as the literal meaning thereof shall be taken into account in determining whether or not the representation is false or misleading in a material respect.

Offence and punishment

(5) Any person who contravenes subsection (1) is guilty of an offence and liable

(a) On conviction on indictment, to a fine in the discretion of the court or to imprisonment for a term not exceeding five years or to both; or

(b) On summary conviction, to a fine not exceeding $25,000 or to imprisonment for a term not exceeding one year or to both. R.S., c. C-23, s. 36; 1974–75–76, c. 76, s. 18.

Representation as to reasonable test and publication of testimonials

53. (1) No person shall, for the purpose of promoting, directly or indirectly, the supply or use of any product, or for the purpose of promoting, directly or indirectly, any business interest,

EXHIBIT E *(concluded)*

    (a) Make a representation to the public that a test as to the performance, efficacy or length of life of the product has been made by any person, or

    (b) Publish a testimonial with respect to the product,

unless he can establish that

    (c) The representation or testimonial was previously made or published by the person by whom the test was made or the testimonial was given, as the case may be, or

    (d) The representation or testimonial was, before being made or published, approved and permission to make or publish it was given in writing by the person by whom the test was made or the testimonial was given, as the case may be

and the representation or testimonial accords with the representation or testimonial previously made, published or approved.

Offence and punishment

    (2) Any person who contravenes subsection (1) is guilty of an offence and liable

    (a) On conviction or indictment, to a fine in the discretion of the court or to imprisonment for a term not exceeding five years or to both; or

    (b) On summary conviction, to a fine not exceeding $25,000 or to imprisonment for a term not exceeding one year or to both. 1974–75–76, c. 76, s. 18.2.

FTC Deceptive and Unsubstantiated Claims Policy Protocol

[¶39,059] Deceptive and Unsubstantiated Claims Policy Protocol

Background: In 1975 the FTC adopted a protocol containing a series of questions for FTC staff to answer in determining which advertising cases should be pursued. According to a 1981 statement by FTC Chairman James C. Miller III (before the American Advertising Federation in Washington, D.C.), the protocol provides a workable framework for analysis and subsequent action and "will receive renewed and increased emphasis under my administration." Text of the protocol follows:

Deceptive and Unsubstantiated Claims Policy Protocol

A. Consumer Interpretations of the Claim

1. List the main interpretation that consumers may place on the claim recommended for challenge, including those that might render the claim true/substantiated as well as those that might render the claim false/unsubstantiated.
2. Indicate which of these interpretations would be alleged to be implications of the claim for purposes of substantiation of litigation. For each interpretation so indicated, state the reasons, if any, for believing that the claim so interpreted would be false/unsubstantiated.

B. Scale of the Deception or Lack of Substantiation

3. What is known about the relative proportions of consumers adhering to each of the interpretations listed above in response to Question 1?
4. What was the approximate advertising budget for the claim during the past year or during any other period of time that would reflect the number of consumers actually exposed to the claim? Is there more direct information on the number of consumers exposed to the claim?

C. Materiality

5. If consumers do interpret the claim in the ways that would be alleged to be implications, what reasons are there for supposing that these interpretations would influence purchase decisions?
6. During the past year, approximately how many consumers purchased the product[*] about which the claim was made?
7. Approximately what price did they pay?
8. Estimate, if possible, the proportion of consumers who would have purchased the product only at some price lower than they did pay, if at all, were they informed that the interpretations identified in response to Question 2 were false.
9. Estimate, if possible, what the advertised product would be worth to the consumers identified by Question 8 if they knew that the product did not have the positive (or unique) attributes suggested by the claim. If the claim can cause consumers to disregard some

---

[*] Throughout, "product" refers to the particular brand advertised.

negative attribute, such as a risk to health and safety, to their possible physical or economic injury, so specify. If so, estimate, if possible, the annual number of such injuries attributable to the claim.

D. Adequacy of Corrective Market Forces

10. If the product to which the claim relates is a low-ticket item, can consumers ordinarily determine prior to purchase whether the claim, as interpreted, is true, or invest a small amount in purchase and then by experience with the product determine whether or not the claim is true? Does the claim relate to a credence quality, that is, a quality of the product that consumers ordinarily cannot evaluate during normal use of the product without acquiring costly information from some source other than their own evaluative faculties?

11. Is the product to which the claim relates one that a consumer would typically purchase frequently? Have produce sales increased or decreased substantially since the claim was made?

12. Are there sources of information about the subject matter of the claim in addition to the claim itself? If so, are they likely to be recalled by consumers when they purchase or use the product? Are they likely to be used by consumers who are not aggressive, effective shoppers? If not, why not?

E. Effect on the Flow of Truthful Information

13. Will the standard of truth/substantiation that would be applied to the claim under the recommendation to initiate proceedings make it extremely difficult as to practical matter to make the type of claim? Is this result reasonable?

14. What are the consequences to consumers of an erroneous determination by the Commission that the claim is false/unsubstantiated? What are the consequences to consumers of an erroneous determination by the Commission that the claim is true/substantiated?

F. Deterrence

15. Is there a possibility of getting significant relief with broad product or claim coverage? What relief is possible? Why would it be significant?

16. Do the facts of the matter recommended present an opportunity to elaborate a rule of law that would be applicable to claims or advertisers other than those that would be directly challenged by the recommended action? If so, describe this rule of law as you would wish the advertising community to understand it. If this rule of law would be a significant precedent, explain why.

17. Does the claim violate a Guide or is it inconsistent with relevant principles embodied in a Guide?

18. Is the fact of a violation so evident to other industry members that, if we do not act, our credibility and deterrence might be adversely affected?

19. Is there any aspect of the advertisement—*e.g.*, the nature of the advertiser, the product, the theme, the volume of the advertising, the memorableness of the ad, the blatancy of the violation—which indicates that an enforcement action would have substantial impact on the advertising community?

20. What, if anything, do we know about the role advertising plays (as against other promotion techniques and other sources of information) in the decision to purchase the product?

21. What is the aggregate dollar volume spent on advertising by the advertiser to be joined in the recommended action?

22. What is the aggregate volume of sales of the advertised product and of products the same type?

G. Law Enforcement Efficiency

23. Has another agency taken action or does another agency have expertise with respect to the claim or its subject matter? Are there reasons why the Commission should defer? What is the position of this other agency? If coordination is planned, what form would it take?

24. How difficult would it be to litigate a case challenging the claim? Would the theory of the proceeding recommended place the Commission in a position of resolving issues that are better left to other modes of resolution, for instance, debate among scientists? If so, explain. Is there a substantial possibility of whole or partial summary judgment?

25. Can the problem seen in the ad be handled by way of a rule? Are the violations widepsread? Should they be handled by way of a rule?

H. Additional Considerations

26. What is the ratio of the advertiser's advertising expense to sales revenues? How, if at all, is this ratio relevant to the public interest in proceeding as recommended?

27. Does the claim specially affect a vulnerable group?

28. Does the advertising use deception or unfairness to offend important values or to exploit legitimate concerns of a substantial segment of the population, whether or not there is direct injury to person or pocketbook, *e.g.* minority hiring or environmental protection?

29. Are there additional considerations not elicited by previous questions that would affect the public interest in proceeding?

FTC Statement on "Unfairness"

FEDERAL TRADE COMMISSION
WASHINGTON, D.C. 20580

December 17, 1980

The Honorable Wendell H. Ford
Chairman, Consumer Subcommittee
Committee on Commerce, Science, and Transportation
Room 130 Russell Office Building
Washington, D.C. 20510

The Honorable John C. Danforth
Ranking Minority Member, Consumer Subcommittee
Committee on Commerce, Science, and Transportation
Room 130 Russell Office Building
Washington, D.C. 20510

Dear Senators Ford and Danforth:

This is in response to your letter of June 13, 1980, concerning one aspect of this agency's jurisdiction over "unfair or deceptive acts or practices." You informed us that the Subcommittee was planning to hold oversight hearings on the concept of "unfairness" as it has been applied to consumer transactions. You further informed us that the views of other interested parties were solicited and compiled in a Committee Print earlier this year.[1] Your letter specifically requested the Commission's views on cases under Section 5 "not involving the content of advertising," and its views as to "whether the Commission's authority should be limited to regulating 'false or deceptive' commercial advertising." Our response addresses these and other questions related to the concept of consumer unfairness.

We are pleased to have this opportunity to discuss the future work of the agency. The subject that you have selected appears to be particularly timely. We recognize that the concept of consumer unfairness is one whose precise meaning is not immediately obvious, and also recognize that this uncertainty has been honestly troublesome for some businesses and some members of the legal profession. This result is understandable in [2] light of the general nature of the statutory standard. At the same time, though, we believe we can respond to legitimate concerns of business and the Bar by attempting to delineate in this letter a concrete framework for future application of the Commission's unfairness authority. We are aided in this process by the cumulative decisions of this agency and the federal courts, which, in our opinion, have brought added clarity to the law. Although the administrative and judicial evolution of the consumer unfairness concept has still left some necessary flexibility in the statute, it is possible to provide a reasonable working sense of the conduct that is covered.

---

1. Unfairness: Views on Unfair Acts and Practices in Violation of the Federal Trade Commission Act (1980) (hereinafter referred to as "Committee Print").

In response to your inquiry we have therefore undertaken a review of the decided cases and rules and have synthesized from them the most important principles of general applicability. Rather than merely reciting the law, we have attempted to provide the Committee with a concrete indication of the manner in which the Commission has enforced, and will continue to enforce, its unfairness mandate. In so doing we intend to address the concerns that have been raised about the meaning of consumer unfairness, and thereby attempt to provide a greater sense of certainty about what the Commission would regard as an unfair act or practice under Section 5.

This letter thus delineates the Commission's views of the boundaries of its consumer unfairness jurisdiction and is subscribed to by each Commissioner. In addition, we are enclosing a companion Commission statement that discusses the way in which this body of law differs from, and supplements, the prohibition against consumer deception, and then considers and evaluates some specific criticisms that have been made of our enforcement of the law.[2] Since you have indicated a particular interest in the possible application of First Amendment principles to commercial advertising, the [3] companion statement is designed to respond to the key questions raised about the unfairness doctrine. However, individual Commissioners may not necessarily endorse particular arguments or particular examples of the Commission's exercise of its unfairness authority contained in the companion statement.

Commission Statement of Policy on the Scope of the
Consumer Unfairness Jurisdiction

Section 5 of the FTC Act prohibits, in part, "unfair . . . acts or practices in or affecting commerce."[3] This is commonly referred to as the Commission's consumer unfairness jurisdiction. The Commission's jurisdiction over "unfair methods of competition" is not discussed in this letter.[4] Although we cannot give an exhaustive treatment of the law of consumer unfairness in this short statement, some relatively concrete conclusions can nonetheless be drawn.

The present understanding of the unfairness standard is the result of an evolutionary process. The statute was deliberately framed in general terms since Congress recognized the impossibility of drafting a complete list of unfair trade [4] practices that would not quickly become outdated or leave loopholes for easy evasion.[5] The task of identifying unfair trade practices was therefore assigned to the Commission, subject to judicial review,[6] in the expectation that the underlying criteria would evolve and develop over time. As the Supreme Court observed as early as 1931, the ban on unfairness "belongs to that class of phrases which do not admit of precise definition, but the meaning and application of which must be arrived at by what this court elsewhere has called 'the gradual process of judicial inclusion and exclusion.' "[7]

By 1964 enough cases had been decided to enable the Commission to identify three factors that it considered when applying the prohibition against consumer unfairness. These were: (1) whether the practice injures consumers; (2) whether it violates established public policy; (3) whether it is unethical or unscrupulous.[8] These

---

2. Neither this letter nor the companion statement addresses ongoing proceedings, but the Commission is prepared to discuss those matters separately at an appropriate time.

3. The operative sentence of Section 5 reads in full as follows: "Unfair methods of competition in or effecting commerce, and unfair or deceptive acts or practices in or affecting commerce, are declared unlawful." 15 U.S.C. 45(a)(1).

4. In fulfilling its competition or antitrust mission the Commission looks to the purposes, policies, and spirit of the other antitrust laws and the FTC Act to determine whether a practice affecting competition or competitors is unfair. *See, e.g., FTC v. Brown Shoe Co.*, 384 U.S. 316 (1966). In making this determination the Commission is guided by the extensive legislative histories of those statutes and a considerable body of antitrust case law. The agency's jurisdiction over "deceptive acts or practices" is likewise not discussed in this letter.

5. *See* H.R. Conf. Rep. No. 1142, 63d Cong., 2d Sess., at 19 (1914) (If Congress "were to adopt the method of definition, it would undertake an endless task"). In 1914 the statue was phrased only in terms of "unfair methods of competition," and the reference to "unfair acts or practices" was not added until the Wheeler–Lee Amendment in 1938. The initial language was still understood as reaching most of the conduct now characterized as consumer unfairness, however, and so the original legislative history remains relevant to the construction of that part of the statute.

6. The Supreme Court has stated on many occasions that the definition of "unfairness" is ultimately one for judicial determination. *See, e.g., FTC v. Sperry & Hutchinson Co.*, 405 U.S. 233, 249 (1972); *FTC v. R.F. Keppel & Bro.*, 291 U.S. 304. 314 (1934).

7. *FTC v. Raladam Co., 283 U.S. 643, 648 (1931). See also FTC v. R.F. Keppel & Bro.*, 291 U.S. 304, 310 (1934) ("Neither the language nor the history of the Act suggests that Congress intended to confine the forbidden methods to fixed and unyielding categories").

8. The Commission's actual statement of the criteria was as follows:

   (1) whether the practice, without necessarily having been previously considered unlawful, offends public policy as it has been established by statutes, the common law, or otherwise—whether, in order words, it is within at least the penumbra of some common-law, statutory, or other established concept of unfairness; (2) whether it is immoral, unethical, oppressive, or unscrupulous; (3) whether it causes substantial injury to consumers (or competitors or other businessmen).

Statement of Basis and Purpose. Unfair or Deceptive Advertising and Labeling of Cigarettes in Relation to the Health Hazards of Smoking, 29 Fed. Reg. 8324, 8355 (1964).

factors [5] were later quoted with apparent approval by the Supreme Court in the 1972 case of *Sperry & Hutchinson*.[9] Since then the Commission has continued to refine the standard of unfairness in its cases and rules, and it has now reached a more detailed sense of both the definition and the limits of these criteria.[10]

Consumer injury

Unjustified consumer injury is the primary focus of the FTC Act, and the most important of the three *S&H* criteria. By itself it can be sufficient to warrant a finding of unfairness. The Commission's ability to rely on an independent criterion of consumer injury is consistent with the intent of the statute, which was to " [make] the consumer who may be injured by an unfair trade practice of equal concern before the law with the merchant injured by the unfair methods of a dishonest competitor."[11]

The independent nature of the consumer injury criterion does not mean that every consumer injury is legally "unfair," however. To justify a finding of unfairness the injury must satisfy three tests. It must be substantial; it must not be outweighed by any countervailing benefits to consumers or competition that the practice produces; and it must be an injury that consumers themselves could not reasonably have avoided.

First of all, the injury must be substantial. The Commission is not concerned with trivial or merely speculative harms.[12] In most cases a substantial injury involves monetary harm, as when sellers coerce consumers into purchasing unwanted goods or [6] services[13] or when consumers buy defective goods or services on credit but are unable to assert against the creditor claims or defenses arising from the transaction.[14] Unwarranted health and safety risks may also support a finding of unfairness.[15] Emotional impact and other more subjective types of harm, on the other hand, will not ordinarily make a practice unfair. Thus, for example, the Commission will not seek to ban an advertisement merely because it offends the tastes or social beliefs of some viewers, as has been suggested in some of the comments.[16]

Second, the injury must not be outweighed by any offsetting consumer or competitive benefits that the sales practice also produces. Most business practices entail a mixture of economic and other costs and benefits for purchasers. A seller's failure to present complex technical data on his product may lessen a consumer's ability to choose, for example, but may also reduce the initial price he must pay for the article. The Commission is aware of these tradeoffs and will not find that a practice unfairly injures consumers unless it is injurious in its net effects.[17] The Commission [7] also takes account of the various costs that a remedy would entail. These include not only the costs to the parties directly before the agency, but also the burdens on society in general in the form of increased paperwork, increased regulatory burdens on the flow of information, reduced incentives to innovation and capital formation, and similar matters.[18]

---

9. *FTC v. Sperry & Hutchinson Co.*, 405 U.S. 223, 244–45 n.5 (1972). The circuit courts have concluded that this quotation reflected the Supreme Court's own views. *See Spiegel, Inc. v. FTC*, 540 F.2d 287, 293 n.8 (7th Cir. 1976); *Heater v. FTC*, 503 F.2d 321, 323 (9th Cir. 1974). The application of these factors to antitrust matters is beyond the scope of this letter.

10. These standards for unfairness are generally applicable to both advertising and non-advertising cases.

11. 83 Cong. Rec. 3255 (1938) (remarks of Senator Wheeler).

12. An injury may be sufficiently substantial, however, if it does a small harm to a large number of people, or if it raises a significant risk of concrete harm.

13. *See, e.g., Holland Furnace Co. v. FTC*, 295 F.2d 302 (7th Cir. 1961) (seller's servicemen dismantled home furnaces and then refused to reassemble them until the consumers had agreed to buy services or replacement parts).

14. Statement of Basis and Purpose. Preservation of Consumers' Claims and Defenses, 40 Fed. Reg. 53506, 53522–23 (1975).

15. For an example *see Philip Morris, Inc.*, 82 F.T.C. 16 (1973) (respondent had distributed free-sample razor blades in such a way that they could come into the hands of small children) (consent agreement). Of course, if matters involving health and safety are within the primary jurisdiction of some other agency, Commission action might not be appropriate.

16. *See, e.g.,* comments of Association of National Advertisers, Committee Print at 120. In an extreme case, however, where tangible injury could be clearly demonstrated, emotional effects might possibly be considered as the basis for a finding of unfairness. *Cf.* 15 U.S.C. 1692 *et seq.* (Fair Debt Collection Practices Act) (banning, *e.g.,* harassing, late-night telephone calls).

17. *See Pfizer, Inc.*, 81 F.T.C. 23, 62–63 n.13 (1972); Statement of Basis and Purpose, Disclosure Requirements and Prohibitions Concerning Franchising and Business Opportunity Ventures, 43 Fed. Reg. 59614, 59636 n.95 (1978). When making this determination the Commission may refer to existing public policies for help in ascertaining the existence of consumer injury and the relative weights that should be assigned to various costs and benefits. The role of public policy in unfairness determinations will be discussed more generally below.

18. For example, when the Commission promulgated the Holder Rule it anticipated an overall lowering of economic costs to society because the rule gave creditors the incentive to police sellers, thus increasing the likelihood that those selling defective goods or services would either improve their practices or leave the marketplace when they could not obtain financing. These benefits, in the Commission's judgment, outweighed any costs to creditors and sellers occasioned by the rule. *See* Statement of Basis and Purpose, Preservation of Consumers' Claims and Defenses, 40 Fed. Reg. 53506, 53522–23 (1975).

Finally, the injury must be one which consumers could not reasonably have avoided.[19] Normally we expect the marketplace to be self-correcting, and we rely on consumer choice—the ability of individual consumers to make their own private purchasing decisions without regulatory intervention—to govern the market. We anticipate that consumers will survey the available alternatives, choose those that are most desirable, and avoid those that are inadequate or unsatisfactory. However, it has long been recognized that certain types of sales techniques may prevent consumers from effectively making their own decisions, and that corrective action may then become necessary. Most of the Commission's unfairness matters are brought under these circumstances. They are brought, not to second-guess the wisdom of particular consumer decisions, but rather to halt some form of seller behavior that unreasonably creates or takes advantage of an obstacle to the free exercise of consumer decisionmaking.[20][8]

Sellers may adopt a number of practices that unjustifiably hinder such free market decisions. Some may withhold or fail to generate critical price or performance data, for example, leaving buyers with insufficient information for informed comparisons.[21] Some may engage in overt coercion, as by dismantling a home appliance for "inspection" and refusing to reassemble it until a service contract is signed.[22] And some may exercise undue influence over highly susceptible classes of purchases, as by promoting fraudulent "cures" to seriously ill cancer patients.[23] Each of these practices undermines an essential precondition to a free and informed consumer transaction, and, in turn, to a well-functioning market. Each of them is therefore properly banned as an unfair practice under the FTC Act.[24][9]

Violation of public policy

The second *S&H* standard asks whether the conduct violates public policy as it has been established by statute, common law, industry practice, or otherwise. This criterion may be applied in two different ways. It may be used to test the validity and strength of the evidence of consumer injury, or, less often, it may be cited for a dispositive legislative or judicial determination that such injury is present.

Although public policy was listed by the *S&H* Court as a separate consideration, it is used most frequently by the Commission as a means of providing additional evidence on the degree of consumer injury caused by specific practices. To be sure, most Commission actions are brought to redress relatively clear-cut injuries, and those determinations are based, in large part, on objective economic analysis. As we have indicated before, the Commission believes that considerable attention should be devoted to the analysis of whether substantial net harm has occurred, not only because that is part of the unfairness test, but also because the focus on injury is the best way to ensure that the Commission acts responsibly and uses its resources wisely. Nonetheless, the Commission wishes to emphasize the importance of examining outside statutory policies and established judicial principles for assistance in helping the agency ascertain whether a particular form of conduct does in fact tend to harm consumers. Thus the agency has referred to First Amendment decisions upholding consumers'

---

19. In some sense any injury can be avoided—for example, by hiring independent experts to test all products in advance, or by private legal actions for damages—but these courses may be too expensive to be practicable for individual consumers to pursue.

20. This emphasis on informed consumer choice has commonly been adopted in other statutes as well. *See, e.g.*, Declaration of Policy, Fair Packaging and Labeling Act, 15 U.S.C. 1451 ("Informed consumers are essential to the fair and efficient functioning of a free market economy").

21. *See, e.g.*, Statement of Basis and Purpose, Labeling and Advertising of Home Insulation, 44 Fed. Reg. 50218, 50222–23 (1979); Statement of Basis and Purpose, Posting of Minimum Octane Numbers on Gasoline Dispensing Pumps, 36 Fed. Reg. 23871, 23882 (1971). *See also Virginia State Board of Pharmacy v. Virginia Citizens Consumer Council, Inc.*, 425 U.S. 748 (1976).

22. *See Holland Furnace Co. v. FTC*, 295 F.2d 302 (7th Cir. 1961); *cf. Arthur Murray Studio, Inc. v. FTC*, 458 F.2d 622 (5th Cir. 1972) (emotional high-pressure sales tactics, using teams of salesmen who refused to let the customer leave the room until a contract was signed). *See also* Statement of Basis and Purpose, Cooling-off Period for Door-to-Door Sales, 37 Fed. Reg. 22934, 22937–38 (1972).

23. *See, e.g., Travel King, Inc.*, 86 F.T.C. 715, 774 (1975). The practices in this case primarily involved deception, but the Commission noted the special susceptibilities of such patients as one reason for banning the ads entirely rather than relying on the remedy of fuller disclosure. The Commission recognizes that "undue influence" in advertising and promotion is difficult to define, and therefore exercises its authority here only with respect to substantial coercive-like practices and significant consumer injury.

24. These few examples are not exhaustive, but the general direction they illustrate is clear. As the Commission stated in promulgating its Eyeglasses Rule, the inquiry should begin, at least, by asking "whether the acts or practices at issue inhibit the functioning of the competitive market and whether consumers are harmed thereby." Statement of Basis and Purpose, Advertising of Ophthalmic Goods and Services, 43 Fed. Reg. 23992, 24001 (1978).

rights to receive information, for example, to confirm that restrictions on advertising tend unfairly to hinder the informed exercise of consumer choice.[25][12]

Conversely, statutes or other sources of public policy may affirmatively allow for a practice that the Commission tentatively views as unfair. The existence of such policies will then give the agency reason to reconsider its assessment of whether the practice is actually injurious in its net effects.[26] In other situations there may be no clearly established public policies, or the policies may even be in conflict. While that does not necessarily preclude the commission from taking action if there is strong evidence of net consumer injury, it does underscore the desirability of carefully examining public policies in all instances.[27] In any event, whenever objective evidence of consumer injury is difficult to obtain, the need to identify and assess all relevant public policies assumes increased importance.

Sometimes public policy will independently support a Commission action. This occurs when the policy is so clear that it will entirely determine the question of consumer injury, so there is little need for separate analysis by the Commission. In these cases the legislature or court, in announcing the policy, has already determined that such injury does exist and thus it need not be expressly provided in each instance. An example [11] of this approach arose in a case involving a mail-order firm.[28] There the Commission was persuaded by an analogy to the due-process clause that it was unfair for the firm to bring collection suits in a forum that was unreasonably difficult for the defendants to reach. In a similar case the Commission applied the statutory policies of the Uniform Commercial Code to require that various automobile manufacturers and their distributors refund to their customers any surplus money that was realized after they repossessed and resold their customer's cars.[29] The Commission acts on such a basis only where the public policy is suitable for administrative enforcement by this agency, however. Thus it turned down a petition for a rule to require fuller disclosure of aerosol propellants, reasoning that the subject of fluorocarbon safety was currently under study by other scientific and legislative bodies with more appropriate expertise or jurisdiction over the subject.[30] [12]

To the extent that the Commission relies heavily on public policy to support a finding of unfairness, the policy should be clear and well-established. In other words, the policy should be declared or embodied in formal sources such as statutes, judicial decisions, or the Constitution as interpreted by the courts, rather than being ascertained from the general sense of the national values. The policy should likewise be one that is widely shared, and not the isolated decision of a single state or a single court. If these two tests are not met the policy cannot be considered as an "established" public policy for purposes of the *S&H* criterion. The Commission would then act only on the basis of convincing independent evidence that the practice was distorting the operation of the market and thereby causing unjustified consumer injury.

---

25. *See* Statement of Basis and Purpose, Advertising of Ophthalmic Goods and Services, 43 Fed. Reg. 23992, 24001 (1978), *citing Virginia State Board of Pharmacy v. Virginia Citizens Consumer Council*, 425 U.S. 748 (1976).

26. *Cf.* Statement of Basis and Purpose, Advertising of Ophthalmic Goods and Services, *supra, see also* n.17 *supra.*

27. The analysis of external public policies is extremely valuable but not always definitive. The legislative history of Section 5 recognizes that new forms of unfair business practices may arise which, at the time of the Commission's involvement, have not yet been generally proscribed. *See* page 4, *supra*. Thus a review of public policies established independently of Commission action may not be conclusive in determining whether the challenged practices should be prohibited or otherwise restricted. At the same time, however, we emphasize the importance of examining public policies, since a thorough analysis can serve as an important check on the overall reasonableness of the Commission's actions.

28. *Spiegel, Inc. v. FTC*, 540 F.2d 287 (7th Cir. 1976). In this case the Commission did inquire into the extent of the resulting consumer injury, but under the rationale involved it presumably need not have done so. *See also FTC v. R.F. Keppel & Bro.*, 291 U.S. 304 (1934) (firm had gained a marketing advantage by selling goods through a lottery technique that violated state gambling policies); *cf. Simeon Management Corp.*, 87 F.T.C. 1184, 1231 (1976), *aff'd*, 579 F.2d 1137 (9th Cir. 1978) (firm advertised weight-loss program that used a drug which could not itself be advertised under FDA regulations) (alternative ground). Since these public-policy cases are based on legislative determinations, rather than on a judgment within the Commission's area of special economic expertise, it is appropriate that they can reach a relatively wider range of consumer injuries than just those associated with impaired consumer choice.

29. A surplus occurs when a repossessed car is resold for more than the amount owed by the debtor plus the expenses of repossession and resale. The law of 49 states requires that creditors refund surpluses when they occur, but if creditors systematically refuse to honor this obligation, consumers have no practical way to discover that they have been deprived of money to which they are entitled. *See Ford Motor Co.*, 94 F.T.C. 564, 618 (1979) *appeal pending*. Nos. 79-7649 and 79-7654 (9th Cir.); *Ford Motor Co.*, 93 F.T.C. 402 (1979) (consent decree); *General Motors Corp.*, D. 9074 (Feb. 1980) (consent decree). By these latter two consent agreements the Commission, because of its unfairness jurisdiction, has been able to secure more than $2 million for consumers allegedly deprived of surpluses to which they were entitled.

30. *See* Letter from John F. Dugan, Acting Secretary, to Action on Smoking and Health (January 13, 1977). *See also* Letter from Charles A. Tobin, Secretary, to Prof. Page and Mr. Young (September 17, 1973) (denying petition to exercise § 6(b) subpoena powers to obtain consumer complaint information from cosmetic firms and then to transmit the data to FDA for that agency's enforcement purposes).

Unethical or unscrupulous conduct

Finally, the third *S&H* standard asks whether the conduct was immoral, unethical, oppressive, or unscrupulous. This test was presumably included in order to be sure of reaching all the purposes of the underlying statute, which forbids "unfair" acts or practices. It would therefore allow the Commission to reach conduct that violates generally recognized standards of business ethics. The test has proven, however, to be largely duplicative. Conduct that is truly unethical or unscrupulous will almost always injure consumers or violate public policy as well. The Commission has therefore never relied on the third element of *S&H* as an independent basis for a finding of unfairness, and it will act in the future only on the basis of the first two. [13]

We hope this letter has given you the information that you require. Please do not hesitate to call if we can be of any further assistance. With best regards,

/s/ Michael Pertschuk
Chairman

/s/ Paul Rand Dixon
Commissioner

/s/ David A. Clanton
Commissioner

/s/ Robert Pitofsky
Commissioner

/s/ Patricia P. Bailey
Commissioner

FTC Statement on Advertising Substantiation

FEDERAL TRADE COMMISSION
POLICY STATEMENT REGARDING ADVERTISING SUBSTANTIATION
PROGRAM

Agency: Federal Trade Commission.

Action: The Federal Trade Commission has issued a Policy Statement regarding its advertising substantiation program.

Summary: The Federal Trade Commission has issued a Policy Statement that articulates its policy regarding the legal requirement pursuant to section 5 of the Federal Trade Commission Act that advertisers and ad agencies have a reasonable basis for their objective claims before their initial dissemination. This Policy Statement is a result of a review of the comments filed in response to a public inquiry that the Commission initiated in March 1983. The Policy Statement reaffirms the Commission's commitment to the reasonable basis requirement and at the same time explains several refinements that will lead to a more efficient program of law enforcement with lower costs to the public, the advertising industry, and the agency.

    For further information contact: Collot Guerard, Federal Trade Commission, Bureau of Consumer Protection, Washington, D.C. 20580, 202-376-8648.

FTC POLICY STATEMENT REGARDING ADVERTISING SUBSTANTIATION

Introduction

On March 11, 1983, the Commission published a notice requesting comments on its advertising substantiation program.[1] To facilitate analysis of the program, the notice posed a number of questions concerning the program's procedures, standards, benefits, and costs, and solicited suggestions for making the program more effective. Based on the public comments and the staff's review, the Commission has drawn certain conclusions about how the program is being implemented and how it might be refined to serve better the objective of maintaining a marketplace free of unfair and deceptive acts or practices. This statement articulates the Commission's policy with respect to advertising substantiation.

The Reasonable Basis Requirement

First, we reaffirm our commitment to the underlying legal requirement of advertising substantiation—that advertisers and ad agencies have a reasonable basis for advertising claims before they are disseminated.

    The Commission intends to continue vigorous enforcement of this existing legal require-

---

1. 48 FR 10471, March 11, 1983.

ment that advertisers substantiate express and implied claims, however conveyed, that make objective assertions about the item or service advertised. Objective claims for products or services represent explicitly or by implication that the advertiser has a reasonable basis supporting these claims. These representations of substantiation are material to consumers. That is, consumers would be less likely to rely on claims for products and services if they knew the advertiser did not have a reasonable basis for believing them to be true.[2] Therefore, a firm's failure to possess and rely upon a reasonable basis for objective claims constitutes an unfair and deceptive act or practice in violation of section 5 of the Federal Trade Commission Act.

Standards for Prior Substantiation

Many ads contain express or implied statements regarding the amount of support the advertiser has for the product claim. When the substantiation claim is express (*e.g.,* "tests prove," "doctors recommend," and "studies show"), the Commission expects the firm to have at least the advertised level of substantiation. Of course, an ad may imply more substantiation than it expressly claims or may imply to consumers that the firm has a certain type of support; in such cases, the advertiser must possess the amount and type of substantiation the ad actually communicates to consumers.

Absent an express or implied reference to a certain level of support, and absent other evidence indicating what consumer expectations would be, the Commission assumes that consumers expect a "reasonable basis" for claims. The Commission's determination of what constitutes a reasonable basis depends, as it does in an unfairness analysis, on a number of factors relevant to the benefits and costs of substantiating a particular claim. These factors include: the type of claim, the product, the consequences of a false claim, the benefits of a truthful claim, the cost of developing substantiation for the claim, and the amount of substantiation experts in the field believe is reasonable. Extrinsic evidence, such as expert testimony or consumer surveys, is useful to determine what level of substantiation consumers expect to support a particular product claim and the adequacy of evidence an advertiser possesses.

One issue the Commission examined was substantiation for implied claims. Although firms are unlikely to possess substantiation for implied claims they do not believe the ad makes, they should generally be aware of reasonable interpretations and will be expected to have prior substantiation for such claims. The Commission will take care to assure that it only challenges reasonable interpretations of advertising claims.[3]

Procedures for Obtaining Substantiation

In the past, the Commission has sought substantiation from firms in two different ways: Through industry-wide "rounds" that involved publicized inquiries with identical or substantially similar demands to a number of firms within a targeted industry or to firms in different industries making the same type of claim; and on a case-by-case basis, by sending specific requests to individual companies under investigation. The Commission's review indicates that "rounds" have been costly to both the recipient and to the agency and have produced little or no law enforcement benefit over a case-by-case approach.

The Commission's traditional investigatory procedure allows the staff to investigate a number of firms within an industry at the same time, to develop necessary expertise within the area of investigation, and to announce our activities publicly in circumstances

---

2. Nor presumably would an advertiser have made such claims unless the advertiser thought they would be material to consumers.

3. Individual Commissioners have expressed differing views as to how claims should be interpreted so that advertisers are not held to outlandish or tenuous interpretations. Notwithstanding these variations in approach, the focus of all Commissioners on reasonable interpretations of claims is intended to ensure that advertisers are not required to substantiate claims that were not made.

where public notice or comment is desirable. The Commission intends to continue undertaking such law enforcement efforts when appropriate. However, since substantiation is principally a law enforcement tool and the Commission's concern in such investigations is with the substantiation in the *advertiser's* possession, there is little, if any, information that the public could contribute in such investigations. Therefore, the Commission anticipates that substantiation investigations will rarely be made public before they are completed.

Accordingly, the Commission has determined that in the future it will rely on nonpublic requests for substantiation directed to individual companies via an informal access letter or, if necessary, a formal civil investigative demand. The Commission believes that tailored, firm-specific requests, whether directed to one firm or to several firms within the same industry, are a more efficient law enforcement technique. The Commission cannot presently foresee circumstances under which the past approach of industry-wide rounds would be appropriate in the ad substantiation area.

Relevance of Post-Claim Evidence in Substantiation Cases

The reasonable basis doctrine requires that firms have the substantiation before disseminating a claim. The Commission has on occasion exercised its discretion, however, to consider supporting materials developed after dissemination.[4] The Commission has not previously identified in one document the circumstances in which it may, in its discretion, consider post-claim evidence in substantiation cases.[5] Such guidance can serve to clarify the program's actual operation as well as focus consideration of post-claim evidence on cases in which it is appropriate.

The Commission emphasizes that as a matter of law, firms lacking a reasonable basis before an ad is disseminated violate section 5 of the FTC Act and are subject to prosecution. The goal of the advertising substantiation requirement is to assure that advertising is truthful, however, and the truth of falsity of a claim is always relevant to the Commission's deliberations. Therefore, it is important that the agency retain the discretion and flexibility to consider additional substantiating evidence, not as a substitute for an advertiser's prior substantiation, but rather in the following circumstances:

- When deciding, before issuance of a complaint, whether there is a public interest in proceeding against a firm;
- When assessing the adequacy of the substantiation an advertiser possessed before a claim was made; and
- When deciding the need for or appropriate scope of an order to enter against a firm that lacked a reasonable basis prior to disseminating an advertisement.

First, using post-claim evidence to evaluate the truth of a claim, or otherwise using such evidence in deciding whether there is a public interest in continuing an investigation or issuing a complaint, is appropriate policy. This does not mean that the Commission will postpone action while firms create post-claim substantiation to prove the truthfulness of claims, nor does it mean that subsequent evidence of truthfulness absolves a firm of liability for failing to possess prior substantiation for a claim. The Commission focuses instead on whether existing evidence that claims are true should lead us in the exercise of our prosecutorial discretion to decline to initiate a law enforcement proceeding. If available post-claim evidence proves that the claim is true, issuing a complaint against a firm that may have violated the prior substantiation requirement is often inappropriate, particularly in light of competing demands on the Commission's resources.

---

4. The Commission's evidentiary rule, 16 CFR 3.40, has sometimes been interpreted as precluding introduction of post-claim substantiation. In fact, it does not. Section 3.40 only provides a sanction against the introduction of evidence that should have been produced in response to a subpoena, but was not.

5. The distinction between pre-claim and post-claim evidence is only relevant when the charge is lack of substantiation. For other charges, such as falsity, when evidence was developed is irrelevant to its admissibility at trial.

Second, post-claim evidence may indicate that apparent deficiencies in the pre-claim substantiation materials have no practical significance. In evaluating the adequacy of prior substantiation, the Commission will consider only post-claim substantiation that sheds light on pre-existing substantiation. Thus, advertisers will not be allowed to create entirely new substantiation simply because their prior substantiation was inadequate.

Finally, the Commission may use post-claim evidence in determining the need for or appropriate scope of an order to be entered against a firm that lacked a reasonable basis. Thus, when additional evidence offered for the first time at trial suggests that the claim is true, the Commission may frame a narrower order than if there had been no post-claim evidence.

The Commission remains committed to the prior substantiation requirement and further believes that these discretionary factors will provide necessary flexibility. The Commission will consider post-claim evidence only in the circumstances listed above. But, whether it will do so in any particular case remains within its discretion.

Self-Regulation Groups and Government Agencies

The Commission traditionally has enjoyed a close working relationship with self-regulation groups and government agencies whose regulatory policies have some bearing on our law enforcement initiatives. The Commission will not necessarily defer, however, to a finding by a self-regulation group. An imprimatur from a self-regulation group will not automatically shield a firm from Commission prosecution, and an unfavorable determination will not mean the Commission will automatically take issue, or find liability if it does. Rather the Commission will make its judgment independently, evaluating each case on its merits. We intend to continue our useful relationships with self-regulation groups and to rely on the expertise and findings of other government agencies in our proceedings to the greatest extent possible.

Issued: July 27, 1984.

By direction of the Commission.

Benjamin I. Berman,
  Acting Secretary.

FTC Statement on Deception

FEDERAL TRADE COMMISSION
WASHINGTON, D.C. 20580

October 14, 1983

The Honorable John D. Dingell
Chairman
Committee on Energy and Commerce
U.S. House of Representatives
Washington, D.C. 20515

Dear Mr. Chairman:

This letter responds to the Committee's inquiry regarding the Commission's enforcement policy against deceptive acts or practices.[1] We also hope this letter will provide guidance to the public.

Section 5 of the FTC Act declares unfair or deceptive acts or practices unlawful. Section 12 specifically prohibits false ads likely to induce the purchase of food, drugs, devices or cosmetics. Section 15 defines a false ad for purposes of Section 12 as one which is "misleading in a material respect."[2] Numerous Commission and judicial decisions have defined and elaborated on the phrase "deceptive acts or practices" under both Sections 5 and 12. Nowhere, however, is there a single definitive statement of the Commission's view of its authority. The Commission believes that such a statement would be useful to the public, as well as the Committee in its continuing review of our jurisdiction. [2]

We have therefore reviewed the decided cases to synthesize the most important principles of general applicability. We have attempted to provide a concrete indication of the manner in which the Commission will enforce its deception mandate. In so doing, we intend to address the concerns that have been raised about the meaning of deception, and thereby attempt to provide a greater sense of certainty as to how the concept will be applied.[3]

---

1. S. Rep. No. 97–451, 97th Cong., 2d Sess. 16; H.R. Rep. No. 98–156, Part I, 98th Cong., 1st Sess. 6 (1983). The Commission's enforcement policy against unfair acts or practices is set forth in a letter to Senators Ford and Danforth, dated December 17, 1980.

2. In determining whether an ad is misleading, Section 15 requires that the Commission take into account "representations made or suggested" as well as "the extent to which the advertisement fails to reveal facts material in light of such representations or material with respect to consequences which may result from the use of the commodity to which the advertisement relates under the conditions prescribed in said advertisement, or under such conditions as are customary or usual." 15 U.S.C. 55. If an act or practice violates Section 12, it also violates Section 5. *Simeon Management Corp.*, 87 F.T.C. 1184, 1219 (1976), *aff'd*, 579 F.2d 1137 (9th Cir. 1978); *Porter & Dietsch*, 90 F.T.C. 770, 873–74 (1977), *aff'd*, 605 F.2d 294 (7th Cir. 1979), *cert. denied*, 445 U.S. 950 (1980).

3. Chairman Miller has proposed that Section 5 be amended to define deceptive acts. Hearing Before the Subcommittee for Consumers of the Committee on Commerce, Science, and Transportation, United States Senate, 97th Cong., 2d Sess. *FTC's Authority Over Deceptive Advertising*, July 22, 1982, Serial No. 97–134, p. 9. Three Commissioners believe a legislative definition is unnecessary. *Id.* at 45 (Commissioner Clanton), at 51 (Commissioner Bailey) and at 76 (Commissioner Pertschuk). Commissioner Douglas supports a statutory definition of deception. Prepared statement by Commissioner George W. Douglas, Hearing Before the Subcommittee for Consumers of the Committee on Commerce, Science and Transportation, United States Senate, 98th Cong. 1st Sess. (March 15, 1983) p. 2.

I. Summary

Certain elements undergird all deception cases. *First*, there must be a representation, omission or practice that is likely to mislead the consumer.[4] Practices that have been found [3] misleading or deceptive in specific cases include false oral or written representations, misleading price claims, sales of hazardous or systematically defective products or services without adequate disclosures, failure to disclose information regarding pyramid sales, use of bait and switch techniques, failure to perform promised services, and failure to meet warranty obligations.[5]

Second, we examine the practice from the perspective of a consumer acting reasonably in the circumstances. If the representation or practice affects or is directed primarily to a particular group, the Commission examines reasonableness from the perspective of that group.

Third, the representation, omission, or practice must be a "material" one. The basic question is whether the act or practice is likely to affect the consumer's conduct or decision with regard to a product or service. If so, the practice is material, and consumer injury is likely, because consumers are likely to have chosen differently but for the deception. In many instances, materiality, and hence injury, can be presumed from the nature of the practice. In other instances, evidence of materiality may be necessary. [4]

Thus, the Commission will find deception if there is a representation, omission or practice that is likely to mislead the consumer acting reasonably in the circumstances, to the consumer's detriment. We discuss each of these elements below.

II. There Must Be a Representation, Omission, or Practice That Is Likely to Mislead the Consumer

Most deception involves written or oral misrepresentations, or omissions of material information. Deception may also occur in other forms of conduct associated with a sales transaction. The entire advertisement, transaction or course of dealing will be considered. The issue is whether the act or practice is likely to mislead, rather than whether it causes actual deception.[6]

Of course, the Commission must find that a representation, omission, or practice occurred. In cases of express claims, the representation itself established the meaning. In cases of implied claims, the Commission will often be able to determine meaning through an examination of the representation itself, including an evaluation of such factors as the entire document, the juxtaposition of various phrases in the document,

---

4. A misrepresentation is an express or implied statement contrary to fact. A misleading omission occurs when qualifying information necessary to prevent a practice, claim, representation, or reasonable expectation or belief from being misleading is not disclosed. Not all omissions are deceptive, even if providing the information would benefit consumers. As the Commission noted in rejecting a proposed requirement for nutrition disclosures, "In the final analysis, the question whether an advertisement requires affirmative disclosure would depend on the nature and extent of the nutritional claim made in the advertisement." *ITT Continental Baking Co. Inc.*, 83 F.T.C. 865, 965 (1976). In determining whether an omission is deceptive, the Commission will examine the overall impression created by a practice, claim, or representation. For example, the practice of offering a product for sale creates an implied representation it is fit for the purposes for which it is sold. Failure to disclose that the product is not fit constitutes a deceptive omission. [*See* discussion below at 5–6] Omissions may also be deceptive where the representations made are not literally misleading, if those representations create a reasonable expectation or belief among consumers which is misleading, absent the omitted disclosure.

Non-deceptive omissions may still violate Section 5 if they are unfair. For instance, the R-Value Rule, 16 C.F.R. 460.5 (1983), establishes a specific method for testing insulation ability, and requires disclosure of the figure in advertising. The Statement of Basis and Purpose, 44 FR 50,242 (1979), refers to a deception theory to support disclosure requirements when certain misleading claims are made, but the rule's general disclosure requirement is premised on an unfairness theory. Consumers could not reasonably avoid injury in selecting insulation because no standard method of measurement existed.

5. Advertising that lacks a reasonable basis is also deceptive. *Firestone*, 81 F.T.C. 398, 451–52 (1972), *aff'd*, 481 F.2d 246 (6th Cir.), *cert. denied*, 414 U.S. 1112 (1973). *National Dynamics*, 82 F.T.C. 488, 549–50 (1973); *aff'd and remanded on other grounds*, 492 F.2d 1333 (2d Cir.), *cert. denied*, 419 U.S. 993 (1974), *reissued*, 85 F.T.C. 391 (1976). *National Comm'n on Egg Nutrition*, 88 F.T.C. 89, 191 (1976), *aff'd*, 570 F.2d 157 (7th Cir.), *cert. denied*, 439 U.S. 821, *reissued*, 92 F.T.C. 848 (1978). The deception theory is based on the fact that most ads making objective claims imply, and many expressly state, that an advertiser has certain specific grounds for the claims. If the advertiser does not, the consumer is acting under a false impression. The consumer might have perceived the advertising differently had he or she known the advertiser had no basis for the claim. This letter does not address the nuances of the reasonable basis doctrine, which the Commission is currently reviewing. 48 FR 10,471 (March 11, 1983).

6. In *Beneficial Corp. v. FTC*, 542 F.2d 611, 617 (3d Cir. 1976), the court noted "the likelihood or propensity of deception is the criterion by which advertising is measured."

the nature of the claim, and the nature of the transaction.[7] In other [5] situations, the Commission will require extrinsic evidence that reasonable consumers reach the implied claims.[8] In all instances, the Commission will carefully consider any extrinsic evidence that is introduced.

Some cases involve omission of material information, the disclosure of which is necessary to prevent the claim, practice, or sale from being misleading.[9] Information may be omitted from written [10] or oral[11] representations or from the commercial transaction.[12] [6]

In some circumstances, the Commission can presume that consumers are likely to reach false beliefs about the product or service because of an omission. At other times, however, the Commission may require evidence on consumers' expectations.[13]

Marketing and point-of-sales practices that are likely to mislead consumers are also deceptive. For instance, in bait and switch cases, a violation occurs when the offer to sell the product is not a bona fide offer.[14] The Commission has also found deception where a sales representative misrepresented the purpose of the initial contact with customers.[15] When a product is sold, there is an implied representation that the product is fit for the purposes for which it is sold. When it is not, deception occurs.[16] There may be a concern about the way a product or service

---

7. On evaluation of the entire document:

The Commission finds that many of the challenged Anacin advertisements, when viewed in their entirety, did convey the message that the superiority of this product has been proven [footnote omitted]. It is immaterial that the word "established," which was used in the complaint, generally did not appear in the ads; the important consideration is the net impression conveyed to the public. *American Home Products*, 98 F.T.C. 136, 374 (1981), *aff'd*, 695 F.2d (3d Cir. 1982).

On the juxtaposition of phrases:

On this label, the statement "Kills Germs By Millions On Contact" immediately precedes the assertion "For General Oral Hygiene Bad Breath, Colds and Resultant Sore Throats" [footnote omitted]. By placing these two statements in close proximity, respondent has conveyed the message that since Listerine can kill millions of germs, it can *cure*, prevent and ameliorate colds and sore throats [footnote omitted]. *Warner Lambert*, 86 F.T.C. 1398, 1489-90 (1975), *aff'd*, 562 F.2d 749 (D.C. Cir. 1977), *cert. denied*, 435 U.S. 950 (1978) (emphasis in original).

On the nature of the claim, *Firestone* is relevant. There the Commission noted that the alleged misrepresentation concerned the safety of respondent's product, "an issue of great significance to consumers. On this issue, the Commission has required scrupulous accuracy in advertising claims, for obvious reasons." 81 F.T.C. 398, 456 (1972), *aff'd*, 481 F.2d 246 (6th Cir.), *cert. denied*, 414 U.S. 1112 (1973).

In each of these cases, other factors, including in some instances surveys, were in evidence on the meaning of the ad.

8. The evidence can consist of expert opinion, consumer testimony (particularly in cases involving oral representations), copy tests, surveys, or any other reliable evidence of consumer interpretation.

9. As the Commission noted in the Cigarette rule, "The nature, appearance, or intended use of a product may create an impression on the mind of the consumer . . . and if the impression is false, and if the seller does not take adequate steps to correct it, he is responsible for an unlawful deception." Cigarette Rule Statement of Basis and Purpose, 29 FR 8324, 8352 (July 2, 1964).

10. *Porter & Deitsch*, 90 F.T.C. 770, 873-74 (1977), *aff'd*, 605 F.2d 294 (7th Cir. 1979), *cert. denied*, 445 U.S. 950 (1980), *Simeon Management Corp.*, 87 F.T.C. 1184, 1230 (1976), *aff'd*, 579 F.2d 1137 (9th Cir. 1978).

11. *See, e.g., Grolier*, 91 F.T.C. 315, 480 (1978), *remanded on other grounds*, 615 F.2d 1215 (9th Cir. 1980), *modified on other grounds*, 98 FTC 882 (1981), *reissued*, 99 F.T.C. 379 (1982).

12. In *Peacock Buick*, 86 F.T.C. 1532 (1975), *aff'd*, 553 F.2d 97 (4th Cir. 1977), the Commission held that

absent a clear and early disclosure of the prior use of a late model car, deception can result from the setting in which a sale is made and the expectations of the buyer. . . . *Id.* at 1555.

[E]ven in the absence of affirmative misrepresentations, it is misleading for the seller of late model used cars to fail to reveal the particularized uses to which they have been put. . . . When a later model used car is sold at close to list price . . . the assumption likely to be made by some purchasers is that, absent disclosure to the contrary, such car has not previously been used in a way that might substantially impair its value. In such circumstances, failure to disclose a disfavored prior use may tend to mislead. *Id.* at 1557-58.

13. In *Leonard Porter*, the Commission dismissed a complaint alleging that respondents' sale of unmarked products in Alaska led consumers to believe erroneously that they were handmade in Alaska by natives. Complaint counsel had failed to show that consumers of Alaskan craft assumed respondents' products were handmade by Alaskans in Alaska. The Commission was unwilling, absent evidence, to infer from a viewing of the items that the products would tend to mislead consumers.

By requiring such evidence, we do not imply that elaborate proof of consumer beliefs or behavior is necessary, even in a case such as this, to establish the requisite capacity to deceive. However, where visual inspection is inadequate, some extrinsic testimonial evidence must be added. 88 F.T.C. 546, 626, n.5 (1976).

14. *Bait and Switch Policy Protocol*, December 10, 1975; Guides Against Bait Advertising, 16 C.F.R. 238.0 (1967). 32 FR 15,540.

15. *Encyclopedia Britannica*, 87 F.T.C. 421, 497 (1976), *aff'd*, 605 F.2d 964 (7th Cir. 1979), *cert. denied*, 445 U.S. 934 (1980), *modified*, 100 F.T.C. 500 (1982).

16. *See* the complaints in *BayleySuit*, C-3117 (consent agreement) (September 30, 1983) [102 F.T.C. 1285]; *Figgie International, Inc.*, D. 9166 (May 17, 1983).

is marketed, such as where inaccurate or [7] incomplete information is provided.[17] A failure to perform services promised under a warranty or by contract can also be deceptive.[18]

III. The Act or Practice Must Be Considered From the Perspective of the Reasonable Consumer

The Commission believes that to be deceptive the representation, omission or practice must be likely to mislead reasonable consumers under the circumstances.[19] The test is whether the consumer's interpretation or reaction is reasonable.[20] When representations or sales practices are targeted to a specific audience, the Commission determines the effect of the practice on a reasonable member of that group. In evaluating a particular practice, the Commission considers the totality of the practice in determining how reasonable consumers are likely to respond. [8]

A company is not liable for every interpretation or action by a consumer. In an advertising context, this principle has been well-stated:

> An advertiser cannot be charged with liability with respect to every conceivable misconception, however outlandish, to which his representations might be subject among the foolish or feeble-minded. Some people, because of ignorance or incomprehension, may be misled by even a scrupulously honest claim. Perhaps a few misguided souls believe, for example, that all "Danish pastry" is made in Denmark. Is it therefore an actionable deception to advertise "Danish pastry" when it is made in this country? Of course not. A representation does not become "false and deceptive" merely because it will be unreasonably misunderstood by an insignificant and unrepresentative segment of the class of persons to whom the representation is addressed. *Heinz W. Kirchner*, 63 F.T.C. 1282, 1290 (1963).

To be considered reasonable, the interpretation or reaction does not have to be the only one.[21] When a seller's representation conveys more than one meaning to reasonable consumers, one of which is false, the seller is liable for the misleading interpretation.[22] An interpretation will be presumed reasonable if it is the one the respondent intended to convey.

The Commission has used this standard in its past decisions. ". . . The test applied by the Commission is whether the interpretation is reasonable in light of the claim."[23] In the Listerine case, the Commission evaluated the claim from the perspective of the "average listener."[24] In a case involving the sale of encyclopedias, the Commission observed "[i]n determining the meaning of an advertisement, a piece of promotional material [9] or a sales presentation, the important criterion is the net impression that it is likely to make on the general populace."[25] The decisions in *American Home Products, Bristol Myers,* and *Sterling Drug* are

---

17. The Commission's complaints in *Chrysler Corporation*, 99 F.T.C. 347 (1982), and *Volkswagen of America*, 99 F.T.C. 446 (1982), alleged the failure to disclose accurate use and care instructions for replacing oil filters was deceptive. The complaint in *Ford Motor Co.*, D. 9154, 96 F.T.C. 362 (1980), charged Ford with failing to disclose a "piston scuffing" defect to purchasers and owners which was allegedly widespread and costly to repair. *See also General Motors*, D. 9145 (provisionally accepted consent agreement, April 26, 1983). [102 F.T.C. 1741].

18. *See Jay Norris Corp.*, 91 F.T.C. 751 (1978), *aff'd with modified language in order*, 598 F.2d 1244 (2d Cir. 1979), *cert. denied*, 444 U.S. 980 (1979) (failure to consistently meet guarantee claims of "immediate and prompt" delivery as well as money back guarantees); *Southern States Distributing Co.*, 83 F.T.C. 1126 (1973) (failure to honor oral and written product maintenance guarantees, as represented); *Skylark Originals, Inc.*, 80 F.T.C. 337 (1972), *aff'd*, 475 F.2d 1396 (3d Cir. 1973) (failure to promptly honor moneyback guarantee as represented in advertisements and catalogs); *Capitol Manufacturing Corp.*, 73 F.T.C. 872 (1968) (failure to fully, satisfactorily and promptly meet all obligations and requirements under terms of service guarantee certificate).

19. The evidence necessary to determine how reasonable consumers understand a representation is discussed in Section II of this letter.

20. An interpretation may be reasonable even though it is not shared by a majority of consumers in the relevant class, or by particularly sophisticated consumers. A material practice that misleads a significant minority of reasonable consumers is deceptive. *See Heinz W. Kirchner*, 63 F.T.C. 1282 (1963).

21. A secondary message understood by reasonable consumers is actionable if deceptive even though the primary message is accurate. *Sears, Roebuck & Co.*, 95 F.T.C. 406, 511 (1980), *aff'd* 676 F.2d 385, (9th Cir. 1982); *Chrysler*, 87 F.T.C. 749 (1976), *aff'd*, 561 F.2d 357 (D.C. Cir.), *reissued* 90 F.T.C. 606 (1977); *Rhodes Pharmacal Co.*, 208 F.2d 382, 387 (7th Cir. 1953), *aff'd*, 348 U.S. 940 (1955).

22. *National Comm'n on Egg Nutrition*, 88 F.T.C. 89, 185 (1976), *enforced in part*, 570 F.2d 157 (7th Cir. 1977); *Jay Norris Corp.*, 91 F.T.C. 751, 836 (1978), *aff'd*, 598 F.2d 1244 (2d Cir. 1979).

23. *National Dynamics*, 82 F.T.C. 488, 524, 548 (1973), *aff'd*, 492 F.2d 1333 (2d Cir.), *cert. denied*, 419 U.S. 993 (1974), *reissued* 85 F.T.C. 391 (1976).

24. *Warner-Lambert*, 86 F.T.C. 1398, 1415 n.4 (1975), *aff'd*, 562 F.2d 749 (D.C. Cir. 1977), *cert. denied*, 435 U.S. 950 (1978).

25. *Grolier*, 91 F.T.C. 315, 430 (1978), *remanded on other grounds*, 615 F.2d 1215 (9th Cir. 1980), *modified on other grounds*, 98 F.T.C. 882 (1981), *reissued*, 99 F.T.C. 379 (1982).

replete with references to reasonable consumer interpretations.[26] In a land sales case, the Commission evaluated the oral statements and written representations "in light of the sophistication and understanding of the persons to whom they were directed."[27] Omission cases are no different: the Commission examines the failure to disclose in light of expectations and understandings of the typical buyer[28] regarding the claims made.

When representations or sales practices are targeted to a specific audience, such as children, the elderly, or the terminally ill, the Commission determines the effect of the practice on a reasonable member of that group.[29] For instance, if a company markets a cure to the terminally ill, the practice [10] will be evaluated from the perspective of how it affects the ordinary member of that group. Thus, terminally ill consumers might be particularly susceptible to exaggerated cure claims. By the same token, a practice or representation directed to a well-educated group, such as a prescription drug advertisement to doctors, would be judged in light of the knowledge and sophistication of that group.[30]

As it has in the past, the Commission will evaluate the entire advertisement, transaction, or course of dealing in determining how reasonable consumers are likely to respond. Thus, in advertising the Commission will examine "the entire mosaic, rather than each title separately."[31] As explained by a court of appeals in a recent case: [11]

> The Commission's right to scrutinize the visual and aural imagery of advertisements follows from the principle that the Commission looks to the impression made by the advertisements as a whole. Without this mode of examination, the Commission would have limited recourse against crafty advertisers whose deceptive messages were conveyed by means other than, or in addition to, spoken words. *American Home Products*, 695 F.2d 681, 688 (3d Cir. Dec. 3, 1982).[32]

---

26. *American Home Products*, 98 F.T.C. 136 (1981), *aff'd*, 695 F.2d 681 (3d Cir. 1982). ". . . consumers may be led to expect, quite reasonably . . ." (at 386); ". . . consumers may reasonably believe . . ." (*Id.* n.52); ". . . would reasonably have been understood by consumers . . ." (at 371); "The record shows that consumers could reasonably have understood this language . . ." (at 372). *See also*, pp. 373, 374, 375. *Bristol-Myers*, D. 8917 (July 5, 1983), appeal docketed, No. 83–4167 (2nd Cir. Sept. 12, 1983). ". . . ads must be judged by the impression they make on reasonable members of the public . . ." (Slip Op. at 4); ". . .consumers could reasonably have understood . . ." (Slip Op. at 7); ". . . consumers could reasonably infer . . ." (Slip Op. at 11) [102 F.T.C. 21 (1983)]; *Sterling Drug, Inc.*, D. 8919 (July 5, 1983), appeal docketed, No. 83–7700 (9th Cir. Sept. 14, 1983). ". . . consumers could reasonably assume . . ." (Slip Op. at 9); ". . . consumers could reasonably interpret the ads . . ." (Slip Op. at 33) [102 F.T.C. 395 (1983)].

27. *Horizon Corp.*, 97 F.T.C. 464, 810 n.13 (1981).

28. *Simeon Management*, 87 F.T.C. 1184, 1230 (1976).

29. The listed categories are merely examples. Whether children, terminally ill patients, or any other subgroup of the population will be considered a special audience depends on the specific factual context of the claim or the practice.

The Supreme Court has affirmed this approach. "The determination whether an advertisement is misleading requires consideration of the legal sophistication of its audience." *Bates v. Arizona*, 433 U.S. 350, 383 n.37 (1977).

30. In one case, the Commission's complaint focused on seriously ill persons. The ALJ summarized:

According to the complaint, the frustrations and hopes of the seriously ill and their families were exploited, and the representations had the tendency and capacity to induce the seriously ill to forego conventional medical treatment worsening their condition and in some cases hastening death, or to cause them to spend large amounts of money and to undergo the inconvenience of traveling for a non-existent "operation." *Travel King*, 86 F.T.C. 715, 719 (1975).

In a case involving a weight loss product, the Commission observed:

It is obvious that dieting is the conventional method of losing weight. But it is equally obvious that many people who need or want to lose weight regard dieting as bitter medicine. To these corpulent consumers the promises of weight loss without dieting are the Siren's call, and advertising that heralds unrestrained consumption while muting the inevitable need for temperance, if not abstinence, simply does not pass muster. *Porter & Dietsch*, 90 F.T.C. 770, 864–865 (1977), *aff'd*, 605 F.2d 294 (7th Cir. 1979), *cert. denied*, 445 U.S. 950 (1980).

Children have also been the specific target of ads or practices. In *Ideal Toy*, the Commission adopted the Hearing Examiner's conclusion that:

False, misleading and deceptive advertising claims beamed at children tend to exploit unfairly a consumer group unqualified by age or experience to anticipate or appreciate the possibility that representations may be exaggerated or untrue. *Ideal Toy*, 64 F.T.C. 297, 310 (1964).

See also, Avalon Industries Inc., 83 F.T.C. 1728, 1750 (1974).

31 *FTC v. Sterling Drug*, 317 F.2d 669, 674 (2d Cir. 1963).

32. Numerous cases exemplify this point. For instance, in *Pfizer*, the Commission ruled that "the net impression of the advertisement, evaluated from the perspective of the audience to whom the advertisement is directed, is controlling." 81 F.T.C. 23, 58 (1972).

In a subsequent case, the Commission explained that "[i]n evaluating advertising representations, we are required to look at the complete advertisement and formulate our opinions on them on the basis of the net general impression conveyed by them and not on isolated excerpts." *Standard Oil of Calif.*, 84 F.T.C. 1401, 1471 (1974), *aff'd as modified*, 577 F.2d 653 (9th Cir. 1978), *reissued*, 96 F.T.C. 380 (1980).

The Third Circuit stated succinctly the Commission's standard. "The tendency of the advertising to deceive must be judged by viewing it as a whole, without emphasizing isolated words or phrases apart from their context." *Beneficial Corp. v. FTC*, 542 F.2d 611, 617 (3d Cir. 1976), *cert. denied*, 430 U.S. 983 (1977).

Commission cases reveal specific guidelines. Depending on the circumstances, accurate information in the text may not remedy a false headline because reasonable consumers may glance only at the headline.[33] Written disclosures or fine print may be [12] insufficient to correct a misleading representation.[34] Other practices of the company may direct consumers' attention away from the qualifying disclosures.[35] Oral statements, label disclosures or point-of-sale material will not necessarily correct a deceptive misrepresentation or omission.[36] Thus, when the first contact between a seller and a buyer occurs through a [13] deceptive practice, the law may be violated even if the truth is subsequently made known to the purchaser.[37] *Pro forma* statements or disclaimers may not cure otherwise deceptive messages or practices.[38]

Qualifying disclosures must be legible and understandable. In evaluating such disclosures, the Commission recognizes that in many circumstances, reasonable consumers do not read the entirety of an ad or are directed away from the importance of the qualifying phrase by the acts or statements of the seller. Disclosures that conform to the Commission's Statement of Enforcement Policy regarding clear and conspicuous disclosures, which applies to television advertising, are generally adequate, [14] CCH *Trade Regulation Reporter,* ¶ 7569.09 (Oct. 21, 1970). Less elaborate disclosures may also suffice.[39]

Certain practices, however, are unlikely to deceive consumers acting reasonably. Thus, the commission generally will not bring advertising cases based on subjective claims (taste, feel, appearance, smell) or on

---

33. In *Litton Industries*, the Commission held that fine print disclosures that the surveys included only "Litton authorized" agencies were inadequate to remedy the deceptive characterization of the survey population in the headline. 97 F.T.C. 1, 71, n.6 (1981), *aff'd as modified*, 676 F.2d 364 (9th Cir. 1982). Compare the Commission's note in the same case that the fine print disclosure "Litton and one other brand" was reasonable to qualify the claim that independent service technicians had been surveyed. "[F]ine print was a reasonable medium for disclosing a qualification of only limited relevance." 97 F.T.C. 1, 70 n.5 (1981).

In another case, the Commission held that the body of the ad corrected the possibly misleading headline because in order to enter the contest, the consumer had to read the text, and the text would eliminate any false impression stemming from the headline. *D.L. Blair*, 82 F.T.C. 234, 255–256 (1973).

In one case, respondent's expert witness testified that the headline (and accompanying picture) of an ad would be the focal point of the first glance. He also told the administrative law judge that a consumer would spend "[t]ypically a few seconds at most" on the ads at issue. *Crown Central*, 84 F.T.C. 1493, 1543 nn. 14–15 (1974).

34. In *Giant Food*, the Commission agreed with the examiner that the fine-print disclaimer was inadequate to correct a deceptive impression. The Commission quoted from the examiner's finding that "very few if any of the persons who would read Giant's advertisements would take the trouble to, or did, read the fine print disclaimer." 61 F.T.C. 326, 348 (1962).

Cf. Beneficial Corp. v. FTC, 542 F.2d 611, 618 (3d Cir. 1976), where the court reversed the Commission's opinion that no qualifying language could eliminate the deception stemming from use of the slogan "Instant Tax Refund."

35. "Respondents argue that the contracts which consumers signed indicated that credit life insurance was not required for financing, and that this disclosure obviated the possibility of deception. We disagree. It is clear from consumer testimony that oral deception was employed in some instances to cause consumers to ignore the warning in their sales agreement . . ." *Peacock Buick*, 86 F.T.C. 1532, 1558–59 (1974).

36. *Exposition Press*, 295 F.2d 869, 873 (2d Cir. 1961); *Gimbel Bros.*, 61 F.T.C. 1051, 1066 (1962); *Carter Products*, 186 F.2d 821, 824 (1951).

By the same token, money-back guarantees do not eliminate deception. In *Sears*, the Commission observed:

A money-back guarantee is no defense to a charge of deceptive advertising. . . . A money-back guarantee does not compensate the consumer for the often considerable time and expense incident to returning a major-ticket item and obtaining a replacement.

Sears, Roebuck and Co., 95 F.T.C. 406, 518 (1980), *aff'd*, 676 F.2d 385 (9th Cir. 1982). However, the existence of a guarantee, if honored, has a bearing on whether the Commission should exercise its discretion to prosecute. *See* Deceptive and Unsubstantiated Claims Policy Protocol, 1975.

37. *See American Home Products*, 98 F.T.C. 136, 370 (1981), *aff'd*, 695 F.2d 681, 688 (3d Cir. Dec. 3, 1982). Whether a disclosure on the label cures deception in advertising depends on the circumstances:

. . . is is well settled that dishonest advertising is not cured or excused by honest labeling [footnote omitted]. Whether the ill-effects of deceptive nondisclosure can be cured by a disclosure requirement limited to labeling, or whether a further requirement of disclosure in advertising should be imposed, is essentially a question of remedy. As such it is a matter within the sound discretion of the Commission [footnote omitted]. The question of whether in a particular case to require disclosure in advertising cannot be answered by application of any hard-and-fast principle. The test is simple and pragmatic: Is it likely that, unless such disclosure is made, a substantial body of consumers will be misled to their detriment? *Statement of Basis and Purpose for the Cigarette Advertising and Labeling Trade Regulation Rule*, 1965, pp. 89–90. 29 FR 8325 (1964).

Misleading "door openers" have also been found deceptive (*Encyclopedia Britannica*, 87 F.T.C. 421 (1976), *aff'd*, 605 F.2d 964 (7th Cir. 1979), *cert. denied*, 445 U.S. 934 (1980), *as modified*, 100 F.T.C. 500 (1982)), as have offers to sell that are not bona fide offers (*Seekonk Freezer Meats, Inc.*, 82 F.T.C. 1025 (1973)). In each of these instances, the truth is made known prior to purchase.

38. In the Listerine case, the Commission held that *pro forma* statements of no absolute prevention followed by promises of fewer colds did not cure or correct the false message that Listerine will prevent colds. *Warner-Lambert*, 86 F.T.C. 1398, 1414 (1975), *aff'd*, 562 F.2d 749 (D.C. Cir. 1977), *cert. denied*, 435 U.S. 950 (1978).

39. *Chicago Metropolitan Pontiac Dealers' Ass'n*, C 3110 (June 9, 1983) [101 F.T.C. 854 (1983)].

correctly stated opinion claims if consumers understand the source and limitations of the opinion.[40] Claims phrased as opinions are actionable, however, if they are not honestly held, if they misrepresent the qualification of the holder or the basis of his opinion or if the recipient reasonably interprets them as implied statements of fact.[41]

The Commission generally will not pursue cases involving obviously exaggerated or puffing representation, *i.e.*, those that the ordinary consumers do not take seriously.[42] Some exaggerated claims, however, may be taken seriously by consumers and are actionable. For instance, in rejecting a respondent's argument that use of the words "electronic miracle" to describe a television antenna was puffery, the Commission stated:

> Although not insensitive to respondent's concern that the term miracle is commonly used in situations short of changing [15] water into wine, we must conclude that the use of "electronic miracle" in the context of respondent's grossly exaggerated claims would lead consumers to give added credence to the overall suggestion that this device is superior to other types of antennae. *Jay Norris*, 91 F.T.C. 751, 847 n.20 (1978), *aff'd*, 598 F.2d 1244 (2d Cir.), *cert. denied*, 444 U.S. 980 (1979).

Finally, as a matter of policy, when consumers can easily evaluate the product or service, it is inexpensive, and it is frequently purchased, the Commission will examine the practice closely before issuing a complaint based on deception. There is little incentive for sellers to misrepresent (either by an explicit false statement or a deliberate false implied statement) in these circumstances since they normally would seek to encourage repeat purchases. Where, as here, market incentive place strong constraints on the likelihood of deception, the Commission will examine a practice closely before proceeding.

In sum, the Commission will consider many factors in determining the reaction of the ordinary consumer to a claim or practice. As would any trier of fact, the Commission will evaluate the totality of the ad or the practice and ask questions such as: how clear is the representation? how conspicuous is any qualifying information? how important is the omitted information? do other sources for the omitted information exist? how familiar is the public with the product or service?[43]

IV. The Representation, Omission or Practice Must Be Material

The third element of deception is materiality. That is, a representation, omission or practice must be a material one for deception to occur.[44] A "material" misrepresentation or practice is one which is likely to affect a consumer's choice of or [16] conduct regarding a product.[45] In other words, it is information

---

40. An opinion is a representation that expresses only the belief of the maker, without certainty, as to the existence of a fact, or his judgement as to quality, value, authenticity, or other matters of judgment. American Law Institute, *Restatement on Torts, Second* ¶ 538 A.

41. *Id.* ¶ 539. At common law, a consumer can generally rely on an expert opinion. *Id.* ¶ 542(a). For this reason, representations of expert opinion will generally be regarded as representations of fact.

42. "[T]here is a category of advertising themes, in the nature of puffing or other hyperbole, which do not amount to the type of affirmative product claims for which either the Commission or other consumer would expect documentation." *Pfizer, Inc.*, 81 F.T.C. 23, 64 (1972).

The term "puffing" refers generally to an expression of opinion not made as a representation of fact. A seller has some latitude in puffing his goods, but he is not authorized to misrepresent them or to assign to them benefits they do not possess [cite omitted]. Statements made for the purpose of deceiving prospective purchasers cannot properly be characterized as mere puffing. *Wilmington Chemical*, 69 F.T.C. 828, 865 (1966).

43. In *Avalon Industries*, the ALJ observed that the " 'ordinary person with a common degree of familiarity with industrial civilization' would expect a reasonable relationship between the size of package and the size of quantity of the contents. He would have no reason to anticipate slack filling." 83 F.T.C. 1728, 1750 (1974) (I.D.).

44. "A misleading claim or omission in advertising will violate Section 5 or Section 12, however, only if the omitted information would be a material factor in the consumer's decision to purchase the product.". *American Home Products Corp.*, 98 F.T.C. 136, 368 (1981), *aff'd*, 695 F.2d 681 (3d Cir. 1982). A claim is material if it is likely to affect consumer behavior. "Is it likely to affect the average consumer in deciding whether to purchase the advertised product—is there a material deception, in other words?" Statement of Basis and Purpose, *Cigarette Advertising and Labeling Rule*, 1965, pp. 86–87, 29 FR 8325 (1964).

45. Material information may affect conduct other than the decision to purchase a product. The Commission's complaint in *Volkswagen of America*, 99 F.T.C. 446 (1982), for example, was based on provision of inaccurate instructions for oil filter installation. In its *Restatement on Torts, Second*, the American Law Institute defines a material misrepresentation or omission as one which the reasonable person would regard as important in deciding how to act, or one which the maker knows that the recipient, because of his or her own peculiarities, is likely to consider important. Section 538(2). The Restatement explains that a material fact does not necessarily have to affect the finances of a transaction. "There are many more-or-less sentimental considerations that the ordinary man regards as important." Comment on Clause 2(a)(d).

that is important to consumers. If inaccurate or omitted information is material, injury is likely.[46]

The Commission considers certain categories of information presumptively material.[47] First, the Commission presumes that express claims are material.[48] As the Supreme Court stated recently, "[i]n the absence of factors that would distort the decision to advertise, we may assume that the willingness of a business to promote its products reflects a belief that consumers are interested in the advertising."[49] Where the seller knew, or should have known, that an ordinary consumer would need omitted information to evaluate the product or service, or that the claim was false, materiality will be presumed because the manufacturer intended the information or omission to have an effect.[50] [17] Similarly, when evidence exists that a seller intended to make an implied claim, the Commission will infer materiality.[51]

The Commission also considers claims or omissions material if they significantly involve health, safety, or other areas with which the reasonable consumer would be concerned. Depending on the facts, information pertaining to the central characteristics of the product or service will be presumed material. Information has been found material where it concerns the purpose,[52] safety,[53] efficacy,[54] or cost[55] of the product or service. [18] Information is also likely to be material if it concerns durability, performance, warranties or quality. Information pertaining to a finding by another agency regarding the product may also be material.[56]

Where the Commission cannot find materiality based on the above analysis, the Commission may require evidence that the claim or omission is likely to be considered important by consumers. This evidence can be the fact that the product or service with the feature represented costs more than an otherwise comparable product without the feature, a reliable survey of consumers, or credible testimony.[57]

A finding of materiality is also a finding that injury is likely to exist because of the representation, omission, sales practice, or marketing technique. Injury to consumers can take many forms.[58] Injury exists if consumers

---

46. In evaluating materiality, the Commission takes consumer preferences as given. Thus, if consumers prefer one product to another, the Commission need not determine whether that preference is objectively justified. *See Algoma Lumber*, 291 U.S. 54, 78 (1933). Similarly, objective differences among products are not material if the difference is not likely to affect consumer choices.

47. The Commission will always consider relevant and competent evidence offered to rebut presumptions of materiality.

48. Because this presumption is absent for some implied claims, the Commission will take special caution to ensure materiality exists in such cases.

49. *Central Hudson Gas & Electric Co. v. PSC*, 447 U.S. 557, 567 (1980).

50. *Cf. Restatement on Contracts, Second* ¶ 162(1).

51. In *American Home Products*, the evidence was that the company intended to differentiate its products from aspirin. "The very fact that AHP sought to distinguish its products from aspirin strongly implies that knowledge of the true ingredients of those products would be material to purchasers." *American Home Products*, 98 F.T.C. 136, 368 (1981), *aff'd*, 695 F.2d 681 (3d Cir. 1982).

52. In *Fedders*, the ads represented that only Fedders gave the assurance of cooling on extra hot, humid days. "Such a representation is the raison d'etre for an air conditioning unit—it is extremely material representation." 85 F.T.C. 38, 61 (1975) (I.D.), *petition dismissed*, 529 F.2d 1398 (2d Cir.), *cert. denied*, 429 U.S. 818 (1976).

53. "We note at the outset that both alleged misrepresentations go to the issue of the safety of respondent's product, an issue of great significance to consumers." *Firestone*, 81 F.T.C. 398, 456 (1972), *aff'd* 481 F.2d 246 (6th Cir.), *cert. denied*, 414 U.S. 1112 (1973).

54. The Commission found that information that a product was effective in only the small minority of cases where tiredness symptoms are due to an iron deficiency, and that it was of no benefit in all other cases, was material. *J.B. Williams Co.*, 68 F.T.C. 481, 546 (1965), *aff'd*, 381 F.2d 884 (6th Cir. 1967).

55. As the Commission noted in *MacMillan, Inc.*:

> In marketing their courses, respondents failed to adequately disclose the number of lesson assignments to be submitted in a course. These were material facts necessary for the student to calculate his tuition obligation, which was based on the number of lesson assignments he submitted for grading. The nondisclosure of these material facts combined with the confusion arising from LaSalle's inconsistent use of terminology had the capacity to mislead students about the nature and extent of their tuition obligation. *MacMillan, Inc.*, 96 F.T.C. 208, 303–304 (1980).

See also, Peacock Buick, 86 F.T.C. 1532, 1562 (1975), *aff'd*, 553 F.2d 97 (4th Cir. 1977).

56. *Simeon Management Corp.*, 87 F.T.C. 1184 (1976), *aff'd*, 579 F.2d 1137, 1168, n.10 (9th Cir. 1978).

57. In *American Home Products*, the Commission approved the ALJ's finding of materiality from an economic perspective:

> If the record contained evidence of a significant disparity between the prices of Anacin and plain aspirin, it would form a further basis for a finding of materiality. That is, there is a reason to believe consumers are willing to pay a premium for a product believed to contain a special analgesic ingredient, but not for a product whose analgesic is ordinary aspirin. *American Home Products*, 98 F.T.C. 136, 369 (1981), *aff'd*, 695 F.2d 681 (3d Cir. 1982).

58. The prohibitions of Section 5 are intended to prevent injury to competitors as well as to consumers. The Commission regards injury to competitors as identical to injury to consumers. Advertising and legitimate marketing techniques are intended to "injure" competitors by directing business to the advertiser. In fact, vigorous competitive advertising can actually benefit consumers by lowering prices, encouraging product innovation, and increasing the specificity and amount of information available to consumers. Deceptive practices injure both competitors and consumers because consumers who preferred the competitor's product are wrongly diverted.

would have chosen differently but for the deception. If different choices are likely, the claim is material, and injury is likely as well. Thus, injury and materiality are different names for the same concept. [19]

V. Conclusion

The Commission will find an act or practice deceptive if there is a misrepresentation, omission, or other practice, that misleads the consumer acting reasonably in the circumstances, to the consumer's detriment. The Commission will not generally require extrinsic evidence concerning the representations understood by reasonable consumers or the materiality of a challenged claim, but in some instances extrinsic evidence will be necessary.

The Commission intends to enforce the FTC Act vigorously. We will investigate, and prosecute where appropriate, acts or practices that are deceptive. We hope this letter will help provide you and the public with a greater sense of certainty concerning how the Commission will exercise its jurisdiction over deception. Please do not hesitate to call if we can be of any further assistance.

By direction of the Commission, Commissioners Pertschuk and Bailey dissenting, with separate statements attached and with separate response to the Committee's request for a legal analysis to follow.

/s/ James C. Miller III
Chairman

cc: Honorable James T. Broyhill
Honorable James J. Florio
Honorable Norman F. Lent

SPI Voluntary Guidelines: Rigid Plastic Container Material Code System

PETE   HDPE   V   LDPE   PP   PS   OTHER

1.0 PURPOSE

The Society of the Plastic Industry (SPI) has developed a voluntary coding system for plastic containers, to identify material type. The purpose of coding is to assist recyclers in sorting plastic containers by resin composition. The system is intended for voluntary use by bottle and container producers, to be molded or formed or otherwise imprinted onto the bottom surface of plastic containers.

This container coding system has been created and is recommended to the industry to provide a consistent national identification mark that meets the needs of the recycling industry, as defined by the recyclers and collectors themselves. The system is designed to be most convenient for the people who will sort containers and is intended to avoid a complicated system which would require extensive worker training and possibly lead to confusion and/or mis-sorting.

Given today's national marketplace, it is crucial that the coding system be standardized nationally. The use of different code systems by various companies or states could significantly disrupt the flow of commerce.

2.0 BACKGROUND

The problem of solid waste. One of the most pressing environmental issues is the mounting problem of solid waste disposal. Communities across the country are facing the issue of how to dispose of a growing volume of municipal waste efficiently and responsibly. In many areas, the lack of landfill space or proper incineration facilities has created the need to reduce the volume of household waste.

The role of recycling. More and more, recycling is playing a role in solving community waste disposal problems. Many states, counties, cities, and smaller communities are recognizing that recycling can noticeably reduce the volume of waste to be handled by landfills and incinerators. They note recycling can also save on landfill use fees and transportation costs and reuse valuable natural resources. Many new laws, regulations, and public education programs are designed to encourage consumer participation in community recycling projects.

Recycling rigid plastic containers. Rigid plastic containers are injection molded or thermoformed containers used to package and deliver any product to a customer, including food service packages. Rigid plastic containers are not a large part of the waste stream, but their use is growing. Currently, almost a third of municipal waste is paper products, with another third being organic and food waste. Plastic materials of all kinds prepresent about 7 percent of the municipal waste stream, half of which is plastic packaging. Plastic containers, however, are one of the components of household trash which can be recycled, along with newspapers, aluminum cans, and glass containers.

The predominant plastic recycling systems in place today are geared towards handling separate plastic materials, which are primarily polyethylene terephthalate (PET) from soft drink bottle and high density polyethylene (HDPE) from milk bottles. Markets for these materials are well developed, and increased volume should be possible. This has come about because soft drink and milk bottles are readily identified by their size, shape, and color and are easily separated from other plastic containers. They are also available in large volume, together representing more than one-third of all plastic bottles.

The remaining volume of plastic containers are made from a variety of resin materials, including PET and HDPE. These, however, are not readily identifiable by size and style, and are not easily separated for processing by the current recycling systems. The challenge to the plastic industry has been to assist in solving the solid waste disposal problem by finding a way to make these other plastic containers more recyclable.

Needs of recyclers. To determine the most appropriate way to aid recyclers and collectors to separate plastic containers for processing, a survey was conducted among a large portion of the recycling industry. The results indicated that while not all recyclers could handle sorting, a significant portion would benefit by having a system to visually identify container material. After evaluating various methods, this recommended system was determined to be the most practical and the most helpful to recyclers and collectors.

The code system identifies the six most common plastic container materials and applies to large containers representing perhaps 70 percent of all container resin. This is intended to encourage sorting which will result in reasonable volumes of higher value recyclable material. All other large containers, including multi-material, can be grouped with smaller containers and can be recycled as ''mixed'' or ''other'' plastics. A recently developed successful technology which makes use of these mixed plastics is in use in some areas and is expected to grow.

3.0 DESIGN AND USE

The Rigid Plastic Container Material Code System is designed to be easy to read at a glance and distinguishable from existing marks put on rigid plastic containers by manufacturers for use in processing and identification. The basic part of the system uses a triangular-shaped symbol composed of three arrows with a specific number in the center to indicate the material from which the container is made. The number-material equivalents are:

1 = PETE (polyethylene terephthalate) (PET)[1]
2 = HDPE (high density polyethylene)
3 = V (vinyl/polyvinyl chloride) (PVC)
4 = LDPE (low density polyethylene)

5 = PP (polypropylene)
6 = PS (polystyrene)
7 = OTHER

The number code is then supplemented by the common letter identification for the various resins under the symbol, to serve as a constant verification of the material sorted, and for additional identification of actual material when necessary.

---

1. The container code letters for polyethylene terephthalate and polyvinyl chloride are different from the standard industry identification letters in order to avoid confusion with registered trademarks.

3.1 Application

3.1.1. *Containers.*   The material code should be molded, formed, or imprinted on all containers that are large enough to accept the 1/2-inch-minimum size symbol. In any case, the symbol should be applied to all containers of eight-ounce size or larger. The code should be on the bottom of the container, as close to the center as is feasible considering design, other marks, and customer requirements for clear areas. Placing the code in a similar location on all containers will allow those sorting them to quickly locate the code and identify the material.

Containers consisting of more than one resin may carry the code of the basic resin if the combination of materials is known to perform the same as the basic material in current recycling systems and normal reuse applications. Otherwise, use of the code "7 OTHER" is recommended.

3.1.2 *Lids*.   It is recommended that material code be applied to all lids of 50 cubic inches or larger. The code should be applied on the top or the underside of the lid, as close to the center as possible. Producers may voluntarily put the code on lids smaller than 50 cubic inches, so long as the minimum 1/2 inch symbol size is maintained.

3.2 Voluntary Timing

The material identification code is intended to be molded or formed into all rigid plastic containers of appropriate size, including those made from existing molds. To accommodate this procedure without substantial disruption of production schedules, it is suggested that molds can generally be modified to add the code at a time they would be off-line for other reasons. **However, the Rigid Plastic Container Division is recommending that all appropriate container and lid molds be modified by JULY 1, 1990.**

3.3 Symbol Size

3.3.1. *Containers.*   The size of the triangular arrow symbol should be a minimum of 1/2 inch and a maximum size of 2 inches, to which letters under the symbol are added, for ease of reading at a glance and for consistency. Smaller sized symbols may be used on eight-ounce and larger containers with special or restrictive base or bottom designs. This recommendation does not include using smaller sized symbols on containers less than eight ounces. Specific size recommendations are as follows:

PETE

1/2 inch symbol for any container up to 34 fluid ounces

HDPE

3/4 inch symbol for 34 fluid ounces up to one gallon containers

V

1-2 inch symbol for one gallon and larger containers, actual symbol size being proportionate to the size of the container bottom

3.3.2. *Lids.*   Symbol size should be a minimum of 1/2 inch on lids of 50 cubic inches and a maximum size of two inches, the actual size being proportionate to the area of the lid.

4.0 MOLD MODIFICATION

4.1 Containers

New and existing molds used in either injection molding or thermoforming should be marked by one of several methods: stamping, engraving, or sandblasting. The selection of the method depends on the material and the flatness of the mold surface and on the capabilities of the mold shop.

4.1.1. *Mold stamping.*   Some new and existing molds, with mold surfaces which are not hardened, may be marked with a hardened stamp.

Care should be taken to firmly hold the stamp to insure a good impression overall with sufficient depth for satisfactory readability of the molded symbol. The depth may be from .003 to .012 of an inch, depending on the contrast with the surrounding surface.

Hardened stamps may be purchased from a quality engraving shop familiar with stamp fabrication techniques. Alternately, an experienced moldmaker may be consulted for assistance in making or locating stamps. This method will not be satisfactory for molds where the symbol must be applied to a curved surface.

4.1.2. *Mold engraving.*   Molds that have hardened surfaces, or where the surface to be marked is not flat, will not be able to be marked using the stamping method. These will need to have the symbol applied by a different method such as engraving.

Engraving can be done by most moldmakers or by an engraving shop familiar with mold fabrication techniques for injection molding and thermoforming.

Master drawings for the creation of engraving masters are included with this technical bulletin. The symbol drawings and numbers/letters drawings are separate, both at eight times scale for a 1/2 inch symbol. Complete full scale photo masters are also included for convenient use for this or other purposes.

4.1.3. *Sandblasting.*   Sandblasting the symbol onto the mold can be done by most mold shops.

On some molds, particularly thermoforming molds, the mold surface is already sandblasted and the use of a sandblasted symbol would not be readily visible. When modifying existing sandblasted molds, the symbol should be engraved.

For new molds, the symbol should be stamped or engraved prior to sandblasting the mold surface. The symbol should then be masked for protection during the sandblasting operation.

4.2 Container Lids

Most new and existing molds for container lids should be marked by one of two methods, sandblasting or polishing—the selection depending on the surface of the mold. When lids need to be clear for printing or decorating, they should not be stamped or engraved because these methods may have a tendency to disturb the surface. When decorating is not a factor, lid molds may be modified by any appropriate means.

Both sandblasted and polished molds may have a tendency to wear and may require increased maintenance for continued clarity of the symbol.

5.0 IMPRINTING

Under special circumstances where mold modification is not technically feasible, the symbol may be imprinted on the bottom of the container or the top of the lid through the use of appropriate container marking or decorating equipment suitable for logos or special symbols. Care should be taken to use permanent inks, applied to surfaces appropriately conditioned to retain the mark through the entire container handling system to the recycler. This method should not be used to mark the underside of lids where the contents of the container may come in contact with the imprint.

6.0 QUALIFICATIONS

6.1 Implementation

The Society of the Plastics Industry, Inc. (SPI) is promoting a voluntary guideline for a plastic container material code system as a public service. The plastics and packaging industries, recyclers and the general public will be informed of the system through news releases, copies of this technical bulletin, or other appropriate means. The system is available to any company or person to use as appropriate.

However, use of the system is voluntary. SPI is not responsible for implementation of the system by container producers or users. Proper use of the system is the sole responsibility of each manufacturer that chooses to use it.

6.2 Recyclability of Containers

Neither the recommendations of SPI to code container by material, nor the presence of resin code on a container, conveys any guarantee, either expressed or implied, that any particular container is suitable for recycling into any particular product. The suitability of a recycled resin for a particular application will depend upon the demands of the application and the nature of any contamination resulting from prior container use. Furthermore, even within a resin type, virgin materials are manufactured with specific properties to meet the needs of specific applications. It is expected that the initial market for recycled resins will be for those applications that are tolerant of the variations in properties that exist among the various resins of each type that are represented in the waste stream.

6.3 Change in Material

If the resin used to produce a particular style of container is changed, it is the responsibility of the manufacturer to change the code to match the new resin. As noted above, use of the symbols on plastic containers is totally voluntary, and producers are free to change resins for particular containers as they see fit. The code is intended to relate solely to the resin type from which the container is made and does not relate to the contents of the container, its shape, or appearance.

6.4 Legal Status

The plastic container material code system is recommended in the belief that there is no legal impediment to its use by any container producer or user under any existing federal or state laws or regulations. However, SPI does not guarantee that no such impediment exists, nor that any federal, state, or local authority will not in the future ban or otherwise restrict the sale or use of plastic containers carrying the material code or symbol in the manner recommended.

6.5 State Mandatory Coding Requirements

Since this voluntary resin coding system was developed, a number of states have adopted mandatory resin coding requirements. SPI makes no representation, expressed or implied, that the voluntary system will satisfy specific state requirements. Each manufacturer, distributor, and user of rigid plastic containers is responsible for determining the coding requirements and compliance deadlines applicable to it and the containers it makes, distributes, or uses.

7.0 SYMBOL SIZE/LOCATION GUIDELINES

Symbols should be located as close to the center of the container bottom as feasible. This is necessary to achieve national consistency among a large variety of rigid container styles. These illustrations are general guidelines for selection of appropriate sizes and location of symbols on other container styles.

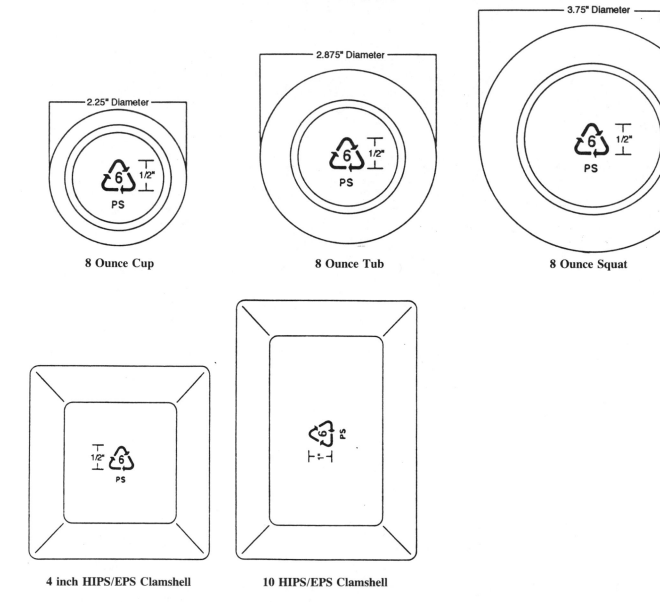

8 Ounce Cup   8 Ounce Tub   8 Ounce Squat

4 inch HIPS/EPS Clamshell   10 HIPS/EPS Clamshell

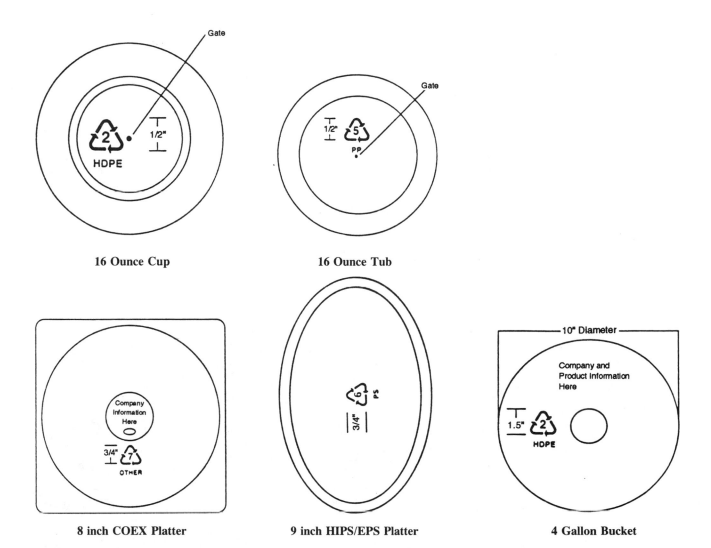

16 Ounce Cup   **16 Ounce Tub**

8 inch COEX Platter   **9 inch HIPS/EPS Platter**   **4 Gallon Bucket**

8.0 PHOTO MASTERS

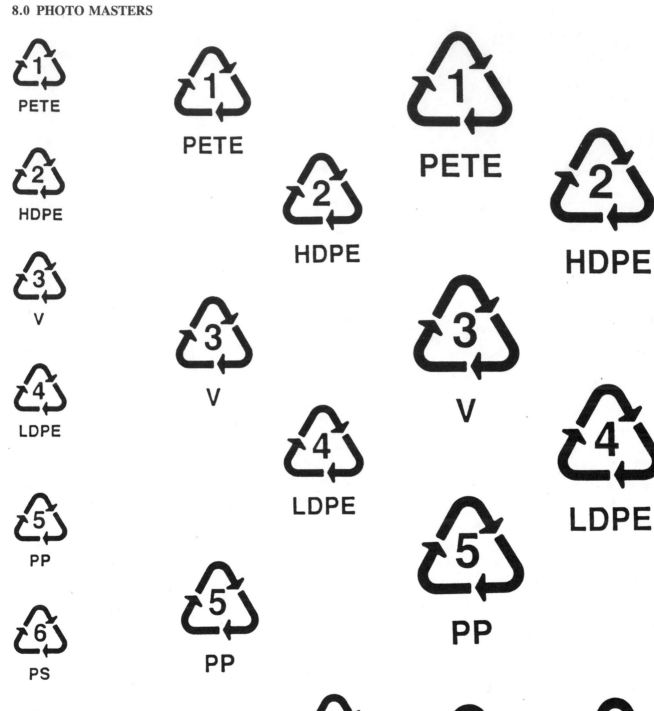

9.0 ENGRAVING MASTERS

9.1 Symbol

ALL CHARACTERS IN HELVETICA BOLD		
Size of Symbol	Approximate Printing Point Size	
	Numbers	Letters
1/2"	13	9
3/4"	20	13
1"	26	17

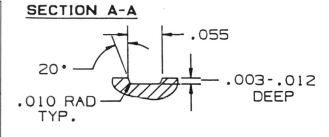

SECTION A-A

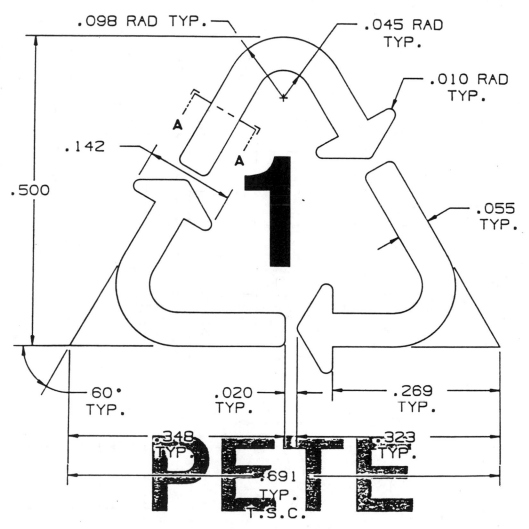

REFERENCE DRAWING FOR CONSTRUCTION OF ENGRAVING MASTER (SCALE 8 × 1/2" SYMBOL)

9.2 Numbers and Letters

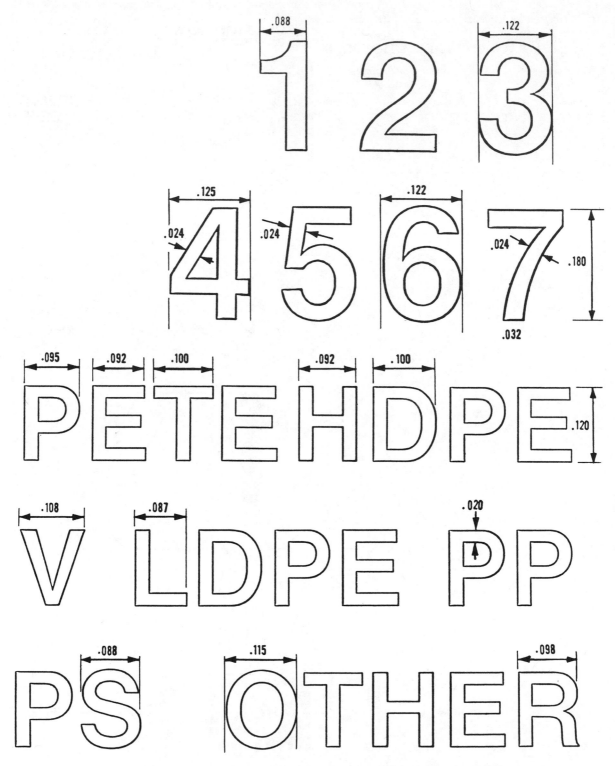

REFERENCE DRAWING FOR CONSTRUCTION OF ENGRAVING MASTER (SCALE 8 × 1/2" SYMBOL)

Application and Rules for Use of New York Recycling Emblem

SUBMITTED TO: NEW YORK STATE DEPARTMENT OF ENVIRONMENTAL CONSERVATION
DIVISION OF SOLID WASTE
BUREAU OF WASTE REDUCTION AND RECYCLING
50 WOLF ROAD—ROOM 200
ALBANY, NEW YORK 12233-4015
PHONE: (518) 457-7337

REQUEST FOR AUTHORIZATION TO USE A RECYCLING EMBLEM
REGULATED UNDER 6 NYCRR PART 368, EFFECTIVE DECEMBER 14, 1990

—PLEASE TYPE OR PRINT—

APPLICANT

Name of Company _____

Address _____

Name of Contact Person _____Telephone (   ) _____

Name of Applicant's Authorized Representative _____
(submit appropriate evidence of authority—e.g., resolution, letter, etc.)

Title of Applicant's Authorized Representative _____

A. 1. Name of package/product _____

     Material _____

  2. Emblem(s) applied for: (check all that apply and provide appropriate information)

    [ ] REUSABLE: Number of times package/product is designed to be refilled or reused _____. Provide a description of the program which the manufacturer, distributor, retailer and/or any contractor thereof, has implemented.

_____

_____

(attach additional 8½″ × 11″ sheets as necessary.)

[ ] RECYCLABLE: Standard being met (check all that apply)

_____paragraph 368.2(k)(1)   _____paragraph 368.2(k)(3)

_____paragraph 368.2(k)(2)   _____paragraph 368.2(k)(4)

Provide the basis of this determination and attach back-up data as appropriate.

[ ] RECYCLED: Minimum percent by weight:

Secondary Material Content _____ Post-Consumer Material Content _____

Provide supportive documentation on feedstocks and outputs of the manufacturing process relating to use of this emblem for accountability in meeting Part 368 standards.

B. 1. Unless otherwise directed by the New York State Department of Environmental Conservation (NYSDEC), provide an appropriate full-sized colored sketch of the package/product displaying the emblem.

2. Upon what package/product is the emblem to be displayed? How will emblem be displayed (e.g., printed, embossed)?

_____

(attach additional 8½" × 11" sheets as necessary.)

3. To what extent is secondary and post-consumer material content limited by technology and what is availability of feedstocks for each?

_____

(attach additional 8½" × 11" sheets as necessary.)

4. Describe each material component of the package/product.

_____

_____

(attach additional 8½" × 11" sheets as necessary.)

5. What marketing outlets are planned for the package/product displaying the emblem?

_____

_____

(attach additional 8½" × 11" sheets as necessary.)

6. What is disposition of packages/products after intended use?

_____

(attach additional 8½" × 11" sheets as necessary.)

C. AFFIDAVIT AND CERTIFICATION
STATE OF _____
                                          ss.:
COUNTY OF _____

I, _____ of _____
the Applicant's Authorized Representative, being duly sworn, depose and say that the package/product, for which authorization to use a recycling emblem is requested, meets or exceeds all applicable standards described in 6 NYCRR Part 368, effective December 14, 1990 and that all representations in this document and supportive material are true and accurate.

It is agreed that the Applicant will supply corroborative data upon request of the New York State Department of Environmental Conservation (NYSDEC) to confirm certification. It is further agreed that representatives of the NYSDEC or the New York State Office of General Services shall have access to the manufacturing plant and purchase/production records at anytime during working hours for the purpose of verifying the certification which has been made for the package/product. If any information submitted as part of the applicant's request is false, authorization for use of such emblem may be denied or revoked by the NYSDEC pursuant to paragraph 368.5(d)(6) and subdivision 368.5(f).

Signature of Applicant's Authorized Representative: _____

On this _____ day of _____ 199____, before me personally came _____, to me known, who being duly sworn, did depose and say that she/he is the Authorized Representative of the Applicant described above, and executed the above instrument on behalf of said Applicant.

_____
Notary Public

Supportive Documentation

Please provide the following documentation for the specific products for which your authorization request has been submitted:

a. Description and quanitification of inventory of post-consumer and pre-consumer material by weight (or volume) on hand at beginning of the period.

b. Summary/description of post-consumer and pre-consumer material by weight (or volume) purchased during the period.

c. Description and quantification of inventory of post-consumer and pre-consumer material by weight (or volume) on hand at the end of the period.

d. Total weight (or volume) of products produced from the pre/post-consumer feedstock for the period.

Invoices and any other supporting documentation must be maintained in a separate file at your place of business. Unless otherwise agreed upon, the period will be a month, weights will be in tons; volume in cubic yards; the minimum justification period will be six months; if a special run is completed for the "recycled" product, the run will be identified; if secondary material is purchased from more than one vendor a summary of purchases will suffice although the actual invoices must be retrievable, if requested. The signature page and the respective pages (showing the values for the documentation requested above) from the annual audit report must be included in this separate file. The meaning of the terms used in this correspondence will be as defined in 6 NYCRR Part 368—Recycling Emblems.

The supportive documentation described above must be approved by this office before authorization is granted for use of an official recycling emblem.

6 NYCRR PART 368: RECYCLING EMBLEMS
(EFFECTIVE DECEMBER 14, 1990)

(Statutory authority: Environmental Conservation Law, Section 27-0717.2)

Section 368.1   Purpose and applicability

(a) Purpose. The purpose of this Part is to establish official recycling emblems and establish standards to be applied for the use of such emblems and terms "recycled," "recyclable," and "reusable." This Part also establishes procedures for petitioning the department for authorization to use an emblem.

Since requests under this Part are not for permits to construct or to operate a solid waste management facility, review of requests for authorization to use an emblem are not subject to Part 621 of this Title. However, the procedures relating to revocation of permits as provided under Part 621 of this Title will apply to revocation of authorizations as provided under this Part.

(b) Applicability. This Part applies to:

(1) Any person who wishes to use an emblem as defined in this Part on a package or product, or in the promotion or advertisement of a package or product, which is sold or offered for sale in this State;

(2) any person who uses the terms "recycled," "recyclable," or "reusable" directly on or in the promotion or advertisement of a package or product which is sold or offered for sale in this State.

Section 368.2   Definitions.

Except as the context indicates otherwise, the following terms have the following meanings when used in this Part:

(a) "Commissioner" means the Commissioner of Environmental Conservation or his duly designated representative.

(b) "Emblem" or "recycling emblem" means:

(1) a symbol illustrated in Section 368.3 of this Part or a facsimile thereof; or the terms "recycled," "recyclable," or "reusable."

(c) "Material" means the elements, constituents, or substances of which something is composed or can be made.

(d) "Material category" means a category of material that can be collected, marketed and utilized in the aggregate as a secondary material feedstock. Such categories include, but are not limited to the following:

(1) clear container glass;
(2) green container glass;
(3) amber container glass;
(4) paper—newspaper;
(5) paper—corrugated containers;
(6) paper—high-grade printing and writing paper;
(7) paper—paperboard;
(8) paper—coated;
(9) paper—kraft bags;
(10) plastic—PET (polyethylene terepthelate);
(11) plastic—HDPE (high density polyethylene);
(12) plastic—PVC (polyvinyl chloride);
(13) plastic—LDPE (low density polyethylene);

(14) plastic—PP (polypropylene);

(15) plastic—PS (polystyrene);

(16) ferrous metal food and beverage containers;

(17) ferrous metal—excluding food and beverage containers;

(18) aluminum food and beverage containers;

(19) aluminum—excluding food and beverage containers;

(20) copper; and

(21) lead-acid batteries.

(e) "Package" or "packaging" means a container, vessel, covering, wrapping, box or device in which a material or product is protected, stored, treated, handled or transported.

(f) "Person" means any individual, public or private corporation, political subdivision, government agency, authority, department or bureau of the State, municipality, industry, co-partnership, association, firm, trust, estate or any other legal entity whatsoever.

(g) "Post-consumer material" means only those products, packages or materials generated by a business or consumer which have served their intended end uses, and which have been separated or diverted from the waste stream for the purposes of collection, recycling and disposition.

(h) "Pre-consumer material" means material and by-products which have not reached a business or consumer for an intended end use and have been recovered or diverted from the waste stream, including but not limited to industrial scrap material, overstock or obsolete inventories from distributors, wholesalers and other companies but such term does not include those materials and by-products generated from, and commonly reused within, an original manufacturing process or separate operations within the same parent company.

(i) "Product" means that which is created as a result of a manufacturing process. For the purposes of this regulation, packages are excluded from this definition.

(j) "Recovered paper material" means paper waste generated after the completion of a papermaking process, such as post-consumer materials, envelope cuttings, bindery trimmings, printing waste, cutting and other converting waste, butt rolls and mill wrappers, obsolete inventories and rejected unused stock. Recovered paper material, however, shall not include fibrous waste generated during the manufacturing process such as fibers recovered from wastewater or trimmings of paper machine rolls (mill broke) regardless of whether such materials are used by the same or another company, and shall also not include fibrous by-products of harvesting, extractive or woodcutting processes or forest residues such as bark. Paper waste generated and reused within operation of the same parent company are excluded from this definition.

(k) "Recyclable" means a material for which any of the following standards are met:

(1) Access to community recyclable recovery programs for that material is available to no less than seventy-five (75) percent of the population of the State; or

(2) A Statewide recycling rate of fifty (50) percent has been achieved within the material category; or

(3) A manufacturer, distributor, or retailer achieves a Statewide recycling rate of fifty (50) percent for the product or package sold within the State; or

(4) A product or package may be recyclable within the jurisdiction of a municipality where an ongoing source separation and recycling program provides the opportunity for recycling of the product or package.

(l) "Recycled" means a package or product containing a specified minimum percentage by weight of secondary material content and minimum percentage by weight of post-consumer material as described in subdivision 368.4(a). The percentage of secondary material content shall be that portion of a package or product weight that is composed of secondary material as demonstrated by an annual mass balance of all feedstocks and outputs of the manufacturing process. The weight of secondary material use in any month shall be no less than eighty (80) percent of the average monthly secondary material usage during the corresponding calendar year.

(m) "Recycling rate" means the percentage by weight of a given material category sold or distributed for sale in the State that would otherwise be destined for the waste stream, including post-consumer and pre-consumer materials, that is collected or otherwise returned for processing or refabrication into marketable end products other than fuel for producing heat or power by combustion.

(n) ''Reusable'' means that the original package or product can be returned for refilling or reuse a minimum of five times as demonstrated by an annual accounting in a program established by a manufacturer, distributor or retailer for refilling or reuse of the manufactured product or package.

(o) ''Secondary material'' or ''recovered material'' means material recovered from or otherwise destined for the waste stream, including pre-consumer material and post-consumer material but such term does not include those materials and by-products generated from, and commonly reused within, an original manufacturing process or separate operations within the same parent company. For paper and paper products, ''secondary material'' shall mean ''recovered paper material.''

Section 368.3 Emblem illustrations.

(a) The recycled emblem is illustrated in Appendix A (Exhibit L–4).

(b) The recyclable emblem is illustrated in Appendix B (Exhibit L–5).

(c) The reusable emblem is illustrated in Appendix C (Exhibit L–6).

Section 368.4 Standards for emblems.

(a) Recycled emblem.

Authorization to use the recycled emblem will not be granted unless:

(1) the package or product is recycled as defined in this Part; and

(2) the package or product meets or exceeds the standards in Table 1 of this section, pertaining to:

(i) minimum percentage by weight of secondary material content; and

(ii) where specified, the minimum percentage by weight of post-consumer material content of a material to which the emblem relates.

(b) Recyclable emblem. Authorization to use the recyclable emblem will not be granted unless the package or product is recyclable as defined in this Part and the package or product to which the emblem pertains can be used in its entirety, excluding labels, stickers, adhesives and closures, as a feedstock at the beginning of a manufacturing process. Where use of the emblem is justified solely upon the existence of a source separation program in the community within which the product or package is sold, authorization to use the recyclable emblem will be limited to the placement of labels on shelves or displays containing the products or packages.

(c) Reusable emblem. Authorization to use the reusable emblem will not be granted unless the package or product is reusable as defined in this Part and, subsequent to cleaning if appropriate, is intended to be reused or refilled a minimum of five times in a program developed and implemented by a manufacturer, distributor or retailer.

(d) Recycled package or product for which standards have not been established in this Part.

(1) All persons selling a package or product in this State, except as specifically prohibited in this Part, may request authorization for use of an emblem. For those cases where standards have not been established, the department will evaluate the request and make a determination using the following considerations:

(i) the extent to which secondary material is limited by technology;

(ii) the supply limitations or availability of secondary material feedstock;

(iii) the current recycling practice within the industry; and

(iv) the necessity for furthering uniformity of standards, particularly with those of the United States Government and other states.

(2) Authorization to use a recycled emblem will not be given for packages or products containing less than 25 percent of secondary material content by weight.

(e) A person that accepts the return of a package or product to the place of purchase for purposes of recycling, whether or not authorized to use an emblem or the terms ''recycled,'' ''recyclable,'' or ''reusable'' may provide written text on or with the package

TABLE L–1

Material	Standard for Minimum Percentage by Weight of Secondary Material Content	Standard for Minimum Percentage by Weight of Post-Consumer Material Content
1. Building Insulation:		
(i) Cellulose; loose-fill and spray-on	75 percent	75 percent
(ii) Rock wool	50 percent	Not Applicable
2. Paper and Paper Products:		
(i) Newsprint	40 percent	40 percent
(ii) High-grade printing and writing papers	50 percent	*
(iii) Tissue products:		
Paper towels	80 percent	40 percent
Paper napkins	80 percent	30 percent
Other	80 percent	20 percent
(iv) Bleached and unbleached packaging:		
Kraft bags	50 percent	20 percent
Other	80 percent	35 percent
(v) Paperboard	90 percent	35 percent
3. Lubricating Oils	50 percent	50 percent
4. Steel		
(i) Packaging	30 percent	15 percent
(ii) Products	75 percent	15 percent
5. Aluminum		
(i) Packaging	30 percent	15 percent
(ii) Products	85 percent	15 percent
6. Copper	50 percent	50 percent
7. Plastics		
(i) Packaging	30 percent	15 percent
(ii) Products	50 percent	15 percent
8. Glass		
(i) Packaging	35 percent	35 percent
(ii) Products	50 percent	35 percent
9. Solvents	75 percent	75 percent
10. Rubber Products	50 percent	25 percent
11. Multi-Material		
(i) Packaging	**	**
(ii) Products	50 percent	50 percent

*Effective January 1, 1994, the minimum percentage by weight of post-consumer material content shall be ten (10) percent.

**Each material component of the package must meet the recycled content standards for that material as described in this section.

or product instructing the purchaser to return the package or product to the place of purchase for purposes of recycling.

(f) The department will reevaluate this Part at least every two years to determine if the standards should be revised to more accurately reflect the current recycling technology and recycling needs of the State. A primary goal of such reevaluation will be the increase of minimum post-consumer material content standards.

Section 368.5 Provisions for use of emblems.

(a) General authorization. An emblem may be used for the following purposes without having first received written approval from the department for such use:

(1) promotion to increase public awareness of an emblem and to encourage consumers to purchase packages or products bearing an authorized emblem; and

(2) promotion of a package or product after the department has authorized the use of the emblem on or for that material.

(b) Use. Use of an emblem as defined in this Part is voluntary.

(c) Prohibitions. After eighteen (18) months from the effective date of this Part:

(1) No person shall sell or offer for sale in this State any package or product displaying an emblem as defined in this Part indicating that the package or product upon which it is placed or to which it refers is recycled, recyclable or reusable unless authorized under provisions of this Part;

(2) Except as provided elsewhere in this Part, no person shall place an emblem as defined in this Part on a package or product without prior written authorization from the department;

(3) Except as provided elsewhere in this Part, no person shall use an emblem as defined in this Part for the promotion or advertisement of a package or product as recycled, recyclable or reusable unless authorized under provisions of this Part.

(d) Specific authorization.

Except as provided elsewhere in this Part, authorization for the use of an emblem may only be granted by the commissioner or his authorized representative.

(1) Requests for use of an emblem must be made in writing and sent to:

Recycling Emblems
New York State Department of Environmental Conservation
Division of Solid Waste
Bureau of Waste Reduction and Recycling
50 Wolf Road
Albany, New York, 12233-4015

The written request including information relating to the package or product must be accompanied by a full sized, colored sketch (as appropriate) of the emblem and the package or product on which it is to be displayed and a certification that the package or product that will be displaying the emblem meets the applicable standards for use of the emblem set forth in section 368.4 of the Part.

(2) Unless otherwise directed, the following information relating to the package or product on which the emblem will be displayed must be provided with the request for authorization:

(i) upon what package or product and how the emblem will be displayed (e.g., stamped, embossed, sticker, printed, etc.);

(ii) whether the label covers all sides of the container;

(iii) for packages or products with different material components, a description of each component;

(iv) the name, title, address and telephone number of the responsible party making the certification under subdivision (c)(1) of this section;

(v) the outlets to be used for marketing the package or product displaying the emblem;

(vi) a statement that a commitment is made to supply corroborative data at the request of the department to confirm certification;

(vii) the waste implications of the package or product after use;

(viii) the extent to which secondary material and post-consumer material content standards are limited by technology; and

(ix) the availability of secondary material and post-consumer material feedstocks.

(3) The certification with evidence of authority must be made a responsible party as follows:

(i) in the case of corporations, by a duly authorized principal executive officer;

(ii) in the case of a partnership or limited partnership, by a general partner;

(iii) in the case of a sole proprietorship, by the proprietor; or,

(iv) in the case of a joint venture, by a joint venture principal.

The required certification shall be in a format as provided by the department.

(4) Within seventy-five (75) days after receipt (date-stamped received by the de-

partment) of a request for authorization to use an emblem, the department will respond or determine whether or not to authorize that use and will notify the person requesting the authorized use of its determination in writing.

(i) For those cases whose standards have not been set forth in section 368.4 of this Part, the department will evaluate the request to use the emblem and make a determination based on considerations listed in this Part.

(ii) The authorized user of the recycled emblem must label the percentage of secondary material content in the center of the arrows or immediately adjacent thereto.

(iii) Authorization will be granted for a maximum period of five years. A request to renew authorization, which must contain all the information identified in paragraphs 368.5 (d)(1), (2) and (3), must be made at least one hundred twenty (120) days before expiration. Failure to do so will result in automatic termination of the authorization to use the emblem effective on the date of expiration.

(iv) Within fifteen (15) calendar days of the receipt of authorization from the department, a final representation of the emblem and package or product on which it will be displayed must be submitted to the department by the person who originally requested its use. The submittal shall be made to the address provided in paragraph 368.5(d)(1) of this Part.

(5) Within one hundred eighty (180) days from notification by the department of any promulgated revision to section 368.4 of this Part that will result in increased recycled content for a package or product, each person authorized to use a recycled emblem for that package or product must request in writing department authorization as outlined in subdivision 368.5(d) to continue use of that emblem. Failure to certify within such period that the package or product meets the new standard will result in revocation of the authorized use of the emblem.

(6) The department will deny authorization to use an emblem if it determines that such use fails to meet the requirements of this Part for authorized uses. If authorization is denied, that person may resubmit a request without prejudice.

(e) Placement of the emblem.

(1) The emblem must be displayed in a manner that clearly indicates the package or product to which it pertains;

(2) Where the emblem applies only to the content of a package, the emblem must be placed on the label describing the content and if possible on the content itself;

(3) Where the emblem applies only to the package, the emblem must be placed on the package with a statement that the emblem applies to the package only; and

(4) The emblem must be exhibited no smaller than three quarters of an inch in size unless otherwise authorized by the department and must be clearly legible.

(f) Revocation of authorized use of an emblem. The department will revoke a general or specific authorization if it is determined that the emblem's use fails to meet the requirements of this Part or the conditions of the authorization.

(1) If a determination is made to revoke the authorized use of an emblem, the person who has granted the authorization will be notified that further use of the emblem is prohibited. The date of receipt of the department's notification, will be the effective revocation date. A person whose authorization to use an emblem has been revoked may not request reauthorization for a period of one year following the effective revocation date. The procedures relating to revocation are those as provided under Part 621 of this Title pertaining to permits.

(2) A notice of revocation will be published in the State Register to alert persons selling packages or products affected by a revocation. In addition, the department will provide notification to trade organizations that request to be placed upon a mailing list for such notifications.

(3) A period of one hundred twenty (120) days from the effective revocation date will be allowed for retailers, brokers, distributors, and wholesalers to clear shelves and stock of existing packages or products affected by a revocation and also to allow manufacturers to discontinue production.

(4) The department reserves the right to revoke the general authorization for the use of an emblem provided in subdivision (a) of this section or specific authorization for

New York State Department of Environmental Conservation
50 Wolf Road, Albany, New York 12233

To The Regulated Community and The General Public

Re: Proposed Amendments to Part 368 Regulations for Recycling Emblems
    Schedule for Public Meetings and Public Hearings

The New York State Department of Environmental Conservation (NYSDEC) has filed proposed amendments to 6 NYCRR Part 368 for the use of the official New York State recycling emblems (copy enclosed) with the Secretary of State.

In New York State, waste reduction and recycling are the more important elements of our solid waste management program. We have made a commitment to: 1) reduce the amount of solid waste being generated; 2) reduce the ultimate amount of solid waste requiring disposal; and 3) safely manage and dispose of the remaining wastes. Recycling, or reusing discarded products or materials, is a very sensible way to reduce the municipal solid waste stream. Many materials now entering the municipal solid waste stream can be resued or recycled, saving landfill space and reducing waste disposal. Paper, glass, waste oil, plastic, yard waste, metals, rubber, certain sludges and other recyclable materials can be separated from nonrecyclable wastes at the source—the home or business where the waste is created. These materials can be used as raw materials for manufacturing, some after processing or refining, others as collected from the source.

New York State promulgated regulations on December 14, 1990, which established official recycling emblems and regulations governing their proper use. These regulations identified the standards that must be met for products to be called recycled, recyclable, or reusable. In turn, consumer demand for environmentally preferable products has been useful in driving manufacturers to utilize secondary materials and to design for recyclability. To improve upon this effort, the proposed amendments to 6 NYCRR Part 368 were developed by the NYSDEC.

The schedule for the public meetings and the public hearing on these proposed amendments to the Part 368 recycling emblems regulations is listed in the enclosed "Notice of Public Hearing." There will be two sessions held at each location, one starting in the morning and the other starting in the afternoon.

We intend to file our proposed amendment to Part 368 with the New York State Environmental Board for consideration at their June meeting in 1994.

We welcome your participation and comments on these proposed regulations.

Sincerely,

John J. Willson, P.E.
Acting Director
Division of Solid Waste

the use of an emblem provided in subdivision (d) of this section if it deems that its use constitutes an abuse or is offensive to the purpose and intent of this Part.

(5) Upon revocation, it is the responsibility of the person who was granted the authorization to use a recycling emblem, to immediately notify affected parties (e.g., retailers, brokers, distributors, wholesalers, etc.) who may have a supply of the packages or products bearing the emblem, and assure that all such packages or products are made unavailable for sale or that the emblem is removed or obliterated on any package or product remaining for sale no later than one hundred twenty (120) days after the revocation date.

Section 368.6 Severability.

If any provision of this Part or its application to any person or circumstances is held invalid, the remainder of this Part, and the application of those provisions or circumstances other than those to which it is held invalid, shall not be affected thereby.

DRAFT 6 NYCRR PART 368 RECYCLING EMBLEMS

Proposed Amendments to New York State's Recycling Emblem Regulations, December 1993

Proposed changes:
(a) Within brackets—to be deleted
(b) Underlined—to be added

368 Proposed Amendments
6 NYCRR Part 368

RECYCLING EMBLEMS

(Statutory authority: Environmental Conservation Law, Section 27-0717.2)

Section 368.1—Purpose [**and**] applicability, and transition
   368.2—Definitions
   368.3—Emblem illustrations
   368.4—Standards for emblems
   368.5—Provisions for use of emblems
   368.6—Severability

Section 368.1   Purpose, [and] applicability, and transition.

(a) Purpose. The purpose of this Part is to establish official recycling emblems and establish standards to be applied for the use of such emblems and the terms "recycled," "recyclable," and "reusable." This Part also established procedures for petitioning the department for authorization to use an emblem. Since requests under this Part are not for permits to construct or to operate a solid waste management facility, review of requests for authorization to use an emblem are not subject to part 621 of this Title. However, the procedures relating to revocation of permits as provided under Part 621 of this Title will apply to revocation of authorizations as provided under this Part.

(b) Applicability. This Part applies to:

(1) any person who wishes to use an emblem as defined in this Part on a package or product, or in the promotion or advertisement of a package or product, which is sold or offered for sale in this State; or

(2) any person who uses the terms "recycled," "recyclable," or "reusable" directly on or in the promotion or advertisement of a package or product which is sold or offered for sale in this State.

(c) Transition.

(1) Authorization by the Department to use an official State recycled, recyclable or reusable emblem which was granted before the effective date of these regulations shall continue under the terms of and for the period specified in such authorization. At the expiration of such period the requirements of this Part shall apply.

(2) Applications for authorization to use an official State recycled, recyclable or reusable emblem which are pending before the department on the effective date of this Part shall comply with the requirements of this Part.

Section 368.2   Definitions.

Except as the context indicates otherwise, the following terms have the following meanings when used in this Part:

(a) "Commissioner" means the Commissioner of Environmental Conservation or [his] a duly designated representative.

(b) "Emblem" or "recycling emblem" means:

(1) a symbol [illustrated] described in Section 368.3 of this Part [or a facsimile thereof]; or

(2) the term[s] "recycled," "recyclable," or "reusable."

(c) "Filing/cover/binding" grades means kraft, file folders; red wallet; colored file folders; data, ring or grip binders, report cover or similar products.

(d) [(c)] "Material" means the elements, constituents, or substances of which something is composed or can be made.

(e) [(d)] "Material category" means [a category of material] that which can be collected, marketed and utilized in the aggregate as a secondary material feedstock. [Such categories include, but are not limited to the following:

(1) clear container glass;
(2) green container glass;
(3) amber container glass;
(4) paper—newspaper;
(5) paper—corrugated containers;
(6) paper—high-grade printing and writing paper;
(7) paper—paperboard;
(8) paper—coated;
(9) paper—kraft bags;
(10) plastic—PET (polyethylene terepthelate);
(11) plastic—HDPE (high density polyethylene);
(12) plastic—PVC (polyvinyl chloride);
(13) plastic—LDPE (low density polyethylene);
(14) plastic—PP (polypropylene);
(15) plastic—PS (polystyrene);
(16) ferrous metal food and beverage containers;
(17) ferrous metal—excluding food and beverage containers;
(18) aluminum food and beverage containers;
(19) aluminum—excluding food and beverage containers;
(20) copper; and
(21) lead-acid batteries.]

[(e)] (f) "Package" or "packaging" means a container, vessel, covering, wrapping, box or device in which a material or product is protected, stored, treated, handled or transported.

[(f)] (g) "Person" means any individual, public or private corporation, political subdivision, government agency or subdivision thereof, authority, [department or bureau of the State] municipality, industry, co-partnership, association, firm, trust, estate or any other legal entity whatsoever.

[(g)] (h) "Post-consumer material" means only those products, packages or materials generated by a business or consumer which have served their intended end use[s] as consumer items, and which have been separated or diverted from the waste stream for the purposes of collection, and recycling [and disposition.] as a secondary material feedstock, but shall not include waste material generated after the completion of a manufacturing or converting process.

[(h)] (i) "Pre-consumer material" means material and by-products which have not reached a business or consumer for an intended end use and have been recovered or diverted from the waste stream, including but not limited to industrial scrap material, overstock, or obsolete inventories from distributors, wholesalers and other companies but such term does not include those materials and by-products generated from, and commonly reused within, an original manufacturing process or separate operations within the same parent company.

[(i)] (j) "Product" means that which is created as a result of a manufacturing or printing process. For the purposes of this Part [regulation,] packages are excluded from this definition.

[(j)] (k) "Recovered paper material" means paper waste generated after the completion of a papermaking process or cotton waste, such as post-consumer materials, envelope cuttings, bindery trimmings, printing waste, cutting and other converting waste, butt rolls and mill wrappers, obsolete inventories and rejected unused stock. Recovered paper material, however, shall not include fibrous waste generated during the manufacturing process such as fibers recovered from wastewater or trimmings of paper machine rolls (mill broke) regardless of whether such materials are used by the same or another company, and shall also not include fibrous by-products of harvesting, extractive or woodcutting processes or forest residues such as bark. As an alternative to meeting the standards in section 368.4, for all printing and writing papers, the minimum content standard may be no less than 50 percent recovered materials that are a waste material byproduct of a finished product other than a paper or textile product which would otherwise be disposed of in a landfill, as determined by the state in which the facility is located. [Paper] Waste generated and reused within operations of the same parent company [are] is excluded from this definition.

[(k)] (l) "Recyclable" means a material for which any of the following standards are met:

(1) For labeling on products or packaging:

(i) [(1)] access to community recyclable recovery programs for that material is available to no less than seventy-five (75) percent of the population of [the] New York State; or

(ii) [(2)] a statewide recycling rate of fifty (50) percent has been achieved within the material category; or

(iii) [(3)] a manufacturer, distributor, or retailer achieves a statewide recycling rate of fifty (50) percent for the product or package sold within the State;

(2) For store shelf or display labeling only: A product or package may be recyclable within the jurisdiction of a municipality where an ongoing source separation and recycling program provides the opportunity for recycling of the product or package.

[(l)] (m) "Recycled" means a package or product containing a specified minimum percentage by weight of secondary material content and minimum percentage by weight of post-consumer material as described in subdivision 368.4(a) of this Part, shown in Table 1 Column A Combined, and approved by the Commissioner; or, alternatively, a percentage by weight of post-consumer material as described in subdivision 368.4(a), of this Part shown in Table 1 Column B: post-consumer only, and approved by the Commissioner. The percentage of secondary material or post-consumer material content shall be that portion of a package or product weight that is composed of secondary material or post-consumer material as demonstrated by an annual mass balance of all feedstocks and outputs of the manufacturing process. The weight of secondary material or post-consumer material use in any month shall be no less than eighty (80) percent of the average monthly secondary material or post-consumer material usage during the corresponding calendar year.

[(m)] (n) "Recycling rate" means the percentage by weight of a given material category sold or distributed for sale in the State that would otherwise be destined for the waste stream, including post-consumer and pre-consumer materials, that is collected or otherwise returned for processing or refabrication into marketable end products other than fuel for producing heat or power by combustion.

[(n)] (o) "Reusable" means that the original package or product can be returned for refilling or reuse a minimum of five times as demonstrated by an annual accounting in a program established by a manufacturer, distributor or retailer for refilling or reuse of the manufactured product or package.

[(o)] (p) "Secondary material" or "recovered material" means material recovered from or otherwise destined for the waste stream, including pre-consumer material and post-consumer material but such term does not include those materials and by-products generated from, and commonly reused within, an original manufacturing process or separate operations within the same parent company. For paper and paper products, "secondary material" shall mean "recovered paper material."

Section 368.3   Emblem illustrations.

(a) The official New York State recycled emblem is illustrated in [Section 368.4 of this Part] Appendix A (Exhibit L–1). The post-consumer percentage must be shown with the recycled emblem as defined in this Part and identified as such. The pre-consumer or secondary material percentage may also be shown. For paper products, the fiber weight may be noted provided it is clearly identified as such and all requirements under this Part have been met.

(b) The official New York State recyclable emblem is illustrated in [Section 368.4 of this Part] Appendix B (Exhibit L–2).

(c) The official New York State reusable emblem is illustrated in [Section 368.4 of this Part] Appendix C (Exhibit L–3).

EXHIBIT L–1

REG. NYSDEC   TM

RECYCLED

EXHIBIT L–2

REG. NYSDEC   TM

RECYCLABLE

EXHIBIT L–3

REG. NYSDEC   TM

REUSABLE

Section 368.4   Standards for emblems.

(a) Recycled emblem.

Authorization to use the recycled emblem will not be granted unless:

(1) the package or product is recycled as defined in this Part; and

(2) the package or product meets or exceeds the standards in either Column A or B of Table 1 of this section, pertaining to:

(i) the minimum percentage by weight of secondary material content shown in Column A and the minimum percentage by weight of post-consumer material shown in Column A; [and] or

[(ii) where specified, the minimum percentage by weight of post-consumer material content of a material to which the emblem relates shown in Column A; or]

(ii) [i] the minimum percentage by weight of post-consumer material content shown in Column B.

NOTE: Table 1 is repealed and a new Table 1 is adopted.

(b) Recyclable emblem.

(1) Authorization to use the recyclable emblem will not be granted unless the package or product is recyclable as defined in this Part and the package or product to which the emblem pertains can be used in its entirety, excluding labels, stickers, adhesives and closures, as a feedback at the beginning of a manufacturing process. Where use of the emblem is justified solely upon the existence of a source separation program in the community within which the product or package is sold, authorization to use the recyclable emblem will be limited to the [placement of labels] use of the emblem on shelves or displays containing the products or packages.

(2) For the purposes of this Part, the department has determined that it may authorize the use of the recyclable emblem on the following materials provided that requests for such authorization are filed with the department pursuant to the requirements for subdivision 368.5(d) of this Part:

(i) Flint glass—(clear container glass);

(ii) Newsprint;

(iii) Corrugated containers;

(iv) Rigid PET plastic containers—blow molded;

(v) Rigid HDPE plastic containers—blow molded;

TABLE L-2

Material	A. Combined		B. Post-Consumer Only
	Standard for Minimum Percentage by Weight of Secondary Material Content	Standard for Minimum Percentage by Weight of Post-Consumer Material Content	Standard for Minimum Percentage by Weight of Post-Consumer Material Content
1. Building Insulation			
(i) Fiberglass	40 percent	40 percent	40 percent
(ii) Other	50 percent	50 percent	50 percent
2. Paper and Paper Products			
(i) Newsprint	40 percent	40 percent	40 percent
(ii) Printing and writing papers—A*	50 percent	10 percent°	20 percent°
Printing and writing papers—B**	50 percent	20 percent°	25 percent°
Printing and writing papers—C***	50 percent	10 percent	15 percent
Filing/Cover/Binding Grades	50 percent	30 percent	40 percent
(iii) Tissue products			
Paper towels	80 percent	40 percent	50 percent
Paper napkins	80 percent	30 percent	40 percent
Other	80 percent	20 percent	40 percent
(iv) Bleached and unbleached packaging			
Kraft bags	50 percent	20 percent	30 percent
Other	55 percent	35 percent	45 percent
(v) Paperboard	85 percent	35 percent	55 percent
Wet Strength Carrier+	30 percent	20 percent	25 percent
3. Steel			
(i) Packaging	30 percent	15 percent	25 percent
(ii) Products	75 percent	15 percent	50 percent
4. Aluminum			
(i) Packaging	55 percent	45 percent	50 percent
(ii) Products	85 percent	20 percent	40 percent
5. Plastics			
(i) Film Packaging	35 percent	15 percent	30 percent
(ii) Rigid Packaging	50 percent	15 percent	30 percent
(iii) Products	50 percent	15 percent	30 percent
6. Glass			
(i) Packaging	45 percent	35 percent	40 percent
(ii) Products	50 percent	35 percent	40 percent
7. Multi-Material			
(i) Packaging++	++	++	++
(ii) Products	50 percent	50 percent	50 percent
8. Other Products or Packaging	50 percent	50 percent	50 percent

  * High speed copier paper, offset paper, forms bond, computer printout paper, carbonless paper, white woven envelopes, and all other printing and writing paper not specified in this subcategory.

 ** Writing and office paper, book paper, cotton fiber paper.

*** Coated paper.

  + Wet strength basket carrier—.021″ caliber or less.

++ Each material component of the package must meet the recycled content standards for that material as described in this section.

  ° These values increase an additional 10 percent of total content beginning on December 31, 1998.

(vi) Ferrous metal food and beverage containers;
(vii) Aluminum food and beverage containers;
(viii) Aluminum foil;
(ix) Lead-acid batteries;
(x) Returnable beverage containers (as defined in 6 NYCRR Part 367); and
(xi) Used engine lubricating oils.
(c) Reusable emblem.

Authorization to use the reusable emblem will not be granted unless the package or product is reusable as defined in this Part and, subsequent to cleaning if appropriate, is intended to be reused or refilled a minimum of five times in a program developed and implemented by a manufacturer, distributor or retailer.

[(d) Recycled package or product for which standards have not been established in this Part.

(1) All persons selling a package or product in this State, except as specifically prohibited in this Part may request authorization for use of an emblem. For those cases where standards have not been established, the department will evaluate the request and make a determination using the following consideration;

(i) the extent to which secondary material content is limited by technology;
(ii) the supply limitations or availability of secondary material feedstock;
(iii) the current recycling practice within the industry; and
(iv) the necessity for furthering uniformity of standards, particularly with those of the United States Government and other states.

(2) Authorization to use a recycled emblem will not be given for packages or products containing less than 25 percent of secondary material content by weight.]

[(e)] (d) A person that accepts the return of a package or product to the place of purchase for purposes of recycling, whether or not authorized to use an emblem or the terms "recycled," "recyclable," or "reusable" may provide written text on or with the package or product instructing the purchaser to return the package or product to the place of purchase for purposes of recycling.

[(f)] (e) The department will reevaluate this Part at least every two years to determine if the standards should be [revised] amended to more accurately reflect the current recycling technology and recycling needs of the State. A primary goal of such reevaluations will be the increase of minimum post-consumer material content standards.

Section 368.5   Provisions for use of emblems.

(a) General authorization. An emblem may be used for the following purposes without having first received written approval from the department for such use:

(1) promotion to increase public awareness of an emblem and to encourage consumers to purchase packages or products bearing an authorized emblem; and

(2) promotion of a package or product after the department has authorized the use of the emblem on or for that material.

(b) Use. Use of an emblem as defined in this Part is voluntary.

(c) Prohibitions. [After eighteen (18) months from the effective date of this Part:]

(1) no person shall sell or offer for sale in this State any package or product displaying an emblem as defined in this Part indicating that the package or product upon which it is placed or to which it refers is recycled, recyclable or reusable unless authorized under provisions of this Part:

(2) except as provided elsewhere in this Part, no person shall place an emblem as defined in this Part on a package or product without prior written authorization from the department; or

(3) except as provided elsewhere in this Part, no person shall use an emblem as defined in this Part for the promotion or advertisement of a package or product as recycled, recyclable or reusable unless authorized under provisions of this Part.

(d) Specific authorization.

Except as provided elsewhere in this Part, authorization for the use of an emblem may only be granted by the commissioner. [or his authorized representative]

(1) Requests for <u>authorization to</u> use [of] an emblem must be made in writing and sent to:

> Recycling Emblems
> New York State Department of Environmental Conservation
> Division of Solid Waste
> Bureau of Waste Reduction and Recycling
> 50 Wolf Road
> Albany, New York 12233-4015

The written request <u>for use of an emblem,</u> including information relating to the package or product, must be accompanied by a full sized, color sketch (as appropriate) of the emblem and the package or product on which it is to be displayed <u>unless the department determines that a sketch is not needed or may be submitted at a specified later date</u> and a certification that the package or product <u>for which the request is made</u> [that will be displaying the emblem] meets the applicable standards for use of the emblem set forth in section 368.4 of <u>this</u> [the] Part.

(2) <u>Provide the information requested on an authorization request form prescribed by the department.</u> Unless otherwise directed, the following information [relating to the package or product on which the emblem will be displayed] must be provided with the request for authorization:

[(i) upon what package or product and how the emblem will be displayed (e.g., stamped, embossed, sticker, printed, etc.);

(ii) whether the label covers all sides of the container;

(iii) for packages or products with different material components, a description of each component;]

[(iv)] <u>(i)</u> the name, title, address and telephone number of the responsible party making the certification under subdivision [(c)] <u>(d)</u> (1) of this section;

[(v)] <u>(ii)</u> the outlets to be used for marketing the package or product <u>for which authorization is requested</u> [displaying the emblem];

[(vi)] <u>(iii)</u> a statement that a commitment is made to supply corroborative data at the request of the department to confirm certification; <u>and</u>

[(vii) the waste implications of the package or product after use;

(viii) the extent to which secondary material and post-consumer material content standards are limited by technology; and]

[(ix)] <u>(iv)</u> [the availability] <u>A demonstration to the satisfaction of the department that an adequate feedstock</u> of secondary material and post-consumer material [feedstocks] <u>exists to meet the requirements of this Part.</u>

(3) The <u>affidavit and</u> certification [with evidence of authority] <u>section of the application</u> must be [made] <u>signed</u> by a responsible party [as follows:] <u>with the legal authority to so sign. An example of a responsible party is as follows:</u>

(i) in the case of corporations, by a duly authorized principal executive officer. <u>Unless otherwise approved by the department a duly authorized principal executive officer shall include the President, Vice President, Treasurer or Secretary;</u>

(ii) in the case of a partnership or limited partnership, by a general partner;

(iii) in the case of a sole proprietorship, by the proprietor; or,

(iv) in the case of a joint venture, by a joint venture principal.

The required certification shall be in a format <u>and on a form</u> [as] provided by the department.

(4) Within seventy-five (75) days after receipt (date-stamped received by the department) of a request for authorization to use an emblem, the department will respond or determine whether or not to authorize that use and will notify the person requesting the authorized use of its determination in writing.

[(i) For those cases where standards have not been set forth in section 368.4 of this Part, the department will evaluate the request to use the emblem and make a determination based on considerations listed in this Part.]

[(ii) (i) The authorized user of the recycled emblem must label the percentage of post-consumer [secondary] material content and identify it as such in a manner acceptable to the department and proximate to the emblem. [the center of the arrows or immediately adjacent thereto.] For paper products, the post-consumer percentage by fiber weight may be shown in lieu of showing the post consumer percentage by total weight provided it is clearly identified as such and all other requirements under this Part have been met.

[(iii)] (ii) Authorization will be granted for a maximum period of five years. A request to renew authorization, which must contain all the information identified in paragraphs 368.5 (d) (1), (2) and (3) [of this subdivision], must be made at least one hundred twenty (120) days before expiration. Failure to do so will result in automatic termination of the authorization to use the emblem effective on the date of expiration.

[(iv)] (iii) Within fifteen (15) calendar days of the receipt of authorization from the department, a final representation of the emblem and package or product on which it will be displayed must be submitted to the department by the person who originally requested its use. The submittal shall be made to the address provided in paragraph 368.5(d) (1) of this [subdivision] Part.

[(5) Within one hundred eighty (180) days from notification by the department of any promulgated revision to section 368.4 of this Part that will result in increased recycled content for a package or product, each person authorized to use a recycled emblem for that package or product must request in writing department authorization as outlined in subdivision 368.5(d) to continue use of that emblem. Failure to certify within such period that the package or product meets the new standard will result in revocation of the authorized use of the emblem.]

(6) (5) The department will deny authorization to use an emblem if it determines that such use fails to meet the requirements of this Part for authorized uses. If authorization is denied, that person may resubmit a request without prejudice.

(e) Placement of the emblem.

(1) the emblem must be displayed in a manner that clearly indicates the package or product to which it pertains[.];

(2) where the emblem applies only to the content of a package, the emblem must be placed on the label describing the content and if possible on the content itself[.];

(3) where the emblem applies only to the package, the emblem must be placed on the package with a statement that the emblem applies to the package only[.]; and

(4) the emblem must be exhibited no smaller than three quarters of an inch in size unless otherwise authorized by the department [and must be] provided that the emblem and lettering are clearly legible.

(f) Revocation of authorized use of an emblem. The department will revoke a general or specific authorization if it is determined that the emblem's use fails to meet the requirements of this Part or the conditions of the authorization.

(1) If a determination is made to revoke the authorized use of an emblem, the person who was granted the authorization will be notified in writing that further use of the emblem is prohibited. The date of receipt of the department's notification, will be the effective revocation date. A person whose authorization to use an emblem has been revoked may not request reauthorization for a period of one year following the effective revocation date. The procedures relating to revocation are those as provided under Part 621 of this Title pertaining to permits.

(2) A notice of revocation will be published in the State Register to alert persons selling packages or products affected by a revocation. In addition, the department will provide notification to trade organizations that request to be placed upon a mailing list for such notifications.

(3) A period of one hundred twenty (120) days from the effective revocation date will be allowed for retailers, brokers, distributors, and wholesalers to clear shelves and stock of existing packages or products affected by a revocation and also to allow manufacturers to discontinue production.

(4) The department reserves the right to revoke the general authorization for the use of an emblem provided in subdivision (a) of this section or specific authorization for

the use of an emblem provided in subdivision (d) of this section if it deems that [its] <u>an emblem's</u> use constitutes an abuse or is offensive to the purpose and intent of this Part.

(5) Upon revocation, it is the responsibility of the person who was granted the authorization to use a recycling emblem, to immediately notify affected parties (e.g., retailers, brokers, distributors, wholesalers, etc.) who may have a supply of the packages or products bearing the emblem, and assure that all such packages or products are made unavailable for sale or that the emblem is removed or obliterated on any package or product remaining for sale no later than <u>one hundred twenty</u> (120) days after the revocation date.

Section 368.6   Severability.

If any provision of this Part or its application to any person or circumstance is held invalid, the remainder of this Part, and the application of those provisions to persons or circumstances other than those to which it is held invalid, shall not be affected thereby.

Appendix M

Resources/Contacts

UNITED STATES STATE REGULATION

For updated and more specific information about a particular state's regulation of environmental marketing claims, contact:

OFFICE OF THE ATTORNEY
  GENERAL
State House
11 South Union Street
Montgomery, AL 36130
(205) 242-7300

OFFICE OF THE ATTORNEY
  GENERAL
State Capital
P.O. Box K
Juneau, AK 99811-0300
(907) 465-3600

OFFICE OF THE ATTORNEY
  GENERAL
1275 West Washington
Phoenix, AZ 85007
(602) 542-4266

DEPARTMENT OF
  ENVIRONMENTAL
  QUALITY
3033 N. Central Ave.
Phoenix, AZ 85012
(602) 207-2300

OFFICE OF THE ATTORNEY
  GENERAL
200 Tower Building
323 Center St.
Little Rock, AR 72201
(501) 682-2007

OFFICE OF THE ATTORNEY
  GENERAL
1515 K St., Suite 511
Sacramento, CA 95814
(916) 445-9555

CALIFORNIA INTEGRATED
  WASTE MGMT. BOARD
  (CIWMB)
8800 Cal Center Drive
Sacramento, CA 95826
(916) 255-2200

OFFICE OF THE ATTORNEY
  GENERAL
Department of Law
1525 Sherman St., 7th Floor
Denver, CO 80203
(303) 866-3617

OFFICE OF THE ATTORNEY
  GENERAL
55 Elm St.
Hartford, CT 06106
(203) 566-2026

DEPARTMENT OF
  ENVIRONMENTAL
  PROTECTION
Waste Management Bureau
165 Capitol Ave.
Hartford, CT 06106
(203) 566-2860

OFFICE OF THE ATTORNEY
  GENERAL
820 North French St., 8th Floor
Wilmington, DE 19801
(302) 577-3838

OFFICE OF THE CORPORATION
  COUNSEL
1350 Pennsylvania Ave., N.W.
Washington, D.C. 20004
(202) 727-6248

OFFICE OF THE ATTORNEY
  GENERAL
Department of Legal Affairs
The Capitol, PL01
Tallahassee, FL 32399-1050
(904) 487-1963

OFFICE OF THE ATTORNEY
  GENERAL
Department of Law
132 State Judicial Building
Atlanta, GA 30334
(404) 656-4585

OFFICE OF THE ATTORNEY
  GENERAL
425 Queen St.
Honolulu, HI 96813

OFFICE OF THE ATTORNEY
  GENERAL
State House
Boise, ID 83720
(208) 334-2400

OFFICE OF THE ATTORNEY
  GENERAL
500 S. Second St.
Springfield, IL 62706
(217) 782-1090

DEPARTMENT OF ENERGY &
  NATURAL RESOURCES
Environmental Protection Agency
325 West Adams St., Room 300
Springfield, IL 62704-1892
(217) 785-2800

OFFICE OF THE ATTORNEY
  GENERAL
219 State House
Indianapolis, IN 46204
(317) 232-6201

INDIANA DEPT. OF
  ENVIRONMENTAL
  MGMT (IDEM)
Office of Pollution Prevention
  & Technical Assistance
P.O. Box 6015
105 S. Meridian St.
Indianapolis, IN 46206-6015
(317) 232-8172
(800) 451-6027 (environmental
help-line)

OFFICE OF THE ATTORNEY
  GENERAL
Hoover State Office Building
Second Floor
Des Moines, IA 53019
(514) 281-5164

DEPARTMENT OF NATURAL
  RESOURCES ENVIRONMENTAL
  PROTECTION DIVISION
Wallace State Office Building
Des Moines, IA 53019-0034
(515) 281-8975

OFFICE OF THE ATTORNEY
GENERAL
Judicial Center, 2nd Floor
Topeka, KS 66612
(913) 296-2215

OFFICE OF THE ATTORNEY
GENERAL
Judicial Center
State Capitol, Room 116
Frankfort, KY 40601
(502) 564-7600

OFFICE OF THE ATTORNEY
GENERAL
Department of Justice
2-3-4 Loyola Building
New Orleans, LA 70112-2096
(504) 342-7013

OFFICE OF THE ATTORNEY
GENERAL
State House
Augusta, ME 04333
(207) 289-3661

MAINE WASTE MANAGEMENT
AGENCY
Office of Planning
State House Station 154
Augusta, ME 04333
(207) 289-5300
(800) 662-4545

OFFICE OF THE ATTORNEY
GENERAL
Consumer Protection Division
200 Saint Paul Place
18th Floor
Baltimore, MD 21202
(410) 576-6300

OFFICE OF THE ATTORNEY
GENERAL
One Ashburton Place
Boston, MA 02108
(617) 727-2200

DEPARTMENT OF
ENVIRONMENTAL
PROTECTION
Division of Solid Waste
Management
One Winter St.
Boston, MA 02108
(617) 292-5583

OFFICE OF THE ATTORNEY
GENERAL
Law Building
525 West Ottawa
P.O. Box 30212
Lansing, MI 48909
(517) 373-1110

OFFICE OF THE ATTORNEY
GENERAL
102 State Capitol Building
St. Paul, MN 55155
(612) 296-6196

MINNESOTA POLLUTION
CONTROL AGENCY
Ground Water and Solid Waste
Division
520 Lafayette Road
St. Paul, MN 55155
(612) 296-7395

OFFICE OF THE ATTORNEY
GENERAL
Carroll Gartin Justice Building
P.O. Box 220
Jackson, MS 39205
(601) 359-3680

OFFICE OF THE ATTORNEY
GENERAL
Supreme Court Building
101 High Street, P.O. Box 899
Jefferson City, MO 65102
(314) 751-3321

OFFICE OF THE ATTORNEY
GENERAL
Department of Justice
Justice Building
215 North Sanders
Helena, MT 59620-1430
(406) 444-2026

OFFICE OF THE ATTORNEY
GENERAL
State Capitol
P.O. Box 98920
Lincoln, NE 68509
(402) 471-2682

OFFICE OF THE ATTORNEY
GENERAL
Heroes Memorial Building
Capital Complex
Carson City, NV 89710
(702) 687-4170

OFFICE OF THE ATTORNEY
GENERAL
State House Annex
25 Capitol St., Suite 208
Concord, NH 03301-6379
(603) 271-3658

DEPARTMENT OF
ENVIRONMENTAL
SERVICES
Waste Management Division
6 Hazen Drive
Concord, NH 03301-6509
(603) 271-2900

OFFICE OF THE ATTORNEY
GENERAL
State House Annex
25 Market St., CN080
Trenton, NJ 08625
(609) 292-4925

CONSUMER AFFAIRS DIVISION
1207 Raymond Blvd.
Newark, NJ 07102
(201) 648-4010

OFFICE OF THE ATTORNEY
GENERAL
Consumer Protection Division
Bataan Memorial Building
P.O. Box 1508
Santa Fe, NM 87504-1508
(505) 827-6060

OFFICE OF THE ATTORNEY
GENERAL
120 Broadway, 25th Floor
New York, NY 10271
(212) 416-8000

NY STATE DEPT. OF
ENVIRONMENTAL
CONSERVATION
Division of Solid Waste
Bureau of Waste Reduction and
Recycling
50 Wolf Road
Albany, NY 12233-4015
(518) 457-3966

NEW YORK CONSUMER
PROTECTION BOARD
99 Washington Ave.
Albany, NY 10007
(518) 473-7205

OFFICE OF THE ATTORNEY
GENERAL
Department of Justice
2 East Morgan St.
P.O. Box 629
Raleigh, NC 27602
(919) 733-3377

OFFICE OF THE ATTORNEY
GENERAL
State Capitol
600 East Boulevard Ave.
Bismarck, ND 58505
(701) 224-2210

OFFICE OF THE ATTORNEY
GENERAL
State Office Tower
30 East Broad Ave.
Columbus, OH 43266-0410
(614) 466-3376

OFFICE OF THE ATTORNEY
GENERAL
State Capitol
Oklahoma City, OK 73105
(405) 521-3921

OFFICE OF THE ATTORNEY
GENERAL
100 Justice Building
Salem, OR 93710
(513) 378-6002

OFFICE OF THE ATTORNEY
GENERAL
Strawberry Square
Harrisburg, PA 17120
(717) 787-3391

DEPARTMENT OF
ENVIRONMENTAL
RESOURCES
P.O. Box 2357
Harrisburg, PA 17105-2357
(717) 787-7382

OFFICE OF THE ATTORNEY
GENERAL
72 Pine St.
Providence, RI 02903
(401) 277-4400

DEPARTMENT OF
ENVIRONMENTAL
MANAGEMENT
Office of Environmental
Coordination
Ocean State Cleanup & Recycling
Program (OSCAR)
83 Park St.
Providence, RI 02903-1037
(401) 277-3434

OFFICE OF THE ATTORNEY
GENERAL
Rembert Dennis Office Building
100 Assembly St.
Columbia, SC 29211
(803) 734-3970

OFFICE OF THE ATTORNEY
GENERAL
500 East Capitol St.
Pierre, SD 57501-5070
(605) 773-3215

OFFICE OF THE ATTORNEY
GENERAL
450 James Robertson Parkway
Nashville, TN 37243
(615) 741-3491

OFFICE OF THE ATTORNEY
GENERAL
Capitol Station
P.O. Box 12548
Austin, TX 78711-2548
(512) 463-2100

OFFICE OF THE ATTORNEY
GENERAL
State Capitol, Room 236
Salt Lake City, UT 84114
(801) 538-1015

OFFICE OF THE ATTORNEY
GENERAL
Pavilion Office Building
Montpelier, VT 05602
(802) 828-3171

OFFICE OF THE ATTORNEY
GENERAL
101 North 8th St., 5th Floor
Richmond, VA 23219
(804) 786-2071

OFFICE OF THE ATTORNEY
GENERAL
Consumer and Fair Practices
Division
900 4th Ave., #2000, TB-14
Seattle, WA 98164-1012
(206) 464-7744

OFFICE OF THE ATTORNEY
GENERAL
State Capitol
Charleston, WV 25305
(304) 348-2021

OFFICE OF THE ATTORNEY
GENERAL
114 East, State Capitol
P.O. Box 7857
Madison, WI 53707
(608) 266-1221

WI DEPT. OF AGRICULTURE,
TRADE AND CONSUMER
PROTECTION
801 W. Badger Rd.
P.O. Box 8911
Madison, WI 53708
(608) 267-6851

OFFICE OF THE ATTORNEY
GENERAL
123 State Capitol
Cheyenne, WY 82002
(307) 777-7841

INTERNATIONAL REGULATION

For updated and more specific information about foreign jurisdictions' regulation of environmental marketing claims, contact:

EUROPEAN COMMUNITY
Ingrid Baschab
EC-COMMISSION—DG XI
Breydel 06/285
200, Rue de la Loi
B-1049, Bruxelles
Belgium
Telephone: (32) 2.235.7165
Facsimile: (32) 2.235.0144

GERMANY
Jurgen Staupe
UNWELTLUNDESANT (FEDERAL
ENVIRONMENT AGENCY)
Bismarckplatz 1
1000 Berlin 33
Germany
Telephone: (49) 30.8903.2636
Facsimile: (49) 30.8903.2285

DUALES SYSTEM
DEUTSCHLAND
Rochus Strasse 226
5300 Bonn
Germany
Telephone: (49) 228.97.92175
Facsimile: (49) 228.97.92123

CANADA
Graham Hardman
ENVIRONMENTAL CHOICE
107 Sparks Street
Second Floor
Ottawa, Ontario
Canada K1A 0H3
Telephone: (1) 613.952.9464
Facsimile: (1) 613.952.9465

JAPAN
Toshiro Kojuma
JAPAN ENVIRONMENT
ASSOCIATION
EcoMark Office
1-5-8 Tranomon Minsto-ku
Tokyo, Japan
Telephone: (81) 3.3508.2068
Facsimile: (81) 3.3593.7195

AUSTRALIA
Kerry Smith
WASTE AND RECYCLING
SECTION
DASETT
GPO Box 787
Canberra ACT 2601
Australia
Telephone: (61) 62.74.15.12
Facsimile: (61) 62.74.11.23

MARKET RESEARCH

Green Action Trends, Yankelovich Clancy Shulman.

Green Gauge, The Roper Organization.

The Kaagan Environmental Monitor, Kaagan Research Associates.

The Environmental Report, Environmental Research Associates.

EnvironMonitor, Burke Marketing Research.

RESEARCH ORGANIZATIONS

YANKELOVICH CLANCY
  SHULMAN INC.
8 Wright Street
Westport, CT 06880-3100
203 227-2700

ROPER ORGANIZATION
205 E. 42nd Avenue
17th Floor
New York, NY 10017
212 599-0700

KAAGAN RESEARCH
  ASSOCIATES
200 W. 57th Street
New York, NY 10019-3211
212 246-3551

ENVIRONMENTAL RESEARCH
  ASSOCIATES
707 State Road 102
Princeton, NJ 08540
609 683-0187

BURKE MARKETING RESEARCH
  INC.
800 Broadway Street
Cincinnati, OH 45202-0000
513 852-8585

INDUSTRY/TRADE ASSOCIATIONS

AMERICAN ASSOCIATION OF
  ADVERTISING AGENCIES
  (AAAA)
666 3rd Ave., 13th Floor
New York, NY 10017-4011
212 682-2500

AMERICAN MARKETING
  ASSOCIATION (AMA)
250 S. Wacker Drive, Suite 200
Chicago, IL 60606-5819
312 648-0536

AMERICAN PAPER INSTITUTE
260 Madison Avenue
New York, NY 10016-2439
212 340-0600

ASSOCIATION OF NATIONAL
  ADVERTISERS INC. (ANA)
155 E. 44th St.
33rd Floor
New York, NY 10017-4201
212 697-5950

CHEMICAL SPECIALTY
  MANUFACTURERS
  ASSOCIATION (CSMA)
1913 I Street NW
Washington, DC 20006-2106
202 872-8110

COUNCIL ON PACKAGING IN
  THE ENVIRONMENT (COPE)
(Formerly, Council on Plastics and
Packaging in the Environment)
1001 Connecticut Avenue NW
Suite 401
Washington, DC 20036
202 331-0099

COSMETIC, TOILETRY, &
  FRAGRANCE ASSOCIATION
  (CFTA)
1110 Vermont Avenue NW
Suite 800
Washington, DC 20005-3564
202 331-1770

DEGRADABLE PLASTIC
  COUNCIL (DPC)
1275 K Street NW
Washington, DC 20005
202 371-5200

FLEXIBLE PACKAGING
  ASSOCIATION (FPA)
1090 Vermont Avenue NW
Suite 500
Washington, DC 20005-4960
202 842-3880

FOOD MARKETING INSTITUTE
  (FMI)
1750 K Street NW
Suite 700
Washington, DC 20006-2394
202 452-8444

ASSOCIATION OF THE
  NONWOVEN FABRICS
  INDUSTRY (INDA)
1001 Winstead Dr.
Suite 460
Cary, NC 27513-2117
919 677-0060

INDEPENDENT COSMETIC
  MANUFACTURER AND
  DISTRIBUTORS (ICMD)
1220 W. Northwest Highway
Palatine, IL 60067-1803
800 334-2623

INSTITUTE OF PACKAGING
  PROFESSIONALS (IOPP)
11800 Sunrise Valley Dr.
Reston, VA 22091-5302
703 620-9380

NATIONAL ASSOCIATION OF
  CONVENIENT STORES (NACS)
1605 King Street
Alexandria, VA 22314-2792
703 684-3600

NATIONAL ASSOCIATION OF
  MANUFACTURERS (NAM)
1331 Pennsylvania Avenue NW
Suite 1500W
Washington, DC 20004-1703
202 637-3000

NATIONAL FOOD PROCESSORS
  ASSOCIATION (NFPA)
1401 New York Avenue
Suite 400
Washington, DC 20005-2154
202 639-5900

NATIONAL RETAIL FEDERATION
  (NRF)
100 W 31st Street
New York, NY 10001-3483
212 244-8780

POLYSTYRENE PACKAGING
  COUNCIL (PPC)
1025 Connecticut Avenue NW
Washington, DC 20036
202 822-6424

SOAP & DETERGENT
  ASSOCIATION (SDA)
475 Park Avenue South
27th Floor
New York, NY 10016-6947
212 725-1262

SOCIETY OF THE PLASTICS
  INDUSTRY, INC. (SPI)
1275 K Street NW
Suite 400
Washington, DC 20005-4006
202 371-5200

CONSULTANTS

Bob Rehak
OGILVY AND MATHER
1415 Louisiana Street
Houston, TX 77002
713 659-6688

ABT ASSOCIATES
55 Wheeler Street
Cambridge, MA 02138

NONPROFIT ORGANIZATIONS

Dianne Ward
COUNCIL OF BETTER BUSINESS
  BUREAUS INC
National Advertising Division
4200 Wilson Blvd
Suite 800
Arlington, VA 22203-1838
703 276-0100

RECYCLING ADVISORY
  COUNCIL OF THE NATIONAL
  RECYCLING COALITION
1101 30th Street NW
Suite 305
Washington, DC 20007
202 625-6406

Susan Alexander
GREEN SEAL
1250 23rd Street NW
Suite 275
Washington, DC 20037
202 331-7337

Resa Dimino
ENVIRONMENTAL ACTION
  FOUNDATION
1525 New Hampshire Avenue NW
Washington, DC 20036

Paul Demko
HEARST CORPORATION'S GOOD
  HOUSEKEEPING INSTITUTE
959 8th Avenue
New York, NY 10019
212 649-2000

Chip Foley
COALITION OF NORTHEAST
  GOVERNORS
400 N. Capitol Street NW
Suite 382
Washington, DC 20001
202 624-8450

Mitchell Friedman
GREEN CROSS CERTIFICATION
  COMPANY
Scientific Certification Company
(parent company)
1611 Telegraph Avenue
Oakland, CA 94612
510 832-1415

Connie Saulter
NORTHEAST RECYCLING
  COUNCIL
139 Main Street
Suite 401
Brattlcboro, VT 05301

INDEPENDENT ORGANIZATIONS

NATIONAL ADVERTISING
  REVIEW BOARD (NARB)
845 3rd Avenue
New York, NY 10022
212 832-1320

RECYCLING ADVISORY
  COUNCIL (RAC)
1101 30th Street NW
Suite 305
Washington, DC 20007
202 625-6410

MAGAZINES/BOOKS

ADVERTISING AGE
965 E. Jefferson
Detroit, MI 48207-3185
800 678-9595

Carl Frankel
GREEN MARKETALERT
345 Woodcreek Road
Bethlehem, CT 06751-1014
203 266-7209

Tom Watson
RESOURCE RECYCLING
1206 NW 21st Avenue
Portland, OR 97209-1609
503 227-1319

GREEN CONSUMER
   SUPERMARKET GUIDE
by Joel Makower
Viking-Penguin

ENVIRONMENTAL PACKAGING:
   U.S. GUIDE TO GREEN
   LABELLING, PACKAGING
   AND RECYCLING
Thompson Publishing Group
Subscription Service Center
1725 North Salisbury Blvd.
Salisbury, MD 21801-9848
1-800-879-3169

UNITED STATES
   ENVIRONMENTAL
   PROTECTION AGENCY
   EVALUATION OF
   ENVIRONMENTAL
   MARKETING TERMS IN THE
   UNITED STATES
US EPA Office of Pollution
Prevention and Toxics
401 M Street, SW
Washington, DC 20406

UNITED STATES
   ENVIRONMENTAL
   PROTECTION AGENCY
   ASSESSING THE
   ENVIRONMENTAL CONSUMER
   MARKET (Pm-221)
US EPA Office of Policy,
Planning and Evaluation
401 M Street, SW
Washington, DC 20406

NCMS
Achieving Environmental Excellence
Implementation Guide

IMPLEMENTATION GUIDE

The self-assessment portion of the Achieving Environmental Excellence program consists of 12 areas of excellence, a score sheet, and a comparison bar chart. The 12 areas of excellence are:

1. Policy
2. Management
3. Planning
4. Cost
5. Stakeholders
6. Human Resources
7. Operations and Facilities
8. Suppliers
9. Compliance
10. Technology and Research
11. Auditing
12. Measurement and Continuous Improvement.

Each area is presented in its own section and begins with an overview. The overview describes the particular area and its requirements. A questionnaire follows, containing several Yes–No questions. The last line of each questionnaire is a score box in which you record the subtotal scores for that section. Lastly, each area has a Comment and Recommendation Sheet to record general information regarding that area. After you complete the evaluation, you will use the numbers in each of the 12 subtotal score boxes to fill in the Score Sheet and display the results graphically in the Comparison Bar Chart.

Area Overview

The overview that begins each of the 12 areas of excellence briefly describes the area and explains its significance to environmental quality. The overview also provides a detailed description of the questions and explains their importance.

Question Format

The AEE program contains more than 150 Yes–No questions divided among the 12 areas of environmental excellence. You may also mark a question Not Applicable (N/A) if it truly does not apply to your company or your particular situation. The questions are structured such that a Yes response indicates a strength, while a No response indicates a weakness. Some questions will have an obvious answer while others may require some additional research to answer. The actual process of completing a question that does not have an obvious answer typically provides a greater understanding of the relevant issues surrounding that particular question and its related area. It is very important that you answer all questions accurately. Only in this manner will a true assessment of your environmental quality be possible. It is this true appraisal that facilitates the exposure of weaknesses and the implementation of relevant, meaningful changes.

There are three criteria that you may use to assign an answer to a question. The three criteria are:

1. The answer is based on available physical documentation or evidence that can be referenced and accessed again if necessary.
2. The answer is based on an assertion by you, or the appropriate personnel, that a particular response should be given.
3. The answer is based on your best judgment or the best judgment of the auditing team, if such a group exists.

You should always try to answer a question based on level 1 criterion, and level 3 criterion should be used only as a last resort. In some instances, documentation is necessary to answer a question. For example, to answer Question 1.1 (Do you have a formal and written corporate environmental policy?) with a Yes response, a physical document must be available. If no documentation is available on which to base your answer, then you should contact the appropriate personnel. For instance, Question 1.2 (Are your employees committed to your environmental policy?) will typically require you to interview several employees. If the appropriate personnel cannot be contacted or cannot give a definite response, then the auditor may use his or her best judgment to answer the question. You may list the documentation that was used and the personnel who were contacted on the Comment and Recommendation Sheet. Retain a copy of the documentation if possible. If you use the third criterion, then document the reasons for the particular response.

When you have arrived at an answer, place a check mark in the box under the appropriate Yes, No, or N/A column. When you have completed all of the questions in a particular area of excellence, add the number of checks in each column, and write down the total in the corresponding Subtotal box at the end of that questionnaire. You will use these numbers later to complete the Score Sheet.

Comment and Recommendation Sheet

Complete the Comment and Recommendation Sheet of each area of excellence concurrently with that area's questionnaire. Reference documentation so that it can be easily obtained if someone wishes to access it. Reference personnel who were contacted so that others may contact them if further information is required. Record any general comments you have as well as the strengths and weaknesses that you find. Also, you should record ideas for possible improvements. It is important to document this information while you are performing the self-assessment since a later review of the answers to the question will not always allow you to remember these ideas. After you have completed all 12 areas of excellence, the entire auditing team should meet, discuss them, and add further remarks to the Comment and Recommendation Sheet at that time.

Score Sheet and Comparison Bar Chart

After you have completed all 12 areas of excellence you may complete the Score Sheet and the Comparison Bar Chart. In the Score Sheet, a percentage is tabulated for each area as well as an overall score. The Comparison Bar Chart visually presents the data and outlines weak and strong areas. Specific details for completing the Score Sheet and Comparison Bar Chart are provided in the Scoring section immediately following the 12 areas of excellence.

1. POLICY

The foundation to achieving environmental excellence is the establishment of a strong, proactive corporate environmental policy along with function-specific policies. Your corporate policy provides a focal point for environmental quality for all of your personnel throughout the corporation. Function-specific policies contain realistic, obtainable goals and objectives and identify the necessary standards and practices for implementation.

A strong environmental policy contains several elements. A major element is employee commitment at all levels, especially from senior management. Specific and appropriate operating procedures are outlined in a strong environmental policy and these procedures are backed by the necessary material, human and financial resources. "Environmental policies are widely distributed, easily accessible, and understood throughout the organization."[1] A strong environmental policy regularly measures environmental performance and "secure(s) continual improvement in environmental performance."[2] The goals and objectives that the policy sets forth are measured, and steps are taken to improve environmental quality, especially where performance is unsatisfactory.

A system to continually improve the environmental policy is established and maintained. The effectiveness of the policy is regularly measured and changes are implemented in the policy as circumstances warrant. The environment is an extremely dynamic area that requires quick adaptability if quality is to be maintained. The environmental policy provides for that adaptability and is itself readily adaptable.

1. Policy	Yes	No	N/A
1.1 Do you have a formal and written corporate environmental policy?			
1.2 Are your employees committed to your corporate environmental policy?			
1.3 Do your function-specific environmental policies have detailed operating procedures?			
1.4 Do your environmental policies have sufficient resources available?			
1.5 Are your environmental policies communicated to all relevant internal and external sources?			
1.6 Does your company contribute to and promote public environmental policies and programs?			
1.7 Are these public environmental policies and programs relevant to the goals of your company's policies?			
1.8 Do your environmental policies have proactive improvement objectives?			
1.9 Do you measure the effectiveness of each environmental policy?			
1.10 Do you have a system to update and continually improve the environmental policies?			
1. Subtotal			

1. Policy
Documentation Used and Personnel Contacted:
General Comments:
Strengths:
Weaknesses:
Recommendations for Improvement:

2. MANAGEMENT

Environmental excellence is ultimately the responsibility of top-level management. "Senior management must visibly accept and commit to quality."[3] They are responsible for creating environmental goals and programs, for motivating employees, for allocating necessary resources, and for implementing and continually improving the goals and programs. Management shows its commitment to environmental quality through visible and active leadership. This commitment is communicated to all employees throughout the corporation to provide motivation for environmental goals and initiatives. Leadership is established by management's active role in specific programs and initiatives.

To ensure that environmental quality is maintained, "the management representative should have sufficient knowledge of the activities of the organization, and of environmental issues, to undertake his or her role effectively."[4] Management regularly receives current information since this information is vital in a highly dynamic field such as the environment. This information is necessary to measure environmental performance and to make needed improvements to maintain environmental excellence. Management has a broad knowledge base, since environmental issues are prevalent in every aspect of business. Management also participates in auditing all environmental programs and activities. This participation facilitates a greater understanding of these activities and the ways in which they can be improved.

2. Management	Yes	No	N/A
2.1 Is management visibly committed to environmental quality in all programs and activities?			
2.2 Does management provide leadership for environmental quality in all programs and activities?			
2.3 Do all environmental programs or initiatives have a specific leader or "champion"?			
2.4 Does management play an active role in achieving environmental goals and objectives?			
2.5 Is there top-level accountability for environmental programs, the management process and the correction of problems?			
2.6 Does management provide the resources necessary to meet environmental goals and objectives?			
2.7 Do managers in charge of environmental processes completely understand the processes and how to implement changes to those processes?			
2.8 Is middle management trained to achieve function-specific environmental goals and objectives?			
2.9 Is management responsible for the final approval of all environmental goals and policies?			
2.10 Is management kept current with the latest environmental information?			
2.11 Does management provide periodic, independent reviews of all environmental activities?			
2. Subtotal			

2. Management
Documentation Used and Personnel Contacted:
General Comments:
Strengths:
Weaknesses:
Recommendations for Improvement:

3. PLANNING

The planning process is an integral part of achieving environmental excellence. Environmental plans include:

- Corporate environmental plan
- Specific environmental initiative plans
- Environmental emergency plans
- Plans to gain company-wide support for environmental programs and initiatives

The corporate environmental plan has far-reaching goals such as lessening the overall impact of operations and activities on the environment. These goals are the responsibility of every employee and provide widespread motivation for environmental quality. Specific initiatives include reducing the use of natural resources, reducing energy consumption, practicing recycling, safely disposing of wastes, and practicing waste minimization. These specific initiatives directly protect the environment and provide visible cost savings. Plans to gain corporation-wide support for environmental programs and initiatives provide for a constant two-way communication with employees regarding relevant environmental issues. Without this support, environmental quality is not achieved. All environmental plans are documented and communicated to all relevant personnel.

Environmental concerns are integrated into the planning of non-environmental activities. These concerns are constantly addressed throughout the design stage of all products. Constantly addressing these concerns ensures the most environmentally sound product possible and avoids costly redesign if the product is found to be out of compliance or receives strong consumer or environmental activist opposition. The impact of environmental programs and initiatives on non-environmental activities is evaluated and necessary improvements are implemented. Environmental concerns are not limited to a single person or department and a fixed set of activities. Environmental excellence is only achieved when environmental concerns are raised throughout the corporation and in all activities.

3.	Planning	Yes	No	N/A
3.1	Do you have a formal, written environmental plan?			
3.2	Are environmental concerns a priority in the design and planning of all business activities?			
3.3	Do you identify and evaluate the effects of new environmental programs and initiatives on all of your business activities?			
3.4	Do you have a plan to gain support throughout your company for new environmental initiatives?			
3.5	Do you have a plan to internally communicate the successes and failures of environmental initiatives?			
3.6	Are environmental programs and practices fully integrated with business planning?			
3.7	Is environmental data integrated with other relevant business data?			
3.8	Do you have a written plan to respond to environmental emergencies and disasters?			
3.9	Are environmental emergency plans integrated with relevant business areas and coordinated with the proper external agencies?			
3.10	Do you have a plan to prevent pollution throughout your company?			
3.11	Do you have a plan to minimize, recycle, reclaim and treat waste?			
3.12	Do you have a plan to reduce the company's dependency on natural resources?			
3.13	Do you have a plan to recycle and use recycled materials wherever it is feasible?			
3.14	Do you have a plan to reduce energy consumption throughout your company?			
3.15	Are products designed to be easily and safely disassembled, recycled, reused, or disposed?			
3.16	Do you have a plan to lessen the overall impact of business activities on the environment?			
3.17	Are all environmental plans backed up by contingency plans?			
3.	Subtotal			

3. Planning
Documentation Used and Personnel Contacted:
General Comments:
Strengths:
Weaknesses:
Recommendations for Improvement:

4. COST

Identification and allocation of environmental costs are critical aspects of environmental excellence. If you do not identify and properly allocate these costs, your company "may subsidize some wasteful practices by not charging the [proper] operating unit the full cost that their wastes impose on the company."[5] Many environmental costs have traditionally been lumped together and charged as indirect costs. These costs may include:[6]

- Monitoring and testing equipment
- Environmental training
- Consumer loyalty and acceptance
- Worker morale and union relations
- Corporate image
- Waste storage and handling
- Record keeping
- Reporting
- Permit fees.

Whenever possible, your company should detail these costs and assign them to the products or operating units that generated them. Charging each unit with their environmental costs provides the necessary motivation to reduce these costs. It provides the basis for efficient allocation of your resources to reduce the highest costs. Assigning environmental costs properly will also lead to a product cost that more accurately reflects the environmental impacts associated with the product.

The next step on the path to excellence is to consider the impacts and costs for the complete product life cycle. Manufacturing is only one stage of the cycle. Other stages include raw materials recovery, product distribution, use, maintenance, and disposal. Environmental impacts and costs are associated with each of these stages. Although your company may not pay for all of these costs, the life cycle perspective is useful in that it can prevent you from reducing an environmental impact or cost at the expense of increasing other environmental impacts and costs. Relevant information gained through a life cycle analysis is communicated to customers and other appropriate parties. This information also outlines areas where improvements are necessary and where they may be easily implemented. Life cycle analysis is a relatively new concept that is complex and expensive. As a minimum, the life cycle perspective is limited to those stages for which costs are borne by your company. For environmental excellence, the perspective is expanded, whenever feasible, to include other life cycle stages.

4. Cost	Yes	No	N/A
4.1 Are the environmental costs and savings, including hidden and intangible ones, of your products and services identified throughout their life cycle?			
4.2 Are the costs and savings of all environmental programs and activities known?			
4.3 Are pollution and waste management costs specifically identified?			
4.4 Do you have a system to identify environmental measures that provide immediate cost savings?			
4.5 Have you identified operations and activities where pollution prevention can reduce costs?			
4.6 Do you have a system to identify operations and activities that create the greatest environmental costs?			
4.7 Do you have a system to identify the individual or unit responsible for each environmental cost of all products and services?			
4.8 Are the environmental impacts and costs of all property acquisitions, divestitures, and joint ventures evaluated?			
4.9 Do risk assessments include a listing of the current and potential environmental effects of all products and activities?			
4.10 Are environmental considerations incorporated into the materials and cost accounting systems?			
4.11 Are sufficient financial resources provided for the implementation of environmental activities and initiatives?			
4.12 Are your customers aware of the hidden and less tangible environmental costs that impact the products and services that they purchase?			
4.13 Are the costs and savings of each specific environmental program or initiative tracked?			
4.14 Are the environmental tasks of a specific operating unit tracked and allocated to that unit?			
4.15 Are all environmental programs and activities evaluated to ensure financial soundness?			
4.16 Are cost accounting time scales large enough such that the savings and other benefits of your products and activities are apparent?			
4. Subtotal			

4.   Cost
Documentation Used and Personnel Contacted:
General Comments:
Strengths:
Weaknesses:
Recommendations for Improvement:

5. STAKEHOLDERS

Stakeholders are typically such groups as employees, the community, investors, customers, suppliers, etc. Environmental stakeholders constitute a wider variety of people and organizations than such stakeholders. Now they include such groups as environmental regulatory agencies, environmental interest groups, and the general public. The environmentally excellent company identifies and categorizes these broad groups of environmental stakeholders. Further, there are "processes for determining stakeholders' environmental needs, expectations and requirements to which the organization may need to respond."[7] Robust processes are necessary since these needs, expectations, and requirements are typically dynamic and sometimes difficult to understand. Stakeholder environmental satisfaction is a primary concern and requires the determination of stakeholder requirements and the methods to satisfy those requirements.

Environmental quality includes two-way communication with environmental stakeholders. Advice, concerns, and comments are actively sought to determine their environmental needs and the methods to satisfy those needs. Relevant concerns, comments, and questions receive a direct, thoughtful response. Communication with stakeholders concentrates on outlining the environmental benefits of your company's products and activities. Many of these benefits remain hidden unless a definite effort is made to bring them forth.

Another component of environmental excellence is the support of public programs and activities. These "activities should be relevant to [your] stakeholders"[8] and your company's goals. This support demonstrates the company's commitment to a sound environment and creates a positive environmental image for your company. Participation in government and industrial consortia is also a component of environmental quality. This participation not only communicates a commitment to a sound environment, but it is a valuable source for environmental information.

5. Stakeholders	Yes	No	N/A
5.1 Do you have a system to identify and categorize all environmental stakeholders?			
5.2 Are the stakeholder's environmental needs and requirements understood?			
5.3 Are the measures and methods to satisfy your stakeholder's environmental needs and requirements known?			
5.4 Are formal procedures used to ensure stakeholder environmental satisfaction?			
5.5 Are efforts made to anticipate stakeholder's environmental needs and concerns?			
5.6 Is the perception of your environmental activities to each set of stakeholders known?			
5.7 Is the importance of your environmental activities to each set of stakeholders known?			
5.8 Is environmental advice solicited from your stakeholders?			
5.9 Does communication with stakeholders include direct response to their environmental concerns and questions?			
5.10 Is relevant environmental information of your products and activities communicated to your stakeholders?			
5.11 Are stakeholders provided with information outlining the benefits of your products, activities, and technologies?			
5.12 Does your company support public environmental programs?			
5.13 Do the public environmental programs address concerns relevant to the public and your company?			
5.14 Does your company participate in consortia to gain environmental knowledge?			
5.15 Does your company enjoy a good environmental reputation with its stakeholders?			
5.16 Do you know how your stakeholders want your company to improve environmentally?			
5. Subtotal			

5. Stakeholders
Documentation Used and Personnel Contacted:
General Comments:
Strengths:
Weaknesses:
Recommendations for Improvement:

6. HUMAN RESOURCES

Another vital component of a company's environmental quality is its employees. Your company must "develop and maintain (in consultation with all employees) appropriate effective two-way communication and training programmes on environmental matters."[9] Employees receive basic training, including emergency training, as well as environmental training related specifically to their job. This training is repeated or reinforced periodically. Employee training includes awareness of relevant environmental issues, activities, and programs. The environmentally excellent company fosters two-way communication regarding pertinent environmental information. Employees, like managers, are kept up to date due to the highly dynamic nature of environmental information. Cross-functional teams provide the necessary diversity of expertise to solve the challenges posed by environmental programs and initiatives.

Although employees require support from their company to achieve environmental quality, they are also required to support environmental quality within their company.

Employees are held accountable for the environmental impacts of their actions, creating motivation for environmental quality. Incentive is also provided via a system to recognize and award employees who significantly contribute to environmental quality. Environmental excellence is continued through the use of a system to maintain momentum for programs and initiatives. Further, employees are regularly evaluated so that their environmental performance is measured, tracked, and improved. Environmental excellence must be the responsibility of every employee.

6. Human Resources	Yes	No	N/A
6.1 Are there sufficient human resources to meet current and future environmental challenges?			
6.2 Are employees made aware of environmental issues, programs, etc. affecting your company and the community?			
6.3 Do all individuals contribute to and support environmental quality in all programs and activities?			
6.4 Are all individuals held accountable for the environmental impacts of their actions?			
6.5 Are cross-functional teams used to solve environmental problems and implement improvements?			
6.6 Are employee-led internal and external environmental initiatives supported?			
6.7 Is openness and communication maintained with employees regarding environmental issues and concerns?			
6.8 Are awards and recognition given to individuals for outstanding environmental performance and improvements?			
6.9 Do all employees receive basic environmental training as well as job-specific training?			
6.10 Do all employees receive emergency and safety training that is periodically repeated or reinforced?			
6.11 Do employees receive professional environmental training and continuing education?			
6.12 Is the environmental training of employees recorded and tracked?			
6.13 Is momentum for environmental improvements built and maintained among employees?			
6.14 Is the environmental performance of employees measured and evaluated?			
6. Subtotal			

6.   Human Resources
Documentation Used and Personnel Contacted:
General Comments:
Strengths:
Weaknesses:
Recommendations for Improvement:

7. OPERATIONS AND FACILITIES

Environmental quality is readily visible in a company's operations and facilities. This is an area where that quality is easily measured and improved. All operations and facilities are evaluated to determine their environmental impacts and their associated risks and hazards. Simple operational changes that result in an increase in environmental performance are identified and implemented. These changes not only benefit the environment, they create immediate savings and provide momentum to carry out broader environmental initiatives. Environmentally dangerous materials are identified, tracked and, where possible, reduced or eliminated.

Many environmental programs are implemented to improve the environmental quality of an operation or facility. Such programs include: pollution prevention, waste minimization, and recycling. These three terms are defined as:[10]

- **Pollution Prevention**—The use of materials, processes, or practices that reduce or eliminate the creation of pollutants or wastes at the source. It includes practices that reduce the use of hazardous materials, energy, water, or other resources and practices that protect natural resources through conservation or more efficient use.
- **Waste Minimization**—Source reduction and the following types of recycling: (1) beneficial use/reuse, and (2) reclamation. Waste minimization does not include recycling activities whose uses constitute disposal and burning for energy recovery.
- **Recycling**—Using, reusing, or reclaiming materials/waste, including processes that regenerate a material or recover a usable product from it.

Achieving Environmental Excellence emphasizes the reduction of energy consumption and dependence on raw materials by addressing these two programs as separate from pollution prevention. These programs not only save money and protect the environment, they create a favorable environmental image for your corporation. Finally, all operations and facilities are evaluated to determine their global and long-term detrimental environmental impacts. Modifications are implemented to reduce or eliminate those impacts.

7. Operations and Facilities	Yes	No	N/A
7.1 Is the environmental performance and impact of all operations, products, and activities evaluated?			
7.2 Are formal and explicit operating standards and practices maintained?			
7.3 Are the environmental risks and hazards of all operations and facilities assessed and controlled?			
7.4 Are operational changes that provide an immediate increase in environmental performance identified and implemented?			
7.5 Are the environmental considerations of all operations and facilities integrated into relevant business functions?			
7.6 Are all environmentally dangerous materials tracked from "cradle to grave"?			
7.7 Are environmentally hazardous materials in all operations and products identified and reduced or eliminated?			
7.8 Is there a system to identify and purchase the most environmentally sound materials for all products and activities?			
7.9 Is waste minimization practiced in all operations and facilities?			
7.10 Is pollution prevention and reduction practiced in all operations and facilities?			
7.11 Is recycling and the use of recycled materials practiced throughout the corporation?			
7.12 Is the energy consumption of all operations, facilities and activities identified and reduced?			
7.13 Is the amount of dependence on raw materials and natural resources identified and reduced?			
7.14 Are the long-term effects of operations, facilities, and activities on ecological systems evaluated and improvements implemented?			
7.15 Are products, operations, and activities modified to prevent serious environmental damage?			
7. Subtotal			

7. Operations and Facilities
Documentation Used and Personnel Contacted:
General Comments:
Strengths:
Weaknesses:
Recommendations for Improvement:

8. SUPPLIERS

Environmental excellence requires not only high standards of quality in your company, but also high standards in your suppliers. "Suppliers and contractors should be carefully selected to reduce upstream environmental consequences."[11] As an absolute minimum, your suppliers are required to comply with all applicable environmental laws and regulations. A supplier who is in constant non-compliance is terminated. Your suppliers are encouraged to adopt a set of environmental principles and to implement sound environmental practices such as pollution prevention, waste minimization, and recycling. Although environmental quality requires high standards in your company's suppliers, it also requires the participation of your company in achieving and maintaining those standards. Your company identifies to its suppliers materials that are acceptable and those that are not. Technical assistance is provided to your suppliers wherever necessary.

The environmental quality of your company ultimately depends not only on the quality of its internal activities but its external activities as well. The quality of your suppliers provides a direct reflection of your own quality. The environmentally excellent company employs a robust system to track and document the environmental problems and performance of its suppliers. Supplier evaluations are ''dependent on records of subcontractors' demonstrated environmental awareness and performance.''[12] Further, this dependence is a high priority in supplier evaluations. Supplier environmental quality is critical to your company's environmental quality and to the overall environment.

8. Suppliers	Yes	No	N/A
8.1 Does your supplier selection depend upon their environmental activities and impacts?			
8.2 Are your suppliers encouraged to adopt a set of environmental principles?			
8.3 Are your suppliers required to comply with all applicable laws and regulations?			
8.4 Are the environmental activities of your company that affect your suppliers communicated to them?			
8.5 Are your suppliers encouraged to practice pollution prevention at their own facilities and in their own operations?			
8.6 Are your suppliers encouraged to practice recycling and to use recycled materials?			
8.7 Are acceptable and unacceptable materials identified to your suppliers?			
8.8 Are your suppliers required to notify you of environmentally hazardous materials used in their products and activities?			
8.9 Do you work with your suppliers to help them achieve their environmental goals?			
8.10 Do you document your supplier's environmental problems?			
8.11 Does your supplier evaluation depend upon their environmental activities and impacts?			
8.12 Do you terminate those suppliers that are in constant non-compliance with environmental laws and regulations?			
8. Subtotal			

8. Suppliers
Documentation Used and Personnel Contacted:
General Comments:
Strengths:
Weaknesses:
Recommendations for Improvement:

9. COMPLIANCE

Meeting applicable environmental laws and regulations is a component to achieving environmental excellence. However, compliance is not equivalent to environmental quality; it is the minimum requirement for environmental quality. Your company "establish(es) and maintain(s) programs to assure that laws and regulations applicable to [your] products and operations are known"[13] as well as obeyed. Further, specific reporting requirements are identified and communicated to the appropriate personnel. A strong, proactive relationship with regulatory agencies is established to assist your company in maintaining its compliance record. Maintaining compliance is also facilitated by keeping documented records of appropriate environmental data.

Environmental quality includes measuring, tracking, and continually improving your company's or facility's ability to maintain compliance. Established procedures are developed to correct situations of non-compliance. The environmentally excellent company has a system to track developments in all relevant laws and regulations. Systematic tracking allows a company to efficiently plan for the adjustment of affected operations and activities. Environmental audits are regularly utilized to ensure that full compliance has been and is being maintained. However, these audits go beyond assessing compliance and "focus on system evaluation and identification of improvement opportunities."[14] Environmental audits address such issues as:

- What changes are necessary to maintain compliance if the laws and regulations change?
- Where and how can effective improvements be implemented?
- What changes are necessary to surpass current and anticipated laws and regulations?

Continuous improvement ensures compliance and provides a competitive edge over companies that are satisfied to maintain the status quo.

9. Compliance	Yes	No	N/A
9.1 Are all applicable environmental regulations identified?			
9.2 Are the specific reporting requirements of each applicable environmental regulation known?			
9.3 Is your company in full compliance with all applicable environmental laws and regulations?			
9.4 Is the individual or department responsible for maintaining compliance with each environmental regulation known?			
9.5 Does your company enjoy a strong, proactive relationship with government environmental regulatory agencies?			
9.6 Are there records of relevant environmental data to ensure compliance with each applicable regulation?			
9.7 Does your company's environmental standards and objectives surpass basic regulatory requirements?			
9.8 Are there established procedures to correct all situations where an operation or activity is found to be out of compliance?			
9.9 Are audits conducted to ensure compliance with all applicable environmental requirements and regulations?			
9.10 Do environmental compliance audits address issues beyond meeting requirements and regulations?			
9.11 Is there a system to track regulatory developments?			
9. Subtotal			

9. Compliance
Documentation Used and Personnel Contacted:
General Comments:
Strengths:
Weaknesses:
Recommendations for Improvement:

10. TECHNOLOGY AND RESEARCH

Environmental quality requires the development and support of sound environmental products, activities, and technologies. Environmental concerns and issues are raised during the design stage of all products. Design for the Environment (DFE) rulebases and standards are utilized. Products are designed to be environmentally sound throughout their entire life cycle, including the period after their intended use. All products are designed to be easily and safely disposed, disassembled, reused, or recycled. The environmentally excellent company employs a system to identify new, relevant environmental technologies to be incorporated into existing and planned operations and activities. Environmental technologies are externally transferred. These technologies are licensed, exchanged through a consortium or, where appropriate, contributed for the benefit of another interested party.

Environmental quality requires your company "to extend knowledge by conducting and supporting research on the . . . environmental effects of products, processes, and waste materials.''[15] Other areas where research is conducted and supported include:

- Waste disposal technologies and standards
- Waste minimization technologies
- Environmental impacts of materials that are purchased, processed, and sold
- Direct and indirect impacts of all business activities

This research benefits the environment and improves your company's profits, environmental image, and competitiveness.

10. Technology and Research	Yes	No	N/A
10.1 Are environmental considerations taken into account during the design stage of all products?			
10.2 Are Design for the Environment (DFE) rulebases and standards used to design products?			
10.3 Are products designed so that they can be easily disassembled, reused, recycled, or disposed?			
10.4 Is there a system to integrate new environmental technologies into products and operating practices?			
10.5 Is research supported and conducted on waste disposal technologies and standards?			
10.6 Is research supported and conducted on waste minimization technologies?			
10.7 Is research supported and conducted on the direct and indirect environmental impacts of all materials?			
10.8 Is research conducted on the direct and indirect environmental impacts of all products and services?			
10.9 Is research conducted on the direct and indirect environmental impacts of all operations and activities?			
10.10 Does your company transfer sound environmental technologies to all interested parties?			
10. Subtotal			

10.   **Technology and Research**
Documentation Used and Personnel Contacted:
General Comments:
Strengths:
Weaknesses:
Recommendations for Improvement:

11. AUDITING

Environmental quality is maintained and improved via an efficient and proactive auditing system. "The organization should establish and maintain policies and programs for planning, conducting, reporting, and follow-up of findings of audits of the environmental management system and its component parts."[16] The environmentally excellent company audits all business activities and functions which impact or influence its environmental performance. Audits are performed at all levels within the company. The auditing system includes self-assessments as well as third-party assessments. Lastly, the auditing system itself is audited and improvements are implemented wherever necessary.

Elements of a successful environmental audit include:

- Formal, written procedures
- Knowledgeable, competent audit team
- Sufficient audit team detachment
- Regular, frequent schedule
- Consistent use

- Evaluation of strengths and weaknesses
- Documented findings and recommendations
- Communication of relevant findings
- Quantitative ratings
- Timely, formal correction procedures
- Identification of improvement opportunities.

A regular, frequent environmental audit with formal, written procedures communicates a strong commitment to environment quality. The detachment of the audit team from the company or facility that they are assessing ensures objectivity and an honest report. Consistent use of an audit provides for the meaningful tracking of performance and quality. An operation's strengths are evaluated and documented so that those "best practices" can be easily applied to similar operations and situations. Findings and recommendations that are documented and communicated to appropriate personnel facilitate the improvement process. This process is also enhanced by quantitative ratings and timely, formal correction procedures.

11.	Auditing	Yes	No	N/A
11.1	Is the environmental management system audited?			
11.2	Are audits conducted to ensure adherence to your company's environmental policies and principles?			
11.3	Do all environmental audits have formal, written procedures that are appropriate to their scope?			
11.4	Do competent professionals with sufficient knowledge of the operation or practice conduct the audit?			
11.5	Does the audit team have sufficient detachment from the company or operating unit? (Note: This question does not apply to self-assessments.)			
11.6	Are environmental audits performed on a schedule that provides sufficient frequency for the particular operation or activity?			
11.7	Are the environmental audits used consistently every time they are conducted?			
11.8	Is environmental auditing performed at all company levels?			
11.9	Do environmental audits include self-assessments?			
11.10	Do environmental audits evaluate the strengths of a program or activity as well as its weaknesses?			
11.11	Are environmental audit findings and recommendations documented in written reports?			
11.12	Are audit findings communicated to appropriate internal and external individuals and organizations?			
11.13	Do environmental audits have quantitative ratings that track performance and improvements?			
11.14	Do audits include timely, formal procedures to correct the problems found or those that consistently reoccur?			
11.15	Do audits systematically identify opportunities for environmental improvements?			
11.16	Are the lessons and improvements from an environmental audit systematically applied to similar operations and activities?			
11.17	Are the auditing systems themselves audited and continually improved?			
11.	Subtotal			

11.   Auditing
Documentation Used and Personnel Contacted:
General Comments:
Strengths:
Weaknesses:
Recommendations for Improvement:

12. MEASUREMENT AND CONTINUOUS IMPROVEMENT

Environmental excellence is maintained only when environmental performance is regularly measured and continually improved via a formal system. Total Quality Management (TQM) is used to manage environmental performance. "TQM is a total organizational approach toward the continuous improvement of quality and productivity that involves all employees."[17] Specifically, TQM graphical tools are used to collect and analyze data. These tools are designed to highlight problems and present relevant data in a manner that facilitates the improvement process.

A continuous improvement program begins with the creation of measurable goals and objectives that track environmental performance. The measurements are lean and quantitative. Including unnecessary data is often a source of confusion and qualitative measurements are too subjective to be meaningful. A quality continuous improvement program includes the use of benchmarking data. "Benchmarking is a process whereby you identify an area in which you want to improve or maintain superiority, find others in industry who do it best, compare yourself to their processes and achievements, and take steps to reduce the gap between yourself and the best-in-class.''[18] Emulating the best practices of other corporations or other business units within your own company is an effective way to increase quality. Lastly, a system is required to solicit feedback from all appropriate internal and external sources. The ability to anticipate relevant changes and to make the necessary adjustments is essential to maintaining environmental quality.

12. Measurement and Continuous Improvement	Yes	No	N/A
12.1 Is Total Quality Management (TQM) used in environmental activities and programs?			
12.2 Are graphical TQM tools (e.g., Pareto charts, control charts, etc.) used to collect and analyze environmental data?			
12.3 Are goals and objectives set for all environmental activities and initiatives?			
12.4 Are environmental goals and objectives measurable?			
12.5 Are measurements used to track the continuous improvement of environmental goals and initiatives?			
12.6 Are environmental measurements lean and quantitative?			
12.7 Is benchmarking used to evaluate and continually improve environmental programs and initiatives?			
12.8 Is feedback on environmental programs and activities actively sought from appropriate internal and external sources?			
12.9 Are all environmental activities and initiatives continually improved?			
12.10 Is the overall impact from your company's activities on the environment tracked and improved?			
12. Subtotal			

12.   Measurement and Continuous Improvement
Documentation Used and Personnel Contacted:
General Comments:
Strengths:
Weaknesses:
Recommendations for Improvement:

SCORING

After the self-assessment of the 12 areas of excellence, you can complete the Score Sheet. You will then convert the score you tabulated in each section, as well as the overall score, into a percentage. You can then plot the percentages on the comparison bar chart where the results are readily visible.

Follow the six steps below to complete the score sheet and the comparison bar chart. You will need to refer to the 12 areas of excellence questionaires you have completed previously.

1. **Column C Yes:** Locate the subtotal under the Yes column of each questionnaire. Place that number in the corresponding row under Column C Yes column in the Score Sheet. Add the numbers in Column C and place that summation in the Total row.

2. **Column D No:** Locate the subtotal under the No column of each questionnaire. Place that number in the corresponding row under Column D No in the Score Sheet. Add the numbers in Column D No and place that summation in the Total row.

3. **Column E N/A:** Locate the subtotal under the N/A column in each questionnaire. Place that number in the corresponding row under Column E N/A in the Score Sheet. Add the numbers in Column E N/A and place that summation in the Total row.

4. **Column F Questions Used:** For each row 1 through 12, subtract the number in Column E N/A from the number in Column B Number of Questions. Place that difference in the corresponding row under Column F Questions Used in the Score Sheet. Add the numbers in Column F Questions Used and place that summation in the Total row.

5. **Column G Score (%):** For each row 1 through 12 and Total, divide the number in Column C Yes by the number in Column F Questions Used. Multiply that number by 100 and place the resulting number in the corresponding row under Column G Score (%).

6. **Comparison Bar Chart:** Fill in each bar of the comparison bar chart up to the percentage that is in the corresponding row of Column G SCORE (%) in the score sheet. Follow this procedure for all 12 areas of excellence as well as for the Total row.

Refer to the score sheet and the comparison bar chart to determine the areas where improvements are necessary. Also use them to track improvement with each successive implementation of the self-assessment program.

Score Sheet

Areas of Excellence	B. Number of Questions	C. Yes	D. No	E. N/A	F. Questions Used (F=B−E)	G: Score (%) G=100*(C/F)
1 Policy	10					
2 Management	11					
3 Planning	17					
4 Cost	16					
5 Stakeholders	16					
6 Human Resources	14					
7 Operations and Facilities	15					
8 Suppliers	12					
9 Compliance	11					
10 Technology and Research	10					
11 Auditing	17					
12 Measurement and Continuous Improvement	10					
Total	159					

Comparison Bar Chart

	0%	20%	40%	60%	80%	100%
1. Policy						
2. Management						
3. Planning						
4. Cost						
5. Stakeholders						
6. Human Resources						
7. Operations and Facilities						
8. Suppliers						
9. Compliance						
10. Technology and Research						
11. Auditing						
12. Measurement and Continuous Improvement						
TOTAL						

BIBLIOGRAPHY

l'Association Francaise de Normalisation (AFNOR). April 1993. *Environmental management system.* Experimental Standard No. X 30-200. Paris.

AT&T. 1991. *AT&T environment and safety report: An investment in our future.* Basking Ridge, N.J.

Booz-Allen & Hamilton Inc. 1991. *Strategic environmental management: Harnessing risk and reward.* Bethesda, Md.

British Standards Institute. 1992. *Specification for environmental management systems.* BB 7750: 1992. London.

Brown, George E. 1993. Green technology: An industrial spring. Presented at the Microelectronics and Computer Technology Conference on an Environmental Technology Initiative for the Electronics Industry. March 10. McLeon, Va.

Canadian Standards Association. February 1992. *A draft report on environmental management systems and models.* 1st rev.

_____. April 1993a. *A guideline for a voluntary environmental management system.* 5th rev. Business Council on National Issues. Business principles for a sustainable and competitive future. p. 37.

Canadian Chemical Producers' Association (CCPA). Responsible care. p. 30.

European Petroleum Industry Association. Environmental guiding principles. p. 29.

International Chamber of commerce (ICC) Business Charter for Sustainable Development. Principles for environmental management. pp. 26–27.

Japan Federation of Economic Organizations. Keidanren global environmental charter. pp. 35–36.

National Round Table on the Environment and the Economy (NRTEE). Objectives for sustainable development. p. 28.

————. April 1993b. *A proposed environmental guideline on green procurement.* Z766.

Chemical Manufacturer's Association. April 1992. *Responsible Care® A Public Commitment: Product stewardship code of management practices.*

Coalition for Environmentally Responsible Economics. 1992. *Guide to the CERES principles.* 2d ed. Boston.

Council of the South African Bureau of Standards. 1993. *South African standards: Code of practice—Environmental management systems.* SABS 0251:1993. Pretoria.

Eastman Kodak Company. [1993?]. *Statement of corporate policy: Health, safety and environment.* Rochester, N.Y.

Executive Enterprises Publications Co. Inc. 1993. *Measuring environmental performance.* New York.

Global Environmental Management Initiative. 1991. *Proceedings—Corporate quality/environmental management: The first conference.* January 9–10. Washington, D.C.

————. 1992. *Total quality environment management: The primer.* Washington, D.C.

————. April 1993. *Environmental self-assessment program.* 1st ed., 2d printing. Washington, D.C.

Green, Mark. 1992. *Total quality environmental management: Performance evaluation TQEMPE®.* Presented at the Executive Enterprises Conference on Total Quality Environmental Management, March 9–10. San Francisco.

International Benchmarking Clearinghouse. 1992. *Benchmarking E/H/S: Proposed business practices and performance indicators.* Study.

International Organization for Standardization. 1993. *Report of the fourth meeting of the ISO/IEC strategic advisory group on environment (SAGE).* June 1. Toronto.

Investor Responsibility Research Center. September 1992. *Institutional investor needs for corporate environmental information.* Washington, D.C.

Lowe, Ernest. November 1992. *Discovering industrial ecology.* Oakland, Calif.: Change Management Center. Draft.

————. January 1993. *Applying industrial ecology.* Oakland, Calif.: Change Management Center. Report.

National Standards Authority of Ireland. 1993. *Guiding principles and generic requirements for environmental management systems—Part 1:Guiding principles and definitions.* Dublin.

Research Triangle Institute. 1992a. *Assessment of the characteristics of small-to-medium-sized companies with successful environmental management programs: Preliminary findings.* Report.

————. 1992b. *Environmental management in small-to-medium-sized firms: Your company's viewpoint.* Questionnaire.

Rushton, Brian M. 1993. How protecting the environment impacts R&D in the United States. *Research•Technology Management.* May–June: 1–21.

Texas Instruments. September 24, 1990. Guiding principles for environmental stewardship.

Texas Instruments Incorporated, Defense Systems & Electronics Group. March 1993. *Pollution prevention program.*

Tusa, Wayne. April 1993. *Assessing existing environmental management programs.* New York: Environmental Risk and Loss Control, Inc.

United States Department of Energy Office of Environmental Audit. [1993?]. *Protocols for conducting environmental management assessments of DOE organizations.* DOE/EH-0326. Washington, D.C.

United States Environmental Protection Agency. May 1992. *Facility pollution prevention guide.* EPA/600/R-92/088. Washington, D.C.

————. January 1993. *Life cycle design guidance manual: Environmental requirements and the product system.* EPA/600/R-92/226. Washington, D.C.

Warren, John and Fagg, Brandon F. January 1993. *Excellence in environmental management: Introduction, preliminary findings, and bibliography.* Research Triangle Park, N.C.: Research Triangle Institute.

REFERENCES

1. British Standards Institute. 1992. *Specification for environmental management systems.* 10. BS 7750:1992. London.

2. United States Department of Energy Office of Environmental Audit. [1993?]. *Protocols for conducting environmental management assessments of DOE organizations. 15 DOE/EH-0326.* Washington, D.C.

3. Hopper, Earl. 1991. Implementing environmental quality at IBM. *Proceedings—Corporate quality/environmental management: The first conference*, 51. January 9–10. Washington, D.C.: Global Environmental Management Initiative.

4. British Standards Institute. 1992. *Specification for environmental management systems.* 10. BS 7750:1992. London.

5. Wells, Richard P., et al. 1993. Measuring environmental success. *Measuring environmental performance*, 9. New York: Executive Enterprises Publications Co. Inc.

6. United States Environmental Protection Agency. January 1993. *Life cycle design guidance manual: Environmental requirements and the product system*, 126–28. EPA/600/R-92/226. Washington, D.C.

7. Canadian Standards Association. April 1993. *A guideline for a voluntary environmental management system*, 14. 5th rev.

8. Investor Responsibility Research Center. September 1992. *Institutional investor needs for corporate environmental information*, iv. Washington, D.C.

9. ISO/IEC SAGE Sub-Group 1. May 11, 1993. Standardization of environmental management systems: A model for discussion, 24. Document: ISO./IEC/SAGE N55 (Annex 3 to ISO/IEC SAGE 82). *Report of the fourth meeting of the ISO/IEC strategic advisory group on environment (SAGE).* June 1. Toronto: International Organization for Standardization.

10. United States Environmental Protection Agency. May 1992. *Facility pollution prevention guide*, 141–43. EPA/600/R-92/088. Washington, D.C.

11. United States Environmental Protection Agency. January 1993. *Life cycle design guidance manual: Environmental requirements and the product system*, 92. EPA/600/R-92/226. Washington, D.C.

12. Council of the South African Bureau of Standards. 1993. *South African standards: Code of practice—Environmental management systems*, 11. SABS 0251:1993. Pretoria.

13. Charm, Joel. 1991. TQM-driven audits work. *Proceedings—Corporate quality/environmental management: The first conference*, 71. January 9–10. Washington, D.C.: Global Environmental Management Initiative.

14. Fisher, Michael T. 1993. Building audits into TQEM measurement systems at P&G. *Measuring environmental performance*, 101. New York: Executive Enterprises Publications Co. Inc.

15. Chemical Manufacturer's Association. April 1992. *Responsible Care® A Public Commitment: Product stewardship code of management practices*, i.

16. Canadian Standards Association, April 1993. *A guideline for a voluntary environmental management system*, 24. 5th rev.

17. Neidert, Anita. 1991. Introduction to environmental health and safety problemsolving through total quality management. *Proceedings—Corporate quality/environmental management: The first conference*, 91. January 9–10. Washington, D.C.: Global Environmental Management Initiative.

18. Klafter, Brenda. 1993. Pollution prevention benchmarking: AT&T and Intel work together with the best. *Measuring environmental performance*, 93. New York: Executive Enterprises Publications Co. Inc.

INDEX

Index

Other books of interest from Irwin Professional Publishing . . .

SYNCHROSERVICE!
The Innovative Way to Build a Dynasty of Customers
Richard J. Schonberger and Edward M. Knod, Jr.

Synchroservice—an organization-wide commitment to seamless, consistent, customer-driven service—can lead to higher customer loyalty, stronger supplier relations, and an undeniable competitive edge. This guide provides everything needed to design and execute a synchronized service strategy, including how to: ensure that every behind-the-scenes operation is working to support and enhance frontline customer encounters, use proven manufacturing planning techniques to anticipate and satisfy ever-changing customer service demands, and organize and link internal teams, supplier teams, and customer teams to make improvements in quality.
ISBN: 0-7863-0245-3

FIRING ON ALL CYLINDERS
The Service/Quality System for High-Powered Corporate Performance
Jim Clemmer

This authoritative, crisply-written book outlines Achieve International's Service/Quality System that is used by dozens of public and private sector companies. The Service/Quality System combines three formerly separate organizational performance fields to help you improve your overall business strategy. You'll discover how to build leadership skills, reduce defects and mistakes, and improve responsiveness in customer service.
ISBN: 1-55623-704-9

INTEGRATED PROCESS DESIGN AND DEVELOPMENT
Dan L. Shunk

This no-nonsense, reader-friendly guide will help managers design and develop integrated processes that are consistent with the needs and capabilities of the plant and the employees.
ISBN: 1-55623-556-9

MANUFACTURING PLANNING AND CONTROL SYSTEMS
Third Edition
Thomas E. Vollman, William L. Berry, and D. Clay Whybark

Stay ahead of your competition! This guide illustrates dozens of examples of the latest key concepts for implementing profitable planning, scheduling, production, and distribution strategies.
ISBN: 1-55623-608-5

Available at fine bookstores and libraries everywhere.